Neutron-Transmutation-Doped
SILICON

Neutron-Transmutation-Doped SILICON

Edited by

Jens Guldberg

Topsil A/S
Frederikssund, Denmark

PLENUM PRESS • NEW YORK AND LONDON

Library of Congress Cataloging in Publication Data

International Conference on Neutron Transmutation
Doping of Silicon (3rd : 1980 : Copenhagen, Denmark)
Neutron-transmutation-doped silicon.

"Proceedings of the Third International Conference on Neutron Transmutation Doping of Silicon, held August 27-29, 1980, in Copenhagen, Denmark."
Bibliography: p.
Includes index.
1. Semiconductor doping, Neutron transmutation—Congresses. 2. Silicon—Defects—Congresses. I. Guldberg, Jens. II. Title.
TK7871.85.I5763 1980 621.3815'2 81-7305
ISBN-13: 978-1-4613-3263-3 e-ISBN-13: 978-1-4613-3261-9 AACR2
DOI: 10.1007/ 978-1-4613-3261-9

Proceedings of the Third International Conference on Neutron Transmutation Doping of Silicon, held August 27—29, 1980, in Copenhagen, Denmark

Softcover reprint of the hardcover 1st edition 1981
A Division of Plenum Publishing Corporation
233 Spring Street, New York, N.Y. 10013

PREFACE

This volume contains the papers presented at the Third International Conference on Neutron Transmutation Doping of Silicon held in Copenhagen on August 27-29, 1980.

The first symposium associated with neutron transmutation doping technology as such was arranged in 1976 at Oak Ridge National Laboratory by John Cleland. At this time it had become clear that the technology could be implemented on a commercial scale and that several types of power devices in the electronic industry would benefit from employing neutron transmutation doped silicon in the fabrication process.

Two years later the Second International Conference on Neutron Transmutation Doping of Semiconductors was arranged at the University of Missouri, Columbia, by Jon Meese. On this occasion the various aspects of silicon fabrication were reviewed, including irradiation control, radiation induced defects, device optimization, and possible benefits of irradiating other semiconductor compounds.

In view of the now wide spread acceptance of neutron doped silicon in the power device industry the present conference was largely directed towards the current status of transmutation doping of silicon. Accordingly, the scope of the three day conference was to review developments in the technology which had occurred during the two years which had passed since the previous conference. In addition, brief accounts were given with respect to other semiconducting compounds and emerging irradiation techniques which may impact on device design principles in the future.

It was a pleasure at Topsil to host the conference and to register the growing interest for this technology in the US, in Europe and in Asia. This is evidenced in the series of invited and contributed papers, as well as in the country-wise distribution among the participants. It is also a pleasure to reconfirm that the next conference will be arranged in Gaithersberg in June 1982 by R.B. Larrabee at the National Bureau of Standards.

We would like to thank the members on the Advisory Committee for their helpful assistance and advice:

S. Arimura	Osaka Titanium Co., Ltd.
K. Heydorn	Research Establishment Risoe
R.D. Larrabee	National Bureau of Standards
J.M. Meese	University of Missouri Research Reactor
A. Senes	Thomson - C.S.F.
H. Stein	Sandia Laboratories
C. Varker	Motorola Inc.
F. Vieweg-Gutberlet	Wacker Chemitronic GmbH

We also extend our thanks to the members of the Organizing Committee

Mr. Finn Larsen
Dr. Olof Malmros
Mr. Kaj Lund
Mrs. Jette Gotfredsen

Finally, we sincerely wish to express our gratitude for the financial support we obtained for the conference from the following organisations:

Monsanto Corporation
Danish Technical Science Research Council
National Science Foundation

This support has been essential in making the conference truly international in its coverage of various aspects of neutron transmutation doping of silicon.

Jens Guldberg
Conference Chairman

Frederikssund
Denmark
December, 1980

CONTENTS

NEUTRON-TRANSMUTATION-DOPED SILICON

NEUTRON-DOPED SILICON - A MARKET REVIEW

Heinz Herzer

Wacker-Chemitronic
P.O.Box 1140
Burghausen, Germany

MARKET DEVELOPMENT OF ND-SILICON

The market development of neutron-transmutation-doped silicon in the past can be divided into the periods:

1973	First trials and evaluation quantities for high power thyristors.
1974/75	Conversion of thyristor production and evaluation quantities for transistors and diodes.
1976-80	Large scale production of high and medium power thyristors, power transistors and power diodes.

The original idea to apply nuclear transmutation as a method for the manufacture of a well-defined n-type silicon crystal goes back to the time when a small number of physicists were concerned with nuclear radiation detection which has always been the special field of overlapping nuclear and semiconductor technology /1-5/.

In 1973 when the transmutation technology was applied first by silicon producers /6-9/, it became evident that ND-silicon was the only way out of the problems with resistivity striations in phosphorus doped dislocation-free silicon grown in the 1-1-1 orientation. The comparison between silicon products doped in the conventional way (Fig. 1 and 2) demonstrates that N-111-FZ - the classical material for power application - comes out worst.

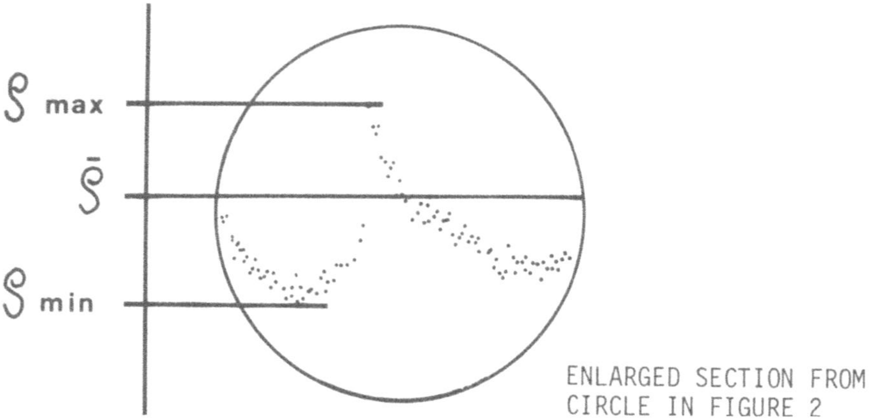

Fig. 1.

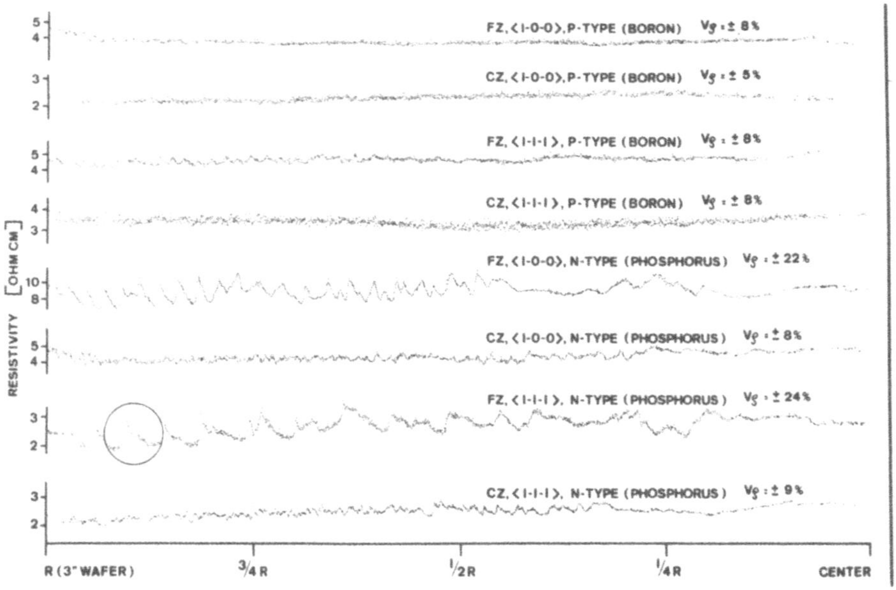

Fig.2. Spreading resistance measurements on silicon wafers for the growth procedures FZ and CZ, the orientations (111) and (100), and the dopants boron and phosphorus /10/.

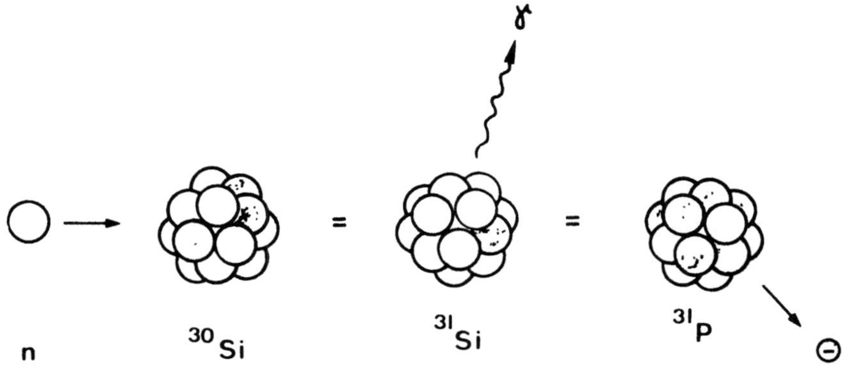

Fig. 3. The transmutation of a ^{30}Si-isotope into a ^{31}P-isotope by neutron capture /1, 2/.

Not only have the microscopic variations been a limiting factor for further progress in thyristor technology, but they have also been one of the reasons for bad yields in the production of power devices at the state of the art those days. In particular the high power thyristor turned out to be the utmost sensitive probe for a group of defects which have been classified by the general expression "hot spot" since in that one chip device a large area of perfect silicon was required. A single tiny imperfection in an area of $20\ cm^2$ in the case of a 2" diameter caused the complete breakdown of the expensive device. Because in general the breakdown is accompanied by a high voltage breakthrough in most cases the defect responsible can no longer be traced and it is difficult to judge whether the reason for the failure was

- a bulk defect due to crystal imperfection, like a swirl defect or a hydrogen hole,
- a bulk imperfection due to resistivity striations,
- a surface defect due to a mechanically introduced disturbance, like a spike, scratch, or crow foot,
- a process induced defect like a slip line or impurities, etc.

Today we know that resistivity striations took a considerable part of the yield decrease and the advent of ND-silicon has been a great help in making improvements in the technology of wafer preparation and device process. The wonderful arrangement of nature allowing silicon atoms themselves to take over the task of doping in such a way that a few of them just capture a thermal neutron being transformed into phosphorus (Fig. 3) provides us with a method to get perfectly doped material, and the additional cost is balanced out by the advantage of tight resistivity tolerances and a better yield in device fabrication. Fig.4 shows the curve of the estimated total quantities of ND-silicon which have been produced until today. 1973 saw the beginning of evaluation quantities only and then - after the excellent material performances had been proven /11, 12/ - a steady increase in production quantities for thyristor application (I) and later the introduction as material for transistors and diodes (II).

The dotted line indicates a possible additional increase due to application in power integrated circuits but from today's point of view it is difficult to judge whether we reach the 100 tons p.a. without ICs or not.

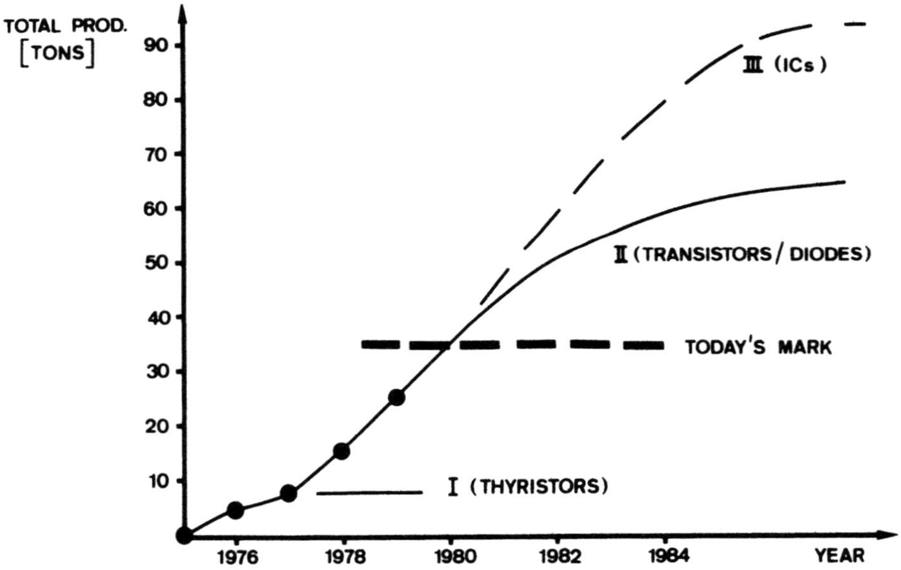

Fig.4. The development of large-scale ND-silicon production and its main application in (I) thyristors and (II) transistors and diodes.
The dotted line indicates a possible future application in integrated circuits.

There are several reasons for the slow start: a) It takes time for the construction and installation of proper irradiation facilities and other equipment to perform the irradiation of silicon ingots, b) a customer will use a new material in his production line and accept additional expenditure only when he is absolutely sure of the technical and commercial benefits and c) of the availability and this means more than just one or two suppliers worldwide.

The last point - the availability from several sources - is an important factor in launching a new product on the market which from a technological point of view is perfectly in line with market conditions.

Today the following companies are suppliers of ND-silicon:

Japan Silicon

Komatsu

Monsanto

Shin-Etsu

Topsil

Wacker

and the applications have been extended to:

high and medium power thyristors

frequency thyristors

power rectifiers

power transistors.

THE PROCEDURE OF ND-SILICON MANUFACTURE

In the manufacture of ND-silicon we have the following steps:

- growth of undoped starting material
- surface cleaning
- neutron irradiation
- radioactivity decay

- surface cleaning
- annealing
- measuring
- wafer production.

These principal steps have not changed since the very beginning /13-15/, and some more details with regard to large scale production in the U.S. will be given by B.D. Stone in the presentation after /16/, and concerning France by P. Breant /17/. From amongst the contributions by reactor operators we will learn how to optimize silicon irradiation by N.W. Crick /18/ and K. Heydorn and K. Anderson /19/, and T.G. Smith /20/ will give us a feeling of security as far as radioactivity is concerned.

Here we would like to make a few comments on the starting material and the quality of the material after annealing. Fig.5 exhibits the spreading resistance profiles measured on three samples which have been irradiated for the same target resistivity of $\rho = 200$ Ohmcm. The resistivity of the starting material differed, amounting to about 400 in the case of F2, 1000 in the case of F5, and 2000 Ohmcm in the case of F10. In the case of the F2 profile we can observe a marked influence of background striations which are of minor importance in the F5 profile and are no longer visible in the F10 profile. This demonstrates that the level of phosphorus concentration in the undoped ingot material has to be considered carefully where we want to produce target resistivities in the higher ranges, i.e. up to 300 Ohmcm. In that respect it is understandable that the additional costs are closer related to the efforts being spent on the starting material than on irradiation and annealing. The range below $\rho = 100$ Ohmcm is obviously not causing any problems in that respect, and the more the target resistivity comes down the more even the resistivity variation will be.

With regard to the specification of ND-silicon, the starting material and the irradiation can be defined precisely whereas the specification of the quality after annealing needs some remarks. The purpose of annealing is to get rid of the irradiation damage induced during neutron bombardment by applying a high temperature treatment. This treatment is limited to the lower side to approximately 630°C which after Nelson /21/ is the recrystallization temperature of an amorphous layer on a single crystalline host

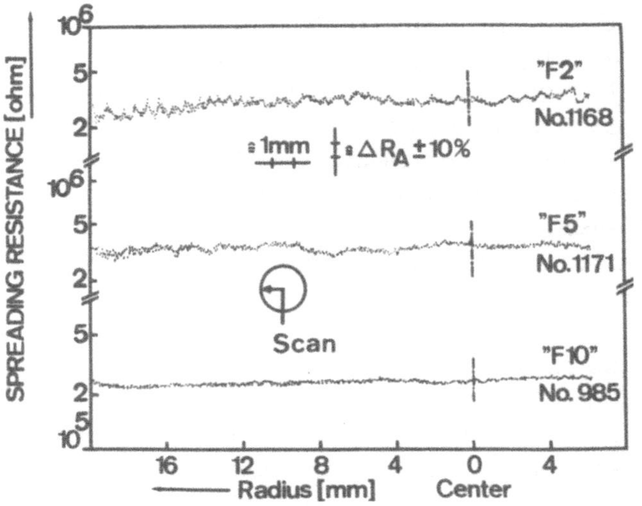

Fig.5. The influence of the doping level in the starting material on the background striations:
Target resistivity $\rho = 200$ Ohmcm;
Resistivity of the starting material (F2) 400 Ohmcm, (F5) 1000 Ohmcm, and (F10) 2000 Ohmcm /14/.

lattice and to the higher side because with increasing temperatures the danger of deteriorating the silicon quality by the introduction of impurities or the generation of slippage increases. Hence, the temperature preferred is in the medium range between 700 and 900°C.

The only disadvantage of this treatment is that depending on the neutron spectrum, a resistivity shift of a few percent can occur to some extent when the material is processed as a wafer in a phosphorus diffusion or in a similar way /14/. Certainly, the resistivity shifts are smaller in the range below $\rho = 150$ Ohmcm and are getting more pronounced above this level. The usual procedure is to compare between target and result and to cut off what is not wihtin specification. But from comparison of the irradiation data with the results we find that the material out of specification would in some cases come within specification as well if it is processed in a phosphorus diffusion, and that concerning the material which is within specification it probably gets to even tighter tolerances as we have specified by four point probe measurement.

An example of how strong such an effect can be is shown in an experiment (Fig.6) where the annealing has been carried through at 650°C: a) is the resistivity variation (RV) of the starting material determined by 4-point probe measurement, b) is the RV on the ND-ingot after annealing - the neutron dose was calculated for a target resistivity of $\rho = 200$ Ohmcm - , c) is the RV of the same ingot after a second annealing with a time factor of about 5, and d) is the RV of a test wafer which was cut from c) and processed in a phosphorus diffusion.

Without trying to give a more or less sophisticated interpretation, it is assumed that the effect has to be seen as the formation and the decay of a donor, possibly a phosphorus-vacancy double minus.

The second parameter - besides resistivity - is the lifetime of minority charge carriers. This parameter is not to be overestimated. From our experience we know that a threshold of $\tau = 100$ µsec is sufficient for all applications and - very important - that the drop of the lifetime from some milliseconds with the starting ingot to 100 microseconds with the ND-ingot does not necessarily have to be in correlation with a decrease in quality. This is best demonstrated in

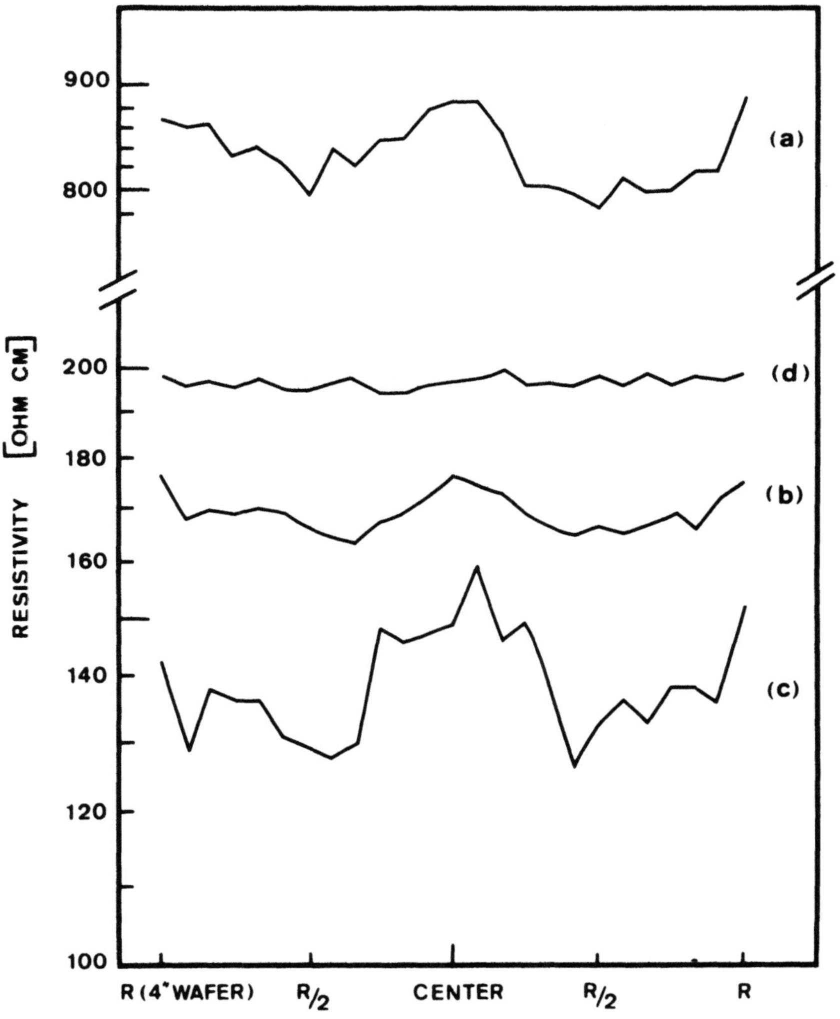

Fig.6. Resistivity variation (RV) on an ingot with strong resistivity shift: Target resistivity $\rho = 200$ Ohmcm; a) starting material, b) after 650° annealing, c) after second annealing with time factor, d) after phosphorus diffusion.

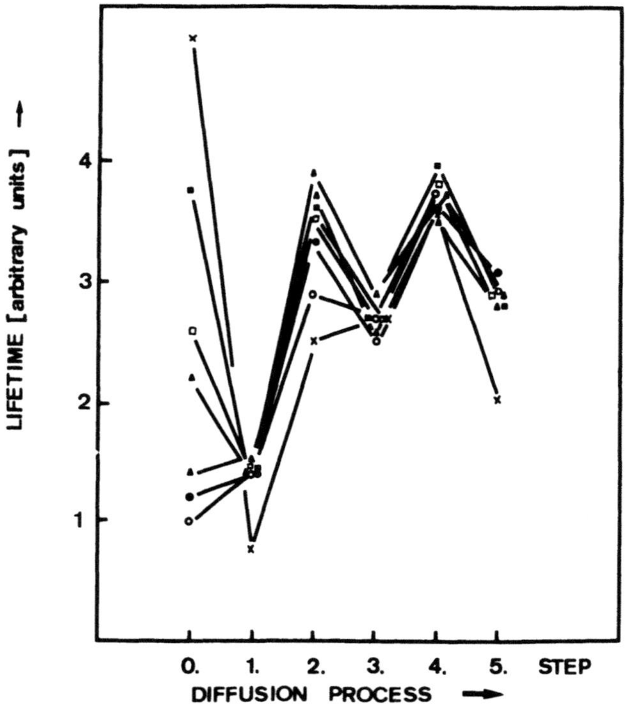

Fig.7. Change of lifetime values of as-delivered ND-silicon (step 0) after various diffusion steps (1 - 5).

Fig. 7 where the behaviour of ND-wafers is shown in a customer's process: the spread in the lifetime values of the as-delivered material vanishes in the first processing step and in all further steps lifetime is only governed by the process conditions.

This does not mean that lifetime is unimportant; we can use it as a

very sensitive parameter in the annealing process, as a control that no failures occur because of contamination, tube breakage and others because in that case lifetime would drop below $\tau = 100\ \mu sec$.
In particular, with the lifetime control on the starting material, a normal decay of the radioactivity and again a lifetime control on the annealed material we have a steady quality assurance of the ND-silicon.

THE ADVANTAGE OF ND-SILICON

When considering the improvements in ND-silicon as compared to conventionally doped FZ-silicon not only the flatness in the microscopic resistivity profile has to be seen, justifying the expression "striation-free", as can be seen by comparing the two methods (Fig.8), but also the advantage of tighter target resistivities which gives an additional effect (Fig.9).

The first point enabled a major step forward resulting in 4" dia. thyristors with blocking voltages above 4000 Volts and several thousand amperes forward current. The second point is important for better yields in nearly all the other power device fabrications which usually apply the dicing technique, producing a number of devices from one wafer. Fig.10 shows the results on the device characteristics of medium power thyristors diced out of a 2" dia. wafer, the material of which was conventionally doped FZ, and as a contrast Fig. 11 demonstrates the device characteristics of the same device and same technique diced out of a 3" dia. ND-wafer. No need to refer to the correlation with the resistivity variation.

The production analysis of the above mentioned example gave the following further data:

Device Results (medium power thyristors)

Standard Float-Zone

Range: 1040 - 1360 Volts
Median is 83% of maximum

Neutron doped silicon

Range: 1200 - 1520 Volts
Median is 99% of maximum

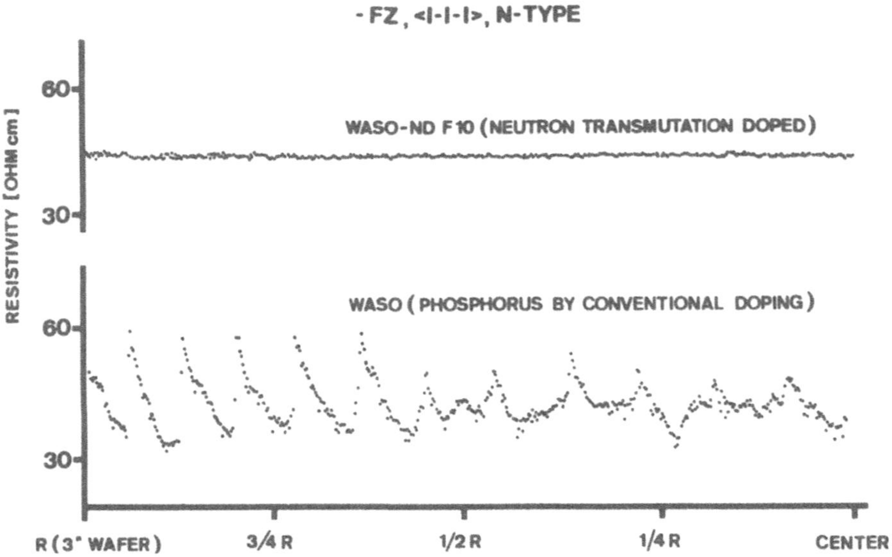

Fig.8. Comparison of the spreading resistance profiles between conventionally doped FZ and ND-silicon /22/.

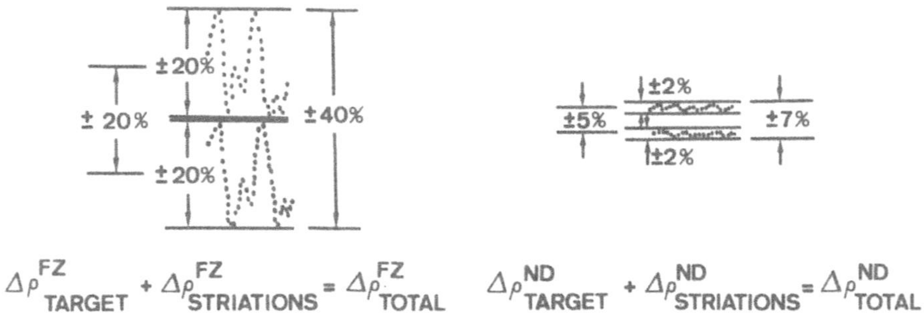

Fig.9. Comparison of the resistivity tolerances - microvariation and target - between conventionally doped FZ and ND-silicon /14/.

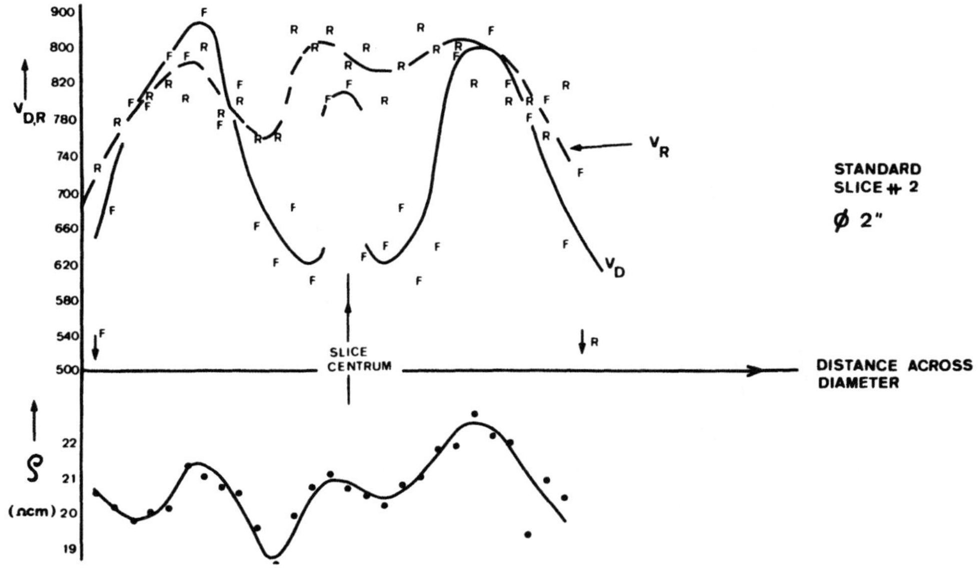

Fig. 10. Results of device characteristics in medium power thyristors produced on a conventionally doped FZ-wafer.

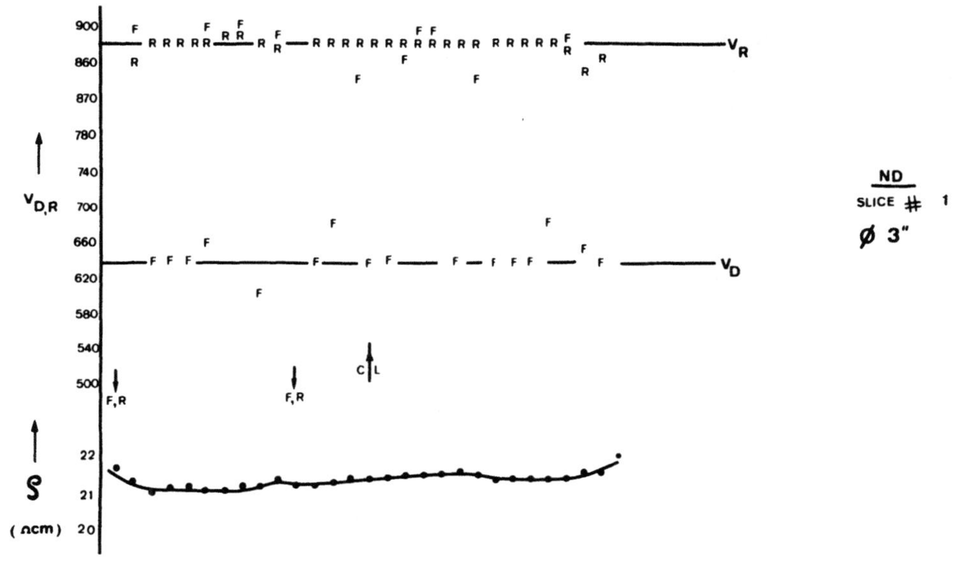

Fig. 11. Results of device characteristics in medium power thyristors produced on an ND-silicon wafer.

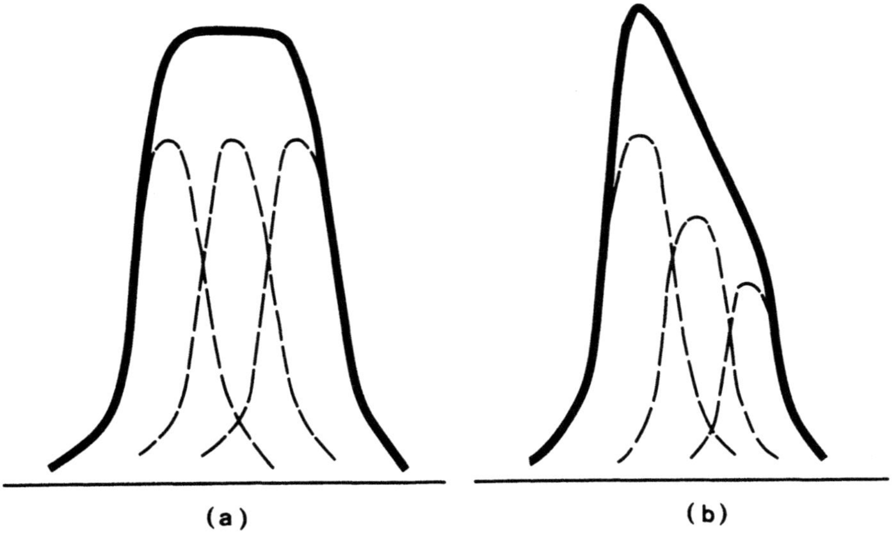

Fig. 12. Profiling of a resistivity specification range: a) equal quantities, b) varying quantities

FUTURE TRENDS

Today neutron transmutation doped silicon is well established and yet we do not make use of all possible advantages. For example, we can profile the resistivity specification range as indicated in Fig. 12: a) dividing into three tight subranges with the additional agreement that a falling out of the middle range into the higher group may be balanced out by an equivalent quantity from the lower group and vice versa - this gives an excellent material specification for stack diodes - , b) profiling with respect to the customer's special requirements. Getting to very tight tolerances - Fig. 13 - could be realized by aiming at a higher target value than the middle of the resistivity specification range and followed by a second irradiation of that part which dropped out of the specification exceeding the upper threshold.

It has been mentioned above that the built-in resistivity after neutron irradiation is even better than we can measure and specify it on the annealed ND-silicon. Here the question arises whether we should go to unannealed material at all. In that case the list of

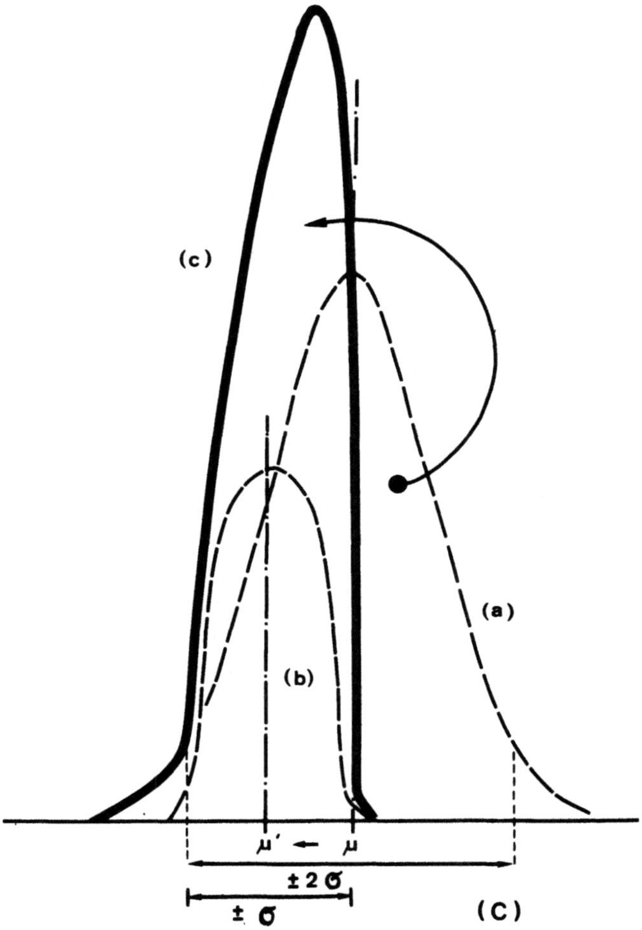

Fig. 13. Distribution model for after-irradiation of ND-silicon for the purpose of tighter target tolerances: a) distribution curve for the first irradiation, b) for the second irradiation of the upper half from a), and c) the sum of the lower half from a) and b).

specification parameters has only to be extended by the point at which the resistivity control should be performed on a test wafer after a phosphorus diffusion and lapping off the n^+-layer.

A final comment on the further development with respect to quantity: can we really go on with unlimited quantities and get involved for example in power ICs? The reactor operators should have the last word. From today's viewpoint there is enough reactor capacity available or will be available with the installation of new equipment. Nevertheless, the silicon suppliers have to handle the matter of where to set the resistivity limit very carefully. If we would go down to $\rho = 5$ Ohmcm and include, just as an example, C-MOS application and p-channel MOS, we would immediately exceed the available capacity. Therefore ND-silicon should be reserved mainly for power application and a lower limit set to $\rho = 10$ Ohmcm. Besides, in the range below $\rho = 10$ Ohmcm CZ-material is the standard product anyway and here striation profiles don't look so bad (Fig.1).

The last question remains: can ND-silicon be replaced by improved conventional doping procedures? It will be the target for everyone working in silicon crystal growth, and we can imagine that there are possible modifications in our procedures resulting in striation-free silicon, too. However, it is difficult to believe that the economical side - first of all the reproducibility and the yield - look as good as in the case of neutron transmutation doping and the best confirmation is the steady increase in quantities the market requires.

REFERENCES

1) K. Lark-Horovitz, Proc.of Conf.at Univ.Reading (Butterworth, London), 47 (1951)
2) M. Tanenbaum and A.D. Mills, J. Electrochem.Soc. 108, 171 (1961)
3) C.N. Klahr and M.S. Cohen, Nucleonics, 22, 62 (1964)
4) J. Messier, Y. Le Coroller and J. Merlo Flores, Proc.of 9th Scintillation and Semiconductor Counter Symposium, Washington, D.C. (1964)
5) G. Dearnaley and D.C. Northrop, "Semiconductor Counters for Nuclear Radiations", Spon Limited, London (1966)
6) M. Schnoeller, IEEE Trans.Electron.Devices ED-21, 313 (1974)
7) E. Haas and M. Schnoeller, Journal of Electronic Materials 5, 57 (1976)

8) H.A. Herrmann and H. Herzer, J.Electrochem.Soc. 122 1568 (1975)
9) H.M.Janus and O. Malmros, IEEE Trans.Electron.Devices ED-23, 805 (1976)
10) Some aspects of FZ and CZ as-grown silicon emphasizing resistivity striations, Technical Information 1/76, Wacker-Chemitronic, Burghausen
11) K. Platzöder and K. Loch, IEEE Trans.Electron.Devices ED-23, 805 (1976)
12) M.J. Hill, P.M. Van Iseghem and W. Zimmermann, IEEE Trans.Electron.Devices ED-23, 809 (1976)
13) H.A. Herrmann, H. Herzer and E. Sirtl, Festkörperprobleme 15: Advances in Solid State Physics, H.J. Queisser, ed., Pergamon Press, 279 (1975)
14) H. Herzer, Proc. of the 3rd International Symposium on Silicon Materials Science and Technology, held at Philadelphia May 1977, H.R. Huff and E. Sirtl, eds., The Electrochem.Soc., Proc.Vol. 77-2 (1977)
15) John M. Meese, ed., Proc. of the 2nd International Conf. on Transmutation Doping in Semiconductors, held at the Univ.of Missouri April 1978, Plenum Press N.Y. (1979)
16) B.D. Stone, this volume
17) P. Breant, this volume
18) N.W. Crick, this volume
19) K. Heydorn and K. Anderson, this volume
20) T.G. Smith, this volume
21) G. Dearnaley, J.H. Freeman, R.S. Nelson and J. Stephen, "Ion Implantation" in Defects in Crystalline Solids 8, 201, North Holland Publishing Company (1973).
22) Waso-ND, Technical Information 1/80, Wacker-Chemitronic, Burghausen

LARGE SCALE PRODUCTION OF NTD SILICON IN THE UNITED STATES

Bobbie D. Stone

Monsanto Company

St. Peters, MO 63376

ABSTRACT

The growth of the demand for NTD float zone silicon in the U.S. is traced in general terms with a discussion of device use patterns. A brief survey of types of irradiation facilities in the U.S. is outlined including an indication of commercial capabilities. Published licensed requirements of the U.S. Nuclear Regulatory Agency are discussed including monitoring of irradiated material for residual radioactivity, personnel protection measures, control of effluent levels, inventory monitoring and associated documentation Product specifications and tolerances are discussed with emphasis on the role played by the starting material. Analysis of the possible sources of inaccuracies is made.

An outline of a zone refining process for producing only NTD product is compared to a conventional one with a discussion of advantages and disadvantages of the two. Trends in the manufacture and usage of NTD float zone silicon are discussed.

INTRODUCTION

The rediscovery of the technique of phosphorus doping of silicon by irradiation with thermal neutrons and the dramatically improved performance of thyristors fabricated from NTD float zone crystal reported by Schnoller[3] in 1974 marked a significant breakthrough in the preparation of silicon crystals for the power device industry. As one who has been involved in the manufacture of float zone silicon almost since its birth in the late fifties, I regard it to be at least as significant as the development of dislocation-free float zone around 1967 and the elimination of swirls of oxygen-vacancy clusters in 1973. The former development released silicon producers from the tedious chore of counting dislocation etch pits and measuring lineage lines and effectively removed the barriers to increasing diameters. While less dramatic, the latter innovation eliminated a yield-reducing feature and simplified processes for several devices such as the Vidicon. In similar fashion, the introduction of the NTD process greatly reduced the concern of the crystal supplier with radial resistivity gradients and microscopic striations and opened the way for more sophisticated devices and better yields for the device houses.

From a silicon producer's standpoint, this development marked the first time that it was necessary to involve a second party, the reactor operator, in the production operation. To some degree, a third party, the Nuclear Regulatory Agency, also became involved. Following the excellent market review that has just been presented here we will attempt to describe how the NTD process and its resulting relationships have changed the float zone silicon industry in the United States. We will review briefly the availability of reactors in the U.S. for large scale irradiations and the criteria for selecting one, the relationship with the regulatory agencies, modifications in the process necessitated by the technique, market trends and some predictions of future development.

A brief historical review at this point may serve to put this newest major development in perspective. The group V element of choice to add to silicon as a dopant to give n-type (electron) conductivity is phosphorus. Unfortunately, phosphorus has a low segregation coefficient in silicon, i.e., it tends to concentrate in the molten phase, which leads to uneven distribution of the dopant in the crystal and thus a slice will have a higher resistivity near the edge than at the center. This so called radial resistivity gradient as well as similar variations called striations which occur on the microscopic scale effectively limits control of voltage characteristics of a device built from the slice.

As early as 1950, Lark-Horowitz[1] recognized that the first nuclear reaction shown in Figure 1 would allow precisely controlled, uniformly distributed introduction of phosphorus in silicon. In the early 1960's Tanenbaum[2] demonstrated the practical feasibility of the technique and the expected improvement in resistivity uniformity and showed that the two reactions shown are the only significant ones when silicon is bombarded with thermal neutrons. (The second reaction results in a low level of radioactivity which must be considered from a health viewpoint.) Finally, in 1974, Schnoller[3] reported a tremendous improvement in the characteristics of thyristors fabricated in this material. At that point, producers of float zone silicon began to distribute so-called NTD crystals and the material gained rapid acceptance due to its extremely uniform resistivity and the tight resistivity tolerances that could be specified.

Characteristics of the NTD product are compared with that from a conventional float zone product in Table I where Monsanto's NTD Mon-x® standard specifications are compared with those for conventional Mon-x®. The ± 10% resistivity tolerance is characteristic of a standard gas zone refining process for the undoped starting crystal. The major part of this variation is caused by non-homogeneity of the starting crystal and even better control is possible using vacuum zone refining for the undoped crystal. Sources of variability in resistivity targeting can be divided into two broad categories--those due to the silicon itself and those due to inaccuracies in the irradiation. Janus and Malmros[11] have derived the quantitative relationship between the resistivity uniformity of the starting crystal, the doping factor (ratio of starting-to-final resistivity) and the uniformity of the product. In addition to these theoretical limitations must be added the error in characterizing the resistivity of the starting crystal (even errors in measuring the type of high resistivity crystal are possible) and in evaluating the final product. Factors affecting the precision of the irradiation include fluence measurement, flux field uniformity,

Figure 1.

$$^{30}Si + N_{thermal} \rightarrow {}^{31}Si \xrightarrow[2.6\ hrs]{t_{1/2} =} {}^{31}P + \beta^-$$

$$^{31}P + N_{thermal} \rightarrow {}^{32}P \xrightarrow[14.3\ days]{t_{1/2} =} {}^{32}S + \beta^-$$

Table I. Comparison of Standard Specifications for NTD Mon-x With Those for Regular Mon-x

	NTD	Regular
Resistivity Tolerance ($<$ 100 Ohm cm)	± 10%	± 25%
Radial Resistivity Gradient (N-type)		
Center-R/2	5%	25%
Center-Edge	5%	18%
Spreading Resistance (Peak/Valley) (Typical)	7-10%	30-50%
Minority Carrier Lifetime Minimum/Typical	100/500	200/6500

Table II
Factors Influencing the Suitability of Reactor Facilities for Volume Irradiation of Silicon

Factor	Desired Level
Thermal Neutron Flux	5×10^{12} to 5×10^{13} neutrons/cm^2/sec.
Length of Uniform Flux Field	At least 30 cm length with ± 5% max. variation
Suitable facilities installed	At least 8 cm diameter with rotation capability and Accurate real time fluence measurement
Operation Schedule	At least 100 hrs/weekly. Preferably continuous.
Proximity to silicon producer	Closer the better
Prices for Services	Consistent with high-volume operation
Research Staff	Helpful but not essential

time involved in manipulation of the crystal, and self shielding by silicon itself. Since the reproducibility of measuring resistivity in silicon with a 4-point probe is generally regarded as about $\pm$ 3%, the radial gradient values of 5% maximum are low indeed. The only parameter for the NTD material which suffers in comparison to conventional material is minority carrier lifetime. The annealing step which is necessary to repair lattice damage and restore resistivity to its proper value can lead to degradation of lifetime. In general, it can be said that a lower distribution of values is obtained, but the 95% point is not much lower.

REACTOR FACILITIES

The organization, ownership and charter of nuclear reactor facilities in the United States differ significantly from those in most other countries. Three distinct classes may be differentiated, i.e., privately owned reactors operated for profit with no government subsidies and no restriction on sales of services; University research reactors whose first priority is to provide research facilities to the University but for whom sales are permitted on a uniform price schedule; and reactors owned by the U.S. Government whose charters allow them to sell research services but not in direct competition with either private or University-owned reactors. In this arrangement, Government-owned reactors could become part of a large-scale commercial operation only if no other vendors of such services were available in the country. While it is not our purpose to identify specific reactors, to my knowledge, one private reactor and at least two University research reactors have engaged in commercial irradiation of silicon and have some capacity for this service. The facilities of one University Research Reactor have been described in considerable detail[4,5]. A second private facility was developing this capability when it was closed temporarily by government order because of an alleged geologic fault in the area[6]. In addition, one government-private partnership and at least one University in Canada offer such services.

In Table II, factors which should be considered in choosing a reactor facility as a source of irradiation services for a volume silicon operation are listed. The flux level is typical for most reactors, but the practical range may be somewhat narrower than that listed. In order to have a rapid turn around of irradiation and still keep the time error in introducing and removing the sample a small part of the total, an irradiation time of a few hours is preferred. Now approximately 3 x 10^{17} neutron/cm^2 total will produce one part per billion atoms of phosphorus in silicon, therefore, at a flux of 1 x 10^{13} neutron/cm^2/second, one part per billion will take about eitht hours. Now usual float zone silicon resistivities require from 0.5 to 10ppb, so a flux much below 1 x 10^{13} would require an unacceptably long time for irradiation. The uniform field in a vertical direction is necessary to maintain axial uniformity in the crystal. Non-uniformity can be compensated for in some cases by flipping the crystal midway through the run or by use of "flux flateners" made of neutron absorbing material placed in the high flux areas. Suitable fixtures and measuring equipment are obvious requirements, but since they represent a sizable investment, their presence indicates the commitment of reactor management to the silicon business. An operating schedule of at least 100 hrs/week is considered the minimum acceptable level of operation. The other factors are "wants" rather than "musts" but neutron price will play a significant role in the use of the method.

RADIOACTIVITY AND REGULATIONS

As mentioned earlier, the secondary reaction of phosphorus with a thermal neutron can produce a detectable level of radioactivity especially at low resistivities which require long irradiations. This is of very little concern to European producers, but must be taken into account by firms in the U.S. who distribute irradiated material.

The difference is in the level of radioactivity from the active isotope, phosphorus-32 that the various regulatory agencies set as exempt from regulations. In most European countries, it is 2×10^{-3} microcuries per gram (This value is also recognized by the International Atomic Energy Commission), while in the U.S. the value is 2×10^{-4} microcuries/gram. In a paper given at the Second International Conference in 1978, we described a technique for calculating the concentration of phosphorus-32 in silicon from counts obtained with conventional beta counting equipment[7]. These data were higher than the value calculated from published data. For example, the highest resistivity calculated to give 2×10^{-4} μ Ci/gram was about 12 ohm cm, but the measured value was about 20 ohm cm.

The measured values would indicate that the 2×10^{-3} μ Ci/g level allowed in Europe would be reached only at resistivities ≤ 8 ohm cm. The net result of these comparisons are that more careful scrutiny is necessary to insure that the regulatory standards are met.

Still another complicating factor is the U.S. NRC regulation that requires anyone who distributes, i.e., sells, materials that has been irradiated must be licensed even though the concentration is below the exempt level. This requirement applies, however, only to the original distributor of the irradiated material at or below the exempt level. In particular it is our understanding that customers of the silicon producer are exempt from licensing even though the irradiated silicon is in the devices they sell. Other regulations require that a licensee maintain detailed records and take certain specific steps to guarantee personnel safety and prevent environmental contamination. Listed in Table III are procedures that must be carried out regularly to conform with the license conditions and guarantee employee safety. A simple mechanical device that is used for counting the surface of rods is shown in Figure 2. The counting head automatically traverses the length of the rod during the counting period while the rod is rotated at a uniform rate. This gives a count averaged over the entire rod surface. The ends are counted separately. The periodic inventory is necessary to insure that the exempt amount of radioactivity which is 5μCi total is not exceeded. This level is the same for the U.S. and Europe.

DEVICES FROM NTD SILICON

The growth of application of NTD float zone to the various types of devices was described very thoroughly by Dr. Janus of Topsil in an excellent paper given at the Second International Conference in 1978[8] and need not be repeated in detail. One point he made which should be emphasized, however, is that while the technique was developed for making high power devices for DC transmission lines and allowed the manufacture of large devices with 4-5 kv/breakdown voltages, it was quickly discovered that the uniform resistivity that allowed the fabrication of higher voltage devices also produced higher yields of median and low-voltage devices. It also gave a narrower distribution of device characteristics. The greater precision in targeting device parameters made the material very popular with those manufacturers. In parallel with this development, the NTD product that had been expected to be a moderately high priced, tight-specification material became a widely-used material in applications where the tight targeting specifications were not necessary. By 1978, even automotive rectifiers and ignition transistors were being made from NTD float zone silicon. It soon became apparent that the NTD product would bring hardly any premium in price and it was encumbent on the silicon producer to recoup the extra charges for neutrons by effecting other economies. At the same time, the increase in device yield reduced the demand for float zone silicon over what it might otherwise have been.

The only quality parameters where NTD float zone appeared to be inferior to standard float zone was in minority carrier lifetime since it typically gave lower values. This effect was quickly traced to the annealing step and most producers are now able to offer distributions of lifetime values that will satisfy their customers.

All of these factors combined to bring NTD float zone from the status of premium speciality material to a standard product with little or no price premium and in some cases, specifications almost as wide as the standard crystal. The customer still benefited from the resistivity uniformity on both the macro and micro scale. In some cases, even the uniformity could be sacrificed to a degree and lower doping ratios used in the interest of minimizing costs. The

Table III
Procedures Used Routinely to Comply with License Requirements, Safety and Prevent Environmental Contamination

Procedure	Frequency
Measure radioactivity on rods and compare concentration to exempt level	Every Rod
Monitor employees with film badges	Each employee. Badges counted monthly
Sample atmosphere in area of slicers and grinders	Weekly
Measure scrap before discarding	All scrap
Measure radioactivity in plant liquid effluent streams	Quarterly
Inventory total radioactivity	Monthly

Table IV
Factors Reported to Affect the Restoration of Minority Carrier Lifetime in NTD Crystals

Ratio of Fast-to-Thermal Neutrons in Reactor

Cleanlines of Crystal Surface in Annealing Step

Means of Preventing Furnace Impurities from Reading Crystal During Annealing

Heating and Cooling Rates in Annealing Step

Annealing Temperature and Time

Figure 2. Mechanical Device for Scanning the Surface of Irradiated Rods for Quantitative Measure of Radioactivity

effect on silicon producers was the requirement that large volumes be produced at essentially no increase in costs over the standard product.

Initially, it was expected that NTD material might prove beneficial for some integrated circuits as well as for discrete devices. For example, a fraction of MOS IC's require relatively tight n-type resistivity tolerances and the low segregration coefficient of phosphorus causes Czochralski grown n-type crystal to have a wide variation in resistivity from one end to another. The NTD process is of course ideal for obtaining a narrow n-type resistivity band. Factors hampering the use of NTD Czochralski for this application are the high neutron costs at the low (5-10 ohm cm) resistivities required and, in the U.S. at least, the long lag time to allow radioactivity to decay to the desired level. Also, a much higher annealing temperature[9,10] is required to restore resistivity and this tends to result in more structural defects.

EFFECT OF NTD ON FLOAT ZONE PROCESSES

The supplier of conventional float zone silicon to the power device industry has traditionally been involved in a job-order business. Small, but significant, differences in specifications for diameter, resistivity range, and crystallographic location marks from customer to customer and sometimes within a given customer, have forced the vendor to treat each order as an individual. This situation, combined with the fact that specifications are constantly changing, makes for maximum lead time as inventorying of finished product is not possible to a significant degree. The situation is complicated further by the relatively poor precision of conventional doping processes which means that tighter-than-normal tolerances required a price premium to compensate for yield losses against the tight specification.

In the paper that I have already referred to from the Second International Conference[8], Dr. Janus described how the introduction of the NTD process had altered the flow of material significantly. I would like to expand on those differences today especially as they relate to the expanding NTD business and to labor practices in the U.S. The stepwise production sequence in the conventional float zone line is shown in Figure 3 compared to that for a standard NTD rod-doping process and a possible slice-doping process which we will discuss in another paper. In the conventional process, we have assumed that doping is done in the crystal-growth rather than in poly-deposition step since this is the most favorable situation from an inventory viewpoint, i.e., poly or slim rod doping complicates the conventional process even further. In each process the horizontal arrow indicates what we refer to as the scheduling step. At this point, a conscious decision is made to produce material against a given specification. All pools labelled inventory prior to this are true inventories where raw material is stockpiled for subsequent steps. All pools after this point are pseudo inventories in that they consist of mainly fall-out material that failed to meet the intended specification. Since inventories are primarily tools for levelling work loads and providing faster response to customer needs, it is apparent that these situations are improved markedly in going from the conventional process to the NTD rod technique and further improved somewhat in going to slice doping.

In the conventional process the only true inventory is polycrystal. Although final diameters vary from customer to customer, there are families of diameters so that inventorying poly is straightforward. In order to maximize the precision of any doping process, it is necessary to determine the accidental or "base level" of impurities in the raw material going into the process. Measurement of resistivity in polycrystal is not precise but some information can be obtained. An alternative method is to maintain poly rods in decomposer run batches and feed back information from the doping results on the first run in the batch. Either method is inferior to the precise information that can be obtained on a crystal before committing it to the NTD process. After the crystal growth step characterization steps are precise and the material moves forward as it passes each specification. The inventories are for material that fails one or more specifications and is held in the hope that it will fit some incoming order. The situation when deliberate inventory is built only at an early step in the process means that the level of activity in the rest of the manufacturing line must fluctuate approximately to the same degree as the level of business. Although there are some exceptions, the general rule is "no orders, no work".

The nature of the float zone process, with its high degree of operator skill, makes it particularly vulnerable to the business fluctuations which are not unusual in the entire electronic field. In the United States, unlike some other countries, it is customary to furlough or lay off workers when business declines. Hopefully, in a large operation this involves only a move to another part of the business. Nevertheless, the effect is the same and the float zone operator, who on the average requires at least a year of on-the-job training to become proficient, is lost to the business. Subsequent steps in the float zone process are common to other parts of the silicon business which are generally more stable, so only the float zone operator is affected.

In the NTD processes, the scheduling operation effectively occurs after the crystal growing step and it should be possible to stablilize the zone refining labor force to grow undoped crystal. As in the case of poly, standard diameters can be grown even though final ground diameter may vary from customer to customer. Since dislocation-free crystals are generally not round and in the <111>

Figure 3. Sequential Steps in Conventional Float Zone Process Compared to NTD Rod-Doping and Slice-Doping Processes.

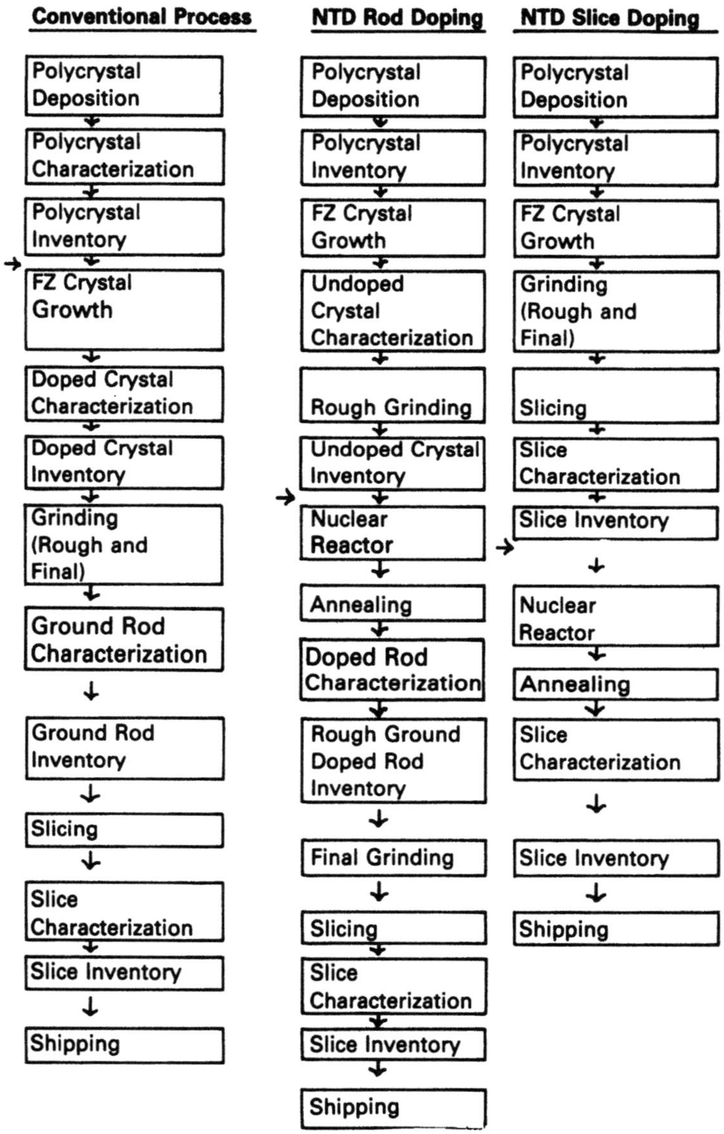

orientation are characterized by large lobes, it is advantageous to rough grind before inventorying. As indicated earlier, resistivity can be measured precisely on the single crystal so that an accurate "base level" value can be entered in the NTD doping equation, adding to the precision of the method.

Moving the inventory downstream in the process reduces lead time which serves to compensate for the trip to the reactor. This trip can be reduced to as little as a week if necessary, but 2-3 weeks is customary. As indicated earlier, the more stringent radioactivity control requirement does not come into play above about 20 ohm cm, but it is as long as 70 days at 5 ohm cm. Overall lead times for the NTD process are comparable to those for conventional float zone. Under certain conditions, the slice doping method could reduce lead time significantly.

In the proposed slice doping process, the inventory point is pushed all the way past the slicing step and this could be the most economical of all. It must be pointed out though, that three important changes must be made in the vendor-customer relationship for this to become possible, i.e.;

Final diameters must be standardized,
Slice thicknesses must be standardized at least to a degree,
And a method of estimating minority carrier lifetime in thin slices must be agreed on.

At this time, it does not appear too likely that any one of these changes are apt to be made although it would appear that there are sufficient economic justifications.

As indicated earlier, if the NTD float zone is to be sold at prices comparable to conventional material for similar specifications, the silicon producer must be able to realize enough economies in the operating efficiency of the float zone operation and increased yields due to the superior precision of the method to recoup the costs of neutrons. (The slice doping process in addition to its efficiencies, seeks to save part of the neutron costs by reducing the volume of reactor occupied.) Somewhat paradoxically, this may be more difficult to achieve for wide specifications than for the tighter ones. The main reason for this is that the poor precision of the conventional process has already established a price differential for tight specifications, which allows some room to recover neutron costs by increased yields. No such room exists for wide-specification material and any hope for recovering those neutron costs must be found in standardization of the line. To be sure, certain losses are entailed in selecting crystal suitable for irradiation to tight specifications for resistivity level and uniformity. For example, a doping factor as defined by Janus and Malmros[11] of at least 10:1 is required to obtain the optimum radial gradients and striations although significant improvement over the conventional product can be made at even lower doping factor levels. At this doping factor and above, it is usually necessary to irradiate crystals that are all of one type so it may be necessary to discard some portion of crystals for the tight specifications. These considerations are the subject of a paper to be given later in the conference by my colleague, Mr. Kramer.

The key to maximum efficiency in operation of the NTD process is clearly in applying it across the board (for everything except p-type, of course) and this is the direction the market is moving.

No discussion on NTD processing would be complete without some attention to the subject of minority carrier lifetime. This aspect of the process has involved more research and has been the subject of more papers than perhaps any other, but the individual process details are one of the best kept manufacturing secrets. The trick is, of course, to carry out the post-irradiation annealing step in such a way that the lifetime of the crystal is restored somewhere near its original level. As a result of the research mentioned, this is generally no problem on a statistical basis, but an occasional crystal with an apparent low lifetime value is lost and detracts from the economics of the process. I believe that there is a need in this area for more work on lifetime measurement methods and more communication on the meaning of the lifetime obtained by different measurement techniques. An American Society of Testing Materials Conference on this subject was held in San Diego, California in early 1979 and I believe such discussions should continue to be held.

Meanwhile, if both producers and users are to enjoy optimum benefits of the NTD process, lifetime specifications and measurement methods should be scrutinized for their applicability to the actual device needs.

Some factors that have been suspected or actually proven to have an effect on lifetime are listed in Table IV.

Although at first thought to be important, the fast/thermal ratio or so-called "Cadmium Ratio" has been proven not to be an important factor in lifetime[9,12]. The other factors are all considered important and their control is probably part of all processes being practiced today.

TRENDS IN NTD UTILIZATION

With the widespread use of NTD float zone silicon in the U.S. approaching its fifth birthday, the material has been almost universally accepted in the power device field. Even device houses that could find no particular advantage for the material in their lines have carried out the laborious tests needed to approve it for purchase. We see no reason why it will not be the material of choice in all power device houses in the U.S. within the very near future. In addition, as the sophistication and packing densities increase in integrated circuits, I expect that some demand will gradually be generated for NTD Czochralski Crystal.

REFERENCES

1. K. Lark-Horowitz "Nuclear-bombarded Semiconductors" in Semiconductor Materials, Conf. University Reading, London, Butterworth's (1951) page 47.
2. M. Tanenbaum and A.D. Mills, J. Electrochemical Society 108,171 (1961).
3. M. S. Schnoller, IEEE Trans Electron Devices ED21, 313 (1974).
4. S. L. Gunn, J. M. Meese and D. M. Alger, "High Precision Irradiation Techniques for NTD Silicon at the University of Missouri Research Reactor", J. M. Meese, Editor, Plenum Press, New York, 1979. page 197.
5. R. Berlinger and S. Wood "A Computer-Controlled Irradiation System for the University of Missouri Research Reactor", Ibid. page 215.
6. J. E. Morrissey, T. Tillingbast, A. P. Ferro and B. J. Berliga, "General Electric Test Reactor NTD Silicon Development Program", Ibid, page 171.
7. B. D. Stone, D. B. Hines, S. L. Gunn and D. McKown "Detection and Identification of Potential Impurities Activated by Neutron Irradiation of Czochralski Silicon", Ibid, page 11.
8. H. M. Janus "Application of NTD Silicon for Power Devices", Ibid, page 37.
9. J. W. Cleland, P. H. Fleming, R. D. Westbrook, R. F. Wood and R. T. Young, "Electrical Properties of Neutron Transmutation Doped Silicon", Ibid page 269.
10. A. M. Liaw and C. J. Varker Semiconductor Silicon/1977 PV 77-2 Electrochemical Society, page 116 (1977).
11. A. M. Janus and O. Malmros, IEEE Transactions of Electron Devices, ED-23, No. 8, 7976 (1976).
12. O. Malmros "The Minority Carrier Lifetime of Neutron Doped Silicon", J. M. Meese, Editor, loc. cit., page 249.

IMPACT OF DEFECTS FORMED IN NEUTRON TRANSMUTATION DOPING OF SILICON ON DEVICE PERFORMANCE

John C. Corelli, James W. Corbett

Department of Nuclear Engineering
Rensselaer Polytechnic Institute
Troy, New York 12181, USA

Department of Physics
State University of New York at Albany
Albany, NY 12222, USA

ABSTRACT

Defects introduced into silicon material as a result of neutron transmutation doping (NTD) cover the entire range of level of complexity in which the defect inventory can include relatively simple point defects to large disordered regions of $\lesssim$500Å size up to <110> rod-like defects 10μm long 200Å diameter. The wide spectrum of defect specie occurs since the dominant defect producer, the fast neutrons, can impart on the average $\sim$50keV of recoil energy to a silicon atom while recoiling silicon atoms subsequent to neutron capture and gamma radiation emission can have energies in the $\sim$400 to 900eV range. In this paper a review will be given on properties of defects introduced in silicon specific to NTD. The approach will be to describe the origin of defects and their effects on extrinsic photoconductivity, optical properties, transmission electron microscopy and electron spin resonance, including defect symmetry properties and the response and alteration of defects with heat treatment. A discussion is given on the nature of unannealable defect microstructure in NTD Si. An analysis of many experiments suggest that it may be advantageous to perform irradiation in NTD Si at temperatures of 500-600°C. Finally, an attempt will be made to show how the defects can effect the operation of devices, eg: high power thyristors, transistors, infrared detecting devices and very large scale integrated circuit systems.

I) INTRODUCTION

The use of reactor irradiation to produce dopant impurities in silicon specifically tailored in terms of concentration and chemical specie type is now a widely used commercial process for the production of devices, eg: high power thyristors, transistors and infrared detecting devices. It would be ideal if one had a purely thermal neutron beam with neutron energies all less than ∿10eV so that then, one would have no defects introduced by fast neutrons (E>0.1MeV) which give rise to high energy recoil silicon atoms. However, since one does not have purely thermal neutron beams to produce the neutron transmutation doping via the $^{30}Si(n,\gamma)\ ^{31}Si \rightarrow ^{31}P + \beta^-$ reaction one has to contend with the defects that are introduced in a mixed spectrum of thermal and fast neutrons where by fast neutrons we include all neutrons of energy greater than say 0.1MeV. Irradiation by typical reactor neutrons introduces defects caused by recoiling Si atoms of ∿150keV down to energies in the 400 to 900eV range.The latter recoil atoms occur subsequent to gamma ray emission by ^{31}Si nuclei. Thus, it is possible to think of the defects introduced in NTD Si as being akin to what would occur if Si were to be irradiated simultaneously with a high fluence of neutrons of E>0.1MeV and a fluence of 10^3-10^4 times higher of 2-3MeV electrons, since it is the latter that can introduce recoil Si atoms of ∿400 to 900eV, and would simulate ^{31}Si recoiling nuclei subsequent to gamma-ray emission. Hence the level of complexity of defect inventory can include a wide spectrum from relatively simple point defects all the way up to large disordered regions of $\lesssim$500Å size and dislocations.

In this paper a discussion of the properties and characteristics of neutron induced defects will be given with emphasis on the studies made using optical, photoconductivity and EPR response of irradiated silicon in addition to microstructure of irradiated and annealed Si. Electrical property studies of NTD silicon have been given in detail elsewhere[1-5] and will not be repeated here except as those recent results aid in understanding the optical and spin resonance measurements. We shall discuss a new approach to study the characteristics of NTD defects in silicon which makes use of electronic transitions of NTD defects which are infrared active.

II) RADIATION-INDUCED EXTRINSIC PHOTOCONDUCTIVITY

The study of radiation-induced <u>extrinsic</u> photoconductivity in NTD Si (before heat treatment) to examine the photoconductive response of radiation-induced defects is difficult due in part to the problem encountered to ensure ohmic contacts, and as will be discussed below the large uncertainty in defining energy level positions. The early work on photoconductivity in neutron irradiated Si has been reported by Fan and Ramdas[6]. Recently Thomas et al[7] and Hobgood et al[8] have reported on photoconductive response of NTD silicon after complete removal of radiation-induced defects by annealing the material to 900°C for 4 hours. The work of Thomas et al[7] and Hobgood et al[8] was directed to producing an infrared detector by introducing controlled amounts of dopant (In) by NTD, and they did not study extrinsic radiation-induced photoconductivity.

Extrinsic photoconductivity studies on electron irradiated Si have been more successful in identifying defects with energy levels[9-11]. In fact Cheng[12], Kalma and Corelli[13], and Young and Corelli[14], were able to obtain microscopic information on simple point defect structures employing uniaxial stress and polarized light to study the photoconductivity in electron-irradiated Si.

The difficulty in determining extrinsic photoconductivity in silicon irradiated by neutrons or other particles capable of imparting high recoil energy to Si atoms (eg: 50MeV electron can impart 196keV to a Si atom) is that the energy levels are broad and effectively appear to be "smeared out" in the forbidden gap. However, low energy (1-2MeV) electron irradiation of silicon gives rise to rather more sharply defined energy levels which yield photoconductivity. The extrinsic photoconductivity induced in n-type Si (before heat treatment) by neutrons[6], 50MeV[13] and 1.5MeV electrons[13,14] are shown in Figure 1. Note that both the neutrons and 50MeV electrons yield energy spectra of what appear to be broad energy levels with no structure whereas the 1.5MeV electron spectrum exhibits distinct well defined energy levels and structure in the photoconductivity spectrum. Thus, both neutrons and 50MeV electrons induce extrinsic photoconductivity in Si dominated by "energy bands", and one cannot ascribe single well defined energy levels to states induced in Si by radiation capable of producing relatively high recoil atoms (∿50-200keV).

Additional experimental evidence on the smeared-out radiation-induced extrinsic photoconductivity of n-and p-type Si after neutron irradiation have been reported by Konopleva et al[15,16]. Vorobeichik et al[17] have reported non-resolved extrinsic photoconductivity levels in 680MeV proton irradiation of both n-and p-type Si. These experimental results imply that the defects induced by more energetic ions give rise to broadened energy states, while the lower energy recoils give rise to well defined energy states of relatively narrow width. In all cases it is found that annealing to 500°C for 1 hr. completely eliminates the radiation-induced extrinsic photoconductivity in Si. We shall return to the idea of defect level width below, but for the moment we can state that Si irradiated by neutrons yields broad non-defined energy states in the forbidden gap making analysis difficult and ambiguous. The interested reader can find additional well defined radiation-induced energy levels which are photoconductive in the papers for which the radiation source was ∿1.5-2MeV electrons[9,14]

III) NEUTRON-INDUCED DEFECTS WHICH ARE INFRARED ACTIVE

1. Impurity and Temperature Effects on Defect Absorption Bands.

Radiation-induced defects in Si can give rise to absorption bands are are rendered infrared active in one of two ways. First, the pairing of an impurity atom with vacancies and/or interstitials

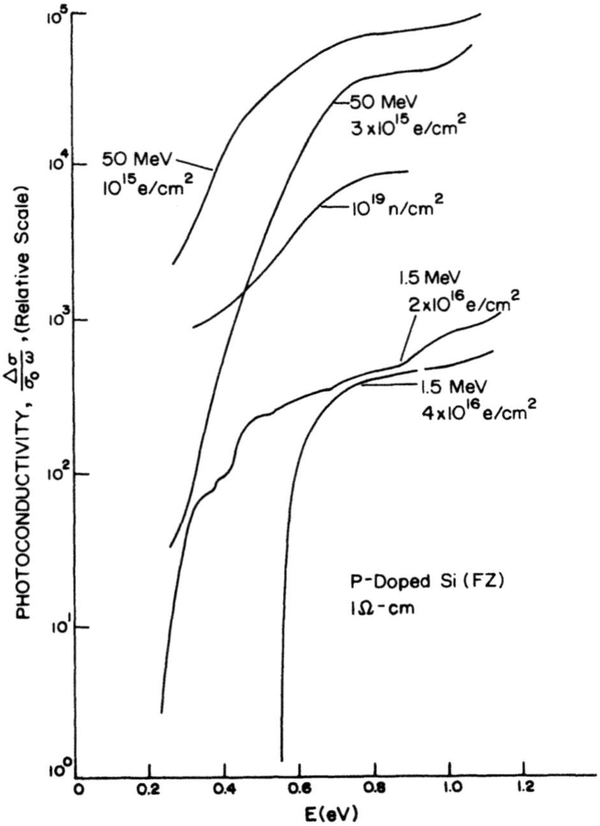

Fig. 1 Radiation-induced extrinsic photoconductivity (78°K) in n-type Si irradiated by 1.5 and 50MeV electrons and reactor neutrons.

or a high degree of lattice disorder, in these cases one observes the predominantly vibrational modes of the defect in the infrared wavelength region. Second, defects that give rise to transitions which are purely electronic in nature and whose energies are in the infrared region of the spectrum. There have been at least 114 different defect infrared absorption bands observed in irradiated silicon which have been documented in the literature during the past twenty years. In Figure 2 are shown the location of all infrared bands known to exist in silicon irradiated by neutrons (n), hydrogen ions (H), and electrons of E$\gtrsim$2MeV. The horizontal lines in the various columns of Figure 2 indicate the location of well known absorption bands in terms of microns (μm) or wavenumber(cm^{-1}). The heading at the top of column 1 labelled V_2 are divacancy-associated bands, and n, n(HOB), and H(HOB) indicate neutron (n), or hydrogen ion (H) as incident particles. The last six columns indicate that the defect which gives rise to the band contains at least one of the impurity atoms (ie: H,P,As,etc.) listed. The results tabulated in Figure 2 have come from various papers in which the most recent are

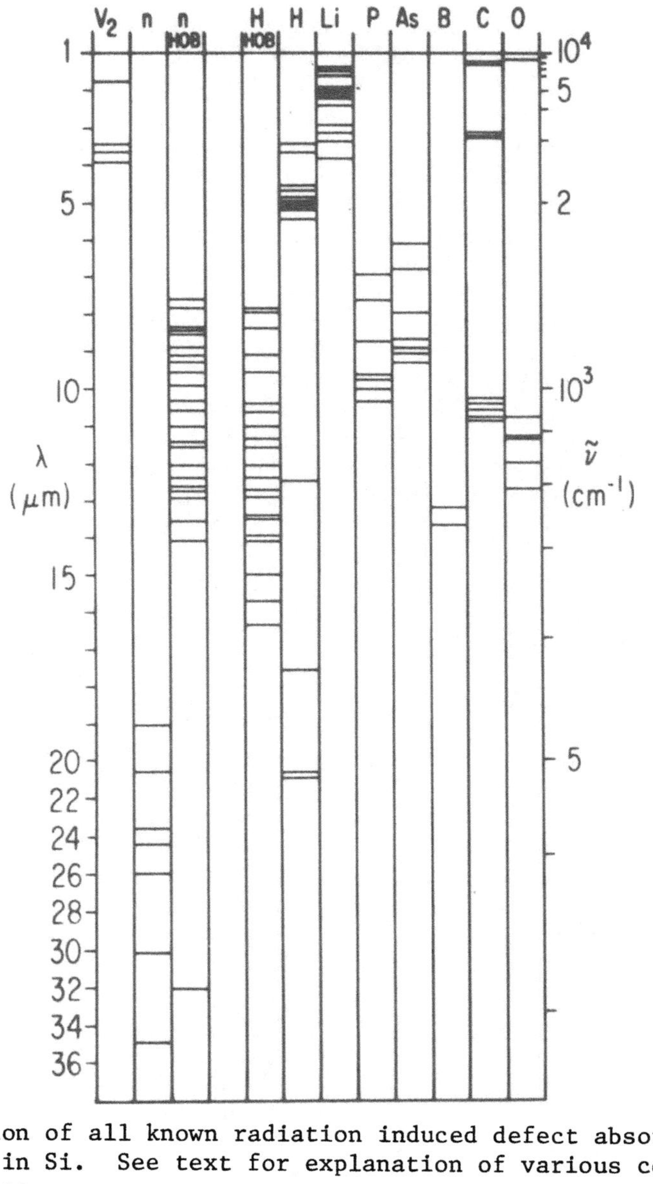

Fig. 2 Location of all known radiation induced defect absorption bands in Si. See text for explanation of various column headings.

by Newman and Totterdell[18], Stein[19], and Corelli et al[20]. References to earlier works are cited in the papers of references[18-20].

We shall focus our attention on the bands shown in Figure 2 and labelled by HOB which is an abreviation for higher order bands. This series of 26 bands arises from defects introduced by reactor neutron irradiation ranging in fluence from $\sim 5X10^{16}n/cm^2$ to $10^{19}n/cm^2$ (E>1MeV). A few of the more intense bands can also be observed in 40-50MeV electron irradiation of Si to relatively high fluences $\gtrsim 10^{18}$e/cm^2. The HOB can only be observed in neutron irradiated Si after heating for ~15 minutes at ~500°C. In order to observe the entire spectrum of bands the sample must be cooled to T<100°K and requires the full energy spectrum of light to be incident on the sample. In Figure 3 we give results on an n-type phosphorus-doped Si sample with initial resistivity of 100 Ω-cm after irradiation to 6X10^{17}n/cm^2 (E>0.1MeV) and 15 minutes anneal at 540°C. The linewidth becomes narrow as the temperature decreases from ~100°K down to ~40°K with no further decrease in linewidth resulting as the temperature is further decreased to 8°K as given in Figure 3. The HOB are independent of initial oxygen concentration[20,21] but are strongly dependent on the phosphorus or boron impurity concentration[20,22] since they are not observed in Si which is chemically doped before irradiation to give a resistivity $\rho \lesssim 1\Omega$-cm (n-type) or 0.1 Ω-cm (p-type). In addition, the bands are not observed if transmutations through the ^{30}Si (n,γ) ^{31}P reaction cause the phosphorus dopant concentration to reach levels of at least 10^{16}cm^{-3}($\rho \cong$ 0.6Ω-cm). All HOB bands disappear upon heat treatment at 600°C for ~20 minutes. Thus, there is clear experimental evidence[20,21] that phosphorus atoms in concentrations $\gtrsim 10^{16}$cm^{-3} must compete with thermally liberated divacancies (anneal at 300°C), vacancies (from vacancy-oxygen complexes which anneal ~350°C), and other defect complexes (eg: - multi-vacancies/interstitials) which give rise to lattice disorder (488 cm^{-1} TO vibrational mode) and exhibit recovery at 300°C. This competition by P atoms for vacancies and/or interstitials apparently prevents the impurity independent defect complexes which give rise to the HOB from forming. Another interesting aspect of the HOB has been reported by Mordkovich et al[23] who find that after neutron irradiation a 10MeV electron fluence of 4X10^{18}e/cm^2 strongly reduces and eliminates the HOB. Newman and Totterdell[18] have attributed the decrease of HOB following electron irradiation to a change of charge state of the defects which appears to be consistent with our charge state model to be given below.

2. Symmetry Properties of Defects Responsible for HOB

In order to study the symmetry of the defects which give rise to the HOB we have performed detailed measurements[20] of the response at 78°K of the bands to externally applied uniaxial compressive stress using polarized light. Typical examples for the response of HOB in the 1000 to 1400 cm^{-1} wavenumber range for three different crystallographic directions under the action of stress are given in Figures 4, 5 and 6. It is clear from the results in Figures 4-6 that the splitting and shifting of the 1048 and 1078 cm^{-1} bands appear to be the same implying similar defect symmetries, whereas the 1101, 1123, 1171 and 1317 cm^{-1} bands have different symmetries. We can summarize all the stress results on the HOB by stating that the bands can be characterized into 8 types each exhibiting a different response to stress and hence a different symmetry, yet all annealing completely after heating to $\gtrsim$600°C.

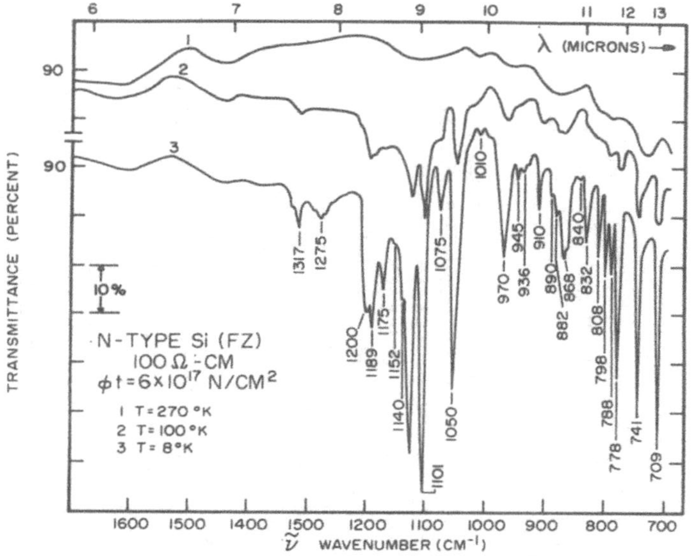

Fig. 3. Infrared spectrum at various temperatures for neutron irradiated silicon after 15 minute anneal at 540°C.

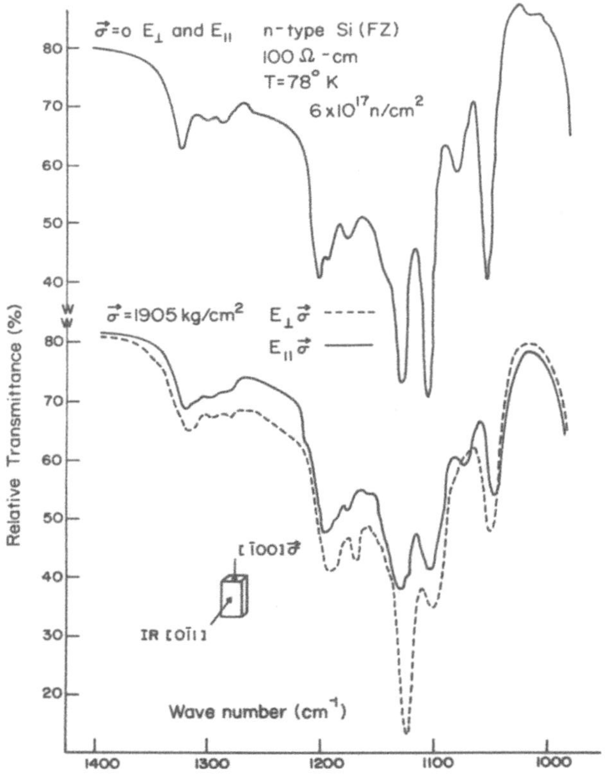

Fig. 4. Response of infrared spectrum (1400-1000 cm^{-1}) with [1̄00] stress and [011] direction of incident polarized light for silicon irradiated to 6 X $10^{17} n/cm^2$,(after 15 minutes anneal at 540°C) measured at 78°K.

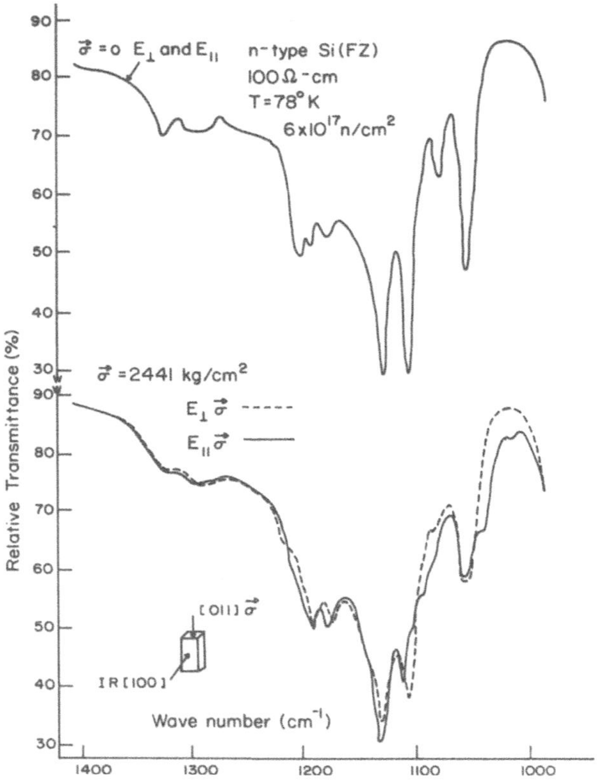

Fig. 5 Response of infrared spectrum (1400-1000 cm^{-1}) with [011] stress and [100] direction of incident polarized light for silicon irradiated to $6X10^{17} n/cm^2$,(after 15 minutes anneal at 540°C) measured at 78°K.

We can obtain some information about the nature of the defects by correlating our IR results with EPR studies in neutron-irradiated ($10^{18} n/cm^2$) intrinsic (FZ) silicon by Lee et al[24], who find a defect consisting of more than six vacancies closely matches our annealing and impurity property of HOB. A defect of six vacancies is a reasonable defect to be expected in heavily damaged silicon as is the case for our studies. Two other defects in neutron-irradiated silicon and studied using EPR correlate with defects responsible for the HOB. Lee et al[24,25], Daly[26] and Brower[27] have reported EPR studies of defects which closely correlate with the 1171 cm^{-1} band and several other HOB. Further probing of the HOB is given below in which a two light beam method is used to extract information about the defects giving rise to the HOB.

3. Excitation of Electrons to Induce Transitions in Defect Absorption Bands - Dual Beam Method

The divacancy associated defect absorption bands[28] at 3.62μm (2762 cm^{-1}) and 3.46 μm (2890 cm^{-1}) and all the HOB discussed above are electronic in nature and are associated with transitions which require that an electron be present in the ground state of the defect center. Thus, the ground state is populated by using all wavelengths of the light source and analyzing the absorption which occurred at a specific wavelength. Recently, a new approach to a study of such bands was used by Mitchell et al[29] and Carton-Merlet et al[30] in which two independent light-beams are incident simultaneously on the sample. One beam is tuned to the wavelength of the defect absorption band under study (eg: the 3.62 μm band, or one of the HOB) called the probe light while the wavelength of the second beam (excitation light) is varied to get electrons excited to the proper ground state energy which initiates the transition. The new method offers the ability to find the energy necessary to populate a defect center and thus observe the absorption band, the time-

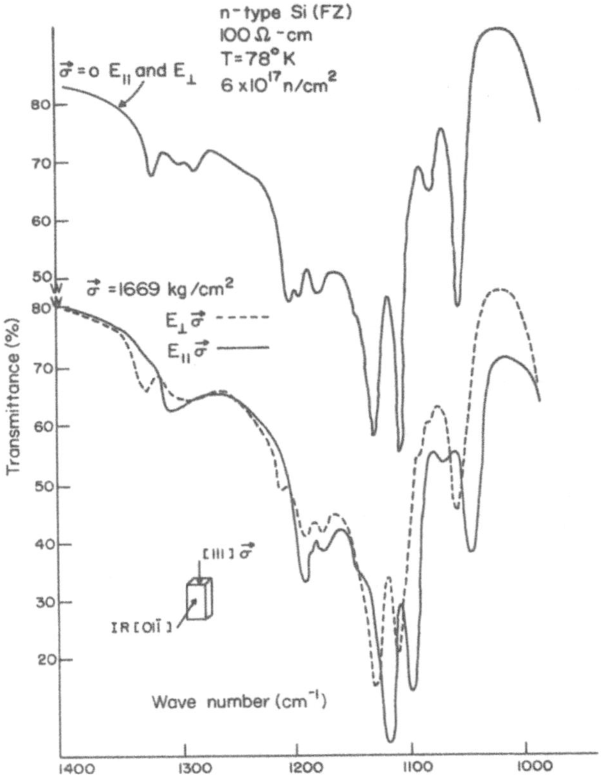

Fig. 6 Response of infrared spectrum (1400-1000 cm^{-1}) with [111] stress and [01$\bar{1}$] direction of incident polarized light for silicon irradiated to 6X10^{17}n/cm^2,(after 15 minutes anneal at 540°C) measured at 540°C.

energy response of the band and the temperature dependence of the photoexcitation cross section of the defect center under investigation. Thus, the method allows one to get new information on the specific defects responsible for the electronic defect absorption bands.

In Figure 7 we show an example of the time response of two HOB (1048 and 1101 cm^{-1}) and the divacancy associated electronic defect absorption band at 3.62 μm (2762 cm^{-1}) for crucible grown 10Ω-cm P-doped Si irradiated to $5X10^{18}n/cm^2$ (E>1MeV). Note that the time to completely populate the state with electrons (decreased transmittance) by the excitation light (E=1.17eV) for the two HOB is different 140 and 210 seconds for the 1048 and 1101 cm^{-1} bands respectively. For the case of the divacancy band at 3.62 μm (2762 cm^{-1}) there is a two-fold process for populating the state, a fast process ∿4 sec followed by a long time of 720 seconds. When the excitation light is turned off the defects responsible for the HOB de-populate

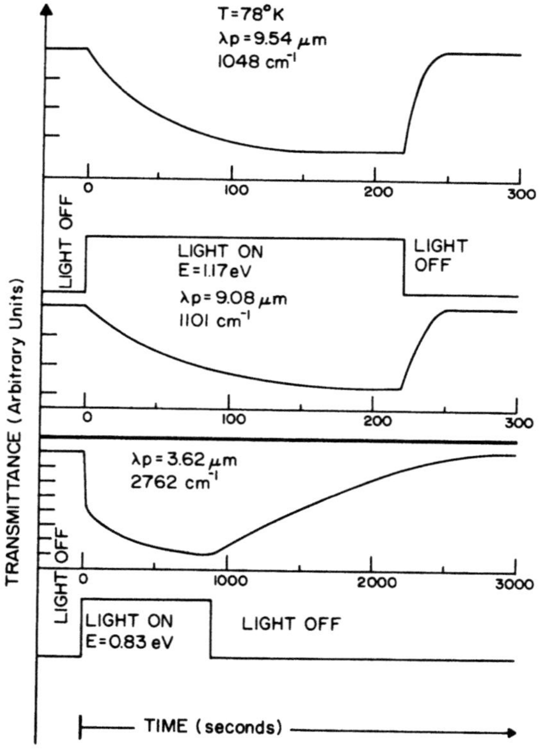

Fig. 7 Photoexcitation and de-excitation time dependence for HOB at 1048 and 1101 cm^{-1} and the divacancy associated band at 2762 cm^{-1} (Excitation energies shown for each band).

in ∿30 seconds, while the de-population time for the 3.62 μm band is ∿1800 seconds. After each excitation of the HOB the absorption response was observed to become faster[29] until both absorption (excitation) and de-population times were identical. Initial signal level was never achieved after the sample was exposed to excitation radiation. After subsequent excitations, the same equilibrium signal level was reached in the light-off condition. Times for absorption and decay for both bands for initial excitation and subsequent excitations are listed in Table I. Note that although both bands (9.54 and 9.08 μm) achieve maximum absorption with the same excitation energy (1.18eV) the times for absorption and decay are different. Subsequent excitation with band gap energy was found to produce the initial response shown in Figure 7 only after waiting a long time or after first exciting the sample with a lower energy. The most effective energy (called the depletion energy) was found to be 0.74eV.

In order to determine the energy dependence of the photoexcitation for these electronic bands the probe light is kept at the absorption band and the energy of excitation light is varied from low energy 0.1eV up to ∿1.8eV. A typical example of the results obtained for two HOB is given in Figure 8, in which maximum absorption occurs for excitation energy of ∿1.17eV which is band gap light.

Using the method described above we can give a model for the response of the defects giving rise to two of the HOB (1101 and 1048 cm^{-1}), and this model is shown in Figure 9.

With initial excitation, electrons are raised into the conduction band from the valence band and then fall into a defect level of charge state T_1. Upon capture of an electron T_1 changes charge state to T_1*. This charge state also has an affinity for capturing electrons from the conduction band and when this occurs T_1* changes to a new charge state T_1**. Once this has been accomplished the probe light (λ_p) excites an electron from T_1** to a new defect level or an excited state of T_1**-T_2. The process, therefore, initially is $e \rightarrow T_1 \rightarrow T_1^*$, $T_1^* + e \rightarrow T_1^{**}$, $T_1^{**} + \lambda_p \rightarrow T_2$ (Fig. 9a). When the excitation light is turned off no more electrons are promoted into T_1 and those that were have changed T_1 to T_1*, also no more electrons are falling into T_1* so that no more T_1** states are being produced. Those T_1** states that were produced are almost totally eliminated by the continuing probe light so what remains are T_1* and the defect level charge state T_1 which exist as quasi stable charge states hindered from recombining due to an energy barrier. When the excitation energy is turned on again (subsequent excitation) the response is faster because T_1* now exists where before it did not; so one has $T_1^* + e \rightarrow T_1^{**}$, $T_1^{**} + \lambda_p \rightarrow T_2$ in addition to the initial response scheme.

The response to the depletion energy is depicted in Figure 9b. The state T_1* is depleted as the depletion energy excites electrons from T_1* to the conduction band. As this occurs, several competing processes take place. First some of the electrons initially removed from T_1* to the conduction band fall back to T_1* and are captured by some of the T_1* charge states not yet depleted. This causes T_1** to be produced giving an initial fast absorption (ie: faster than

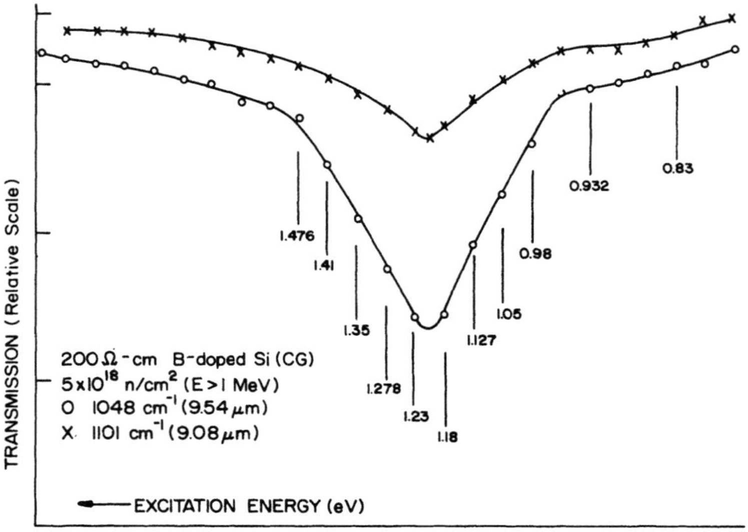

Fig. 8 Dependence of photoexcitation cross section on excitation energy for the 1048 and 1101 cm^{-1} HOB at 78°K.

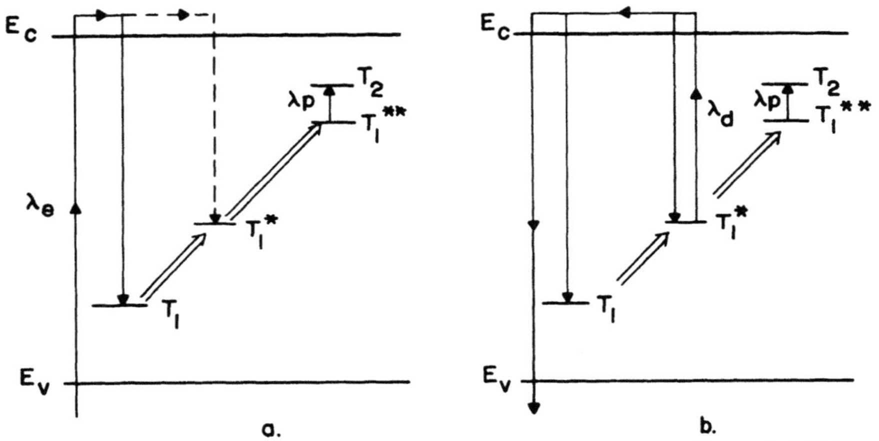

Fig. 9 Model of photoionization process for the 1048 and 1101 cm^{-1} HOB: (a) applies during the excitation process with band gap light; (b) applies to depletion process with light of energy E_e = 0.74eV.

shown in Figure 7 for the HOB) as the probe light makes the transition from T_1** to T_2. Electrons also fall into T_1 and initiate the absorption scheme previously seen for initial excitation. The depletion energy, however, will gradually deplete almost if not all the T_1* charge states eventually as electrons are gradually lost to the valance band. Subsequent excitation with band gap light will then produce the original excitation scheme and slow response as described earlier (see Fig. 7 for HOB) since the number of T_1* charge states will be very low thus minimizing the quick response of capture of an electron by T_1* immediately and forming T_1**. The fact that an exact initial response is not reproduced indicates that some (although relatively small amount) T_1* may exist, further indicating some level of trapping at T_1*. There may be some trapping at T_1** as well. A possible explanation as to why the initial equilibrium level is not reached is that some electrons may be lost to other defects.

All three charge states are necessary to explain the absorption, decay and depletion energy responses. If there was just one charge state T_1 with a defect level T_2, then each subsequent response to excitation would be the same and would be relatively fast since there would be no competition and other charge states would not be produced. This, however, is not the case. The initial excitation is slow and subsequent excitations are faster. If there were just two charge states T_1 and T_1* with a defect level T_2, the response again would be expected to be the same since any T_1* would be depleted by the probe light to T_2 and any electron captured by T_1 would cause the charge state to change to T_1*. The depletion T_1 would not be observed since it would not exist after the light is off. The combination of observed responses to initial excitation, subsequent excitation and depletion energy, therefore dictate the necessity for three charge states and the above model. This model also explains the findings of Koval et al (1973)[21] for sensitivity of charge state of the defects involved as proposed by Newman and Totterdell[18]. Finally, we have yet to investigate the possibility of valence band to T_1* (Fig. 9) transitions which are also explained by the model.

In the above discussion of the two beam method for studying the HOB it is important to recall that before the 500°C heat treatment needed to detect the HOB the band gap in neutron-irradiated Si exhibits a striking decrease (the so-called near-edge absorption) first reported by Fan and Ramdas[6] and subsequently by Cheng et al[28] and later by other workers. The near-edge absorption anneals out completely after heat treatment to 300°C, thus restoring the initial forbidden gap of energy 1.19eV. It is interesting to note that Mitchell and Corelli[31] have used the two beam method to study the divacancy band at 3.62 μm (3762 cm^{-1}) and were able to observe not only indirect band gap transitions but also reported on direct band gap transitions and a decrease of the direct band gap from 2.5eV down to 0.81eV after neutron irradiation of Si to a fluence of $5X10^{18}n/cm^2$ (E>1MeV). Thus the large disorder in the lattice crystal structure of neutron-irradiated Si causes a closure in both the direct and indirect gap, and moreover the direct gap narrows sufficiently (0.81eV) that the cross section for a direct gap transition is larger than for the indirect transition in which the latter necessitates phonon assistance. The details of the model incorporating direct gap transitions can be found elsewhere[31].

IV. Microstructure and EPR Study of Complex Radiation-Induced Defects

It is clear from the foregoing discussion of radiation-induced extrinsic photoconductivity and defect absorption bands that silicon atom recoils in neutron bombardment give rise to defects which depend on the crystal orientation, the Fermi level, temperature of post-irradiation annealing, impurity concentration, energy of incoming particle (neutron, electron, ion, etc.), degree of ionization and the specific change state of the defect. The results on the HOB clearly indicate that they must arise from defects formed after more simple defects (eg: divacancies, interstitials vacancy-impurity pairs, etc.) are redistributed by heat treatment to temperatures above 300°C. In addition to divacancy recovery which occurs after heat treatment to $\sim$300°C there are defects associated with interstitials and detected by EPR[32-35] which anneal below $\sim$200°C. Thus we see that the basic simple defects are precursors of the more complex defects which must be responsible for the HOB in neutron-and ion-irradiated Si. Stein[36] has given an excellent overall picture of the thermal annealing behavior of various neutron-induced defects in Si studied using a variety of probes including optical, EPR, X-ray transmission, transmission electron microscopy and deep level transient spectroscopy. In all cases it is found that the isochronal annealing spectrum shows the disappearance of simpler defect structures with increasing temperature followed by the evolution of new "defects" which ultimately also disappear upon heat treatment to higher temperatures. Finally, after heating to temperatures in the 800-900°C range the removal of neutron induced defects is complete as far as structure sensitive property changes are concerned, ie: apart from (1) the non-annealable neutron transmutation dopant concentrations which yield the final resistivity of the material, and (2) dislocations.

A synthesis of the way in which simple defects can aggregate to form microstructural imperfections observed by transmission electron microscopy (TEM) has been proposed recently by Tan[37]. In his approach Tan makes use of a large body of results from transmission electron microscopy studies on neutron-and ion-bombarded silicon and argues that point defects are the precursors which condense and nucleate to form various types of dislocations (eg:<110>rod-like defects and {113} stacking faults, both are interstitial types). A review of a portion of the TEM research on ion-bombarded Si has been given by Romanov and Smirnov[38], Wu and Washburn[39,40], and a review article amplifying the approach by Tan is currently in preparation by Tan and Corbett[41]. The important element in the model of Tan is he attributes the nucleation of point defects to the number of dangling bonds per point defect rather than the previously accepted view of strain energy associated with the defect.

According to Tan[37] one can envision what he has referred to as the intermediate defect configuration (IDC) extending indefinitely along a <110> direction with their bonding perpendicular to the <110> and viewing the atoms in a plane perpendicular to the <110>.

In Figure 10 is shown a two dimensional projection of the <110>perfect lattice chain with two extra rows of added interstitials. If instead one had a perfect lattice with rows of two atoms removed then the resulting structures in the 110 plane would be as shown in Figure 11, in which the bonding occurs in a 5-8-5 member ring. Note the "tunnel" formed inside the 8-member ring. Tan has also observed that a shear stress in the $(\bar{1}\bar{1}1)$ plane applied in the <112> direction can cause bond rearrangement so as to yield the structure shown in Figure 12 in which there are no dangling bonds and the structure exhibits two 5-7 membered rings side by side. The above are examples of the projections of defects in the (110) plane that are explained in Tan's model. These IDC's subsequently become upon heat treatment <110> elongated dislocation dipoles and correlate with defect rearrangement observed after heat treatment in TEM and infrared studies on the HOB mentioned earlier. The foregoing discussion is only intended to be a brief description of a new intuitively appealing model to understand defects in Si which has very recently been proposed by Tan[37].

In addition to rod-like defects, stacking faults, dislocation loops and vacancy dislocation loops, voids[38] are observed in ion-bombarded Si at temperatures between 400-500°C, and the voids evolve when vacancies condense to form a 3-dimensional configuration. At annealing temperatures of 800-900°C all defects are found to"dissolve into the matrix", however they never are completely eliminated. That this is so can be inferred from the TEM studies of Salisbury and Loretto[42] and Nes and Washburn[43] who use 1MeV and 0.650MeV electron beams from high voltage electron microscopes to induce defect complexes which are analyzed using 100kV and 650kV electron beam energies. It is significant that the fluence levels used in these electron microscope studies are 10^{20}-10^{22}e/cm^2 but of course not much different in damaging power than fast neutron fluences ranging from say 10^{16}-10^{19}n/cm^2 used to obtain NTD Si.

The results obtained by Salisbury and Loretto[42] show that 1MeV electron irradiations cause sufficient lattice disorder to produce rod-like defects consisting of interstitials along <110> directions on {113} planes, and the defects appear at irradiation temperatures of 300 up to 550°C. They[42] conclude that the defects are self interstitials, and the interstitials occupy tetrahedral sites elongated in <110> directions, and, the {113} rather than {111} plane is the normal plane for precipitation of self interstitials. In the work of Nes and Washburn[43] using 0.650MeV electrons to cause disorder, <110> rods are found to be present even after 750°C anneals for ∿1¼ hr. The <110>needles or rod-like defects are composed of interstitials and have sizes of up to 10 μm length, with diameters varying from ∿40Å up to ∿200Å. Thus, we can conclude that simple point defects induced by 1MeV electrons must coalesce in the manner proposed by Tan[37] to form fairly large structures. These structures are apparently never completely eliminated from the crystal matrix with heat treatment. Clearly, they must also be present in NTD

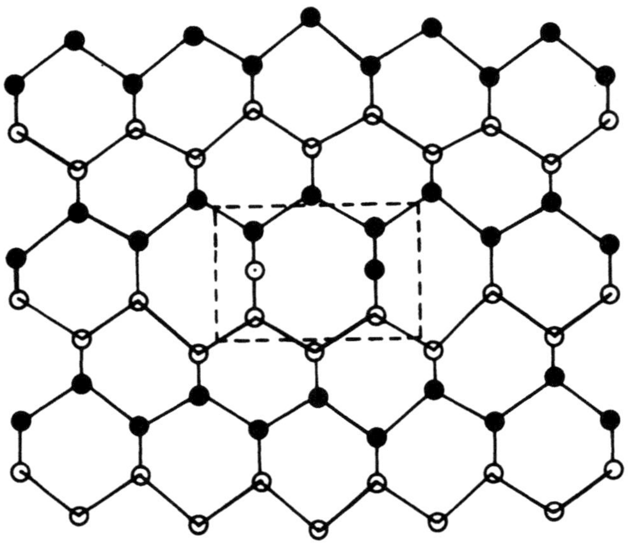

Fig. 10 The projection of the (110) plane of the perfect silicon lattice showing two extra interstitials.

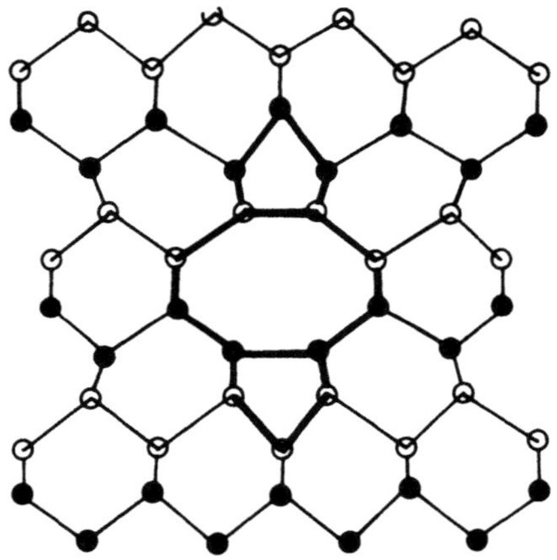

Fig. 11 The projection of the (110) plane showing 5-8-5 member rings bonding arrangement resulting from extracting two atoms from the perfect (110) perfect lattice plane.

silicon after heat treatment even to 700-800°C and probably have a permanent effect on certain structure sensitive properties which depend on minority carrier lifetime. Thus, we must contend with the presence of major damage complexes such as <110> interstitial rods, voids ∿20-40Å diameter, etc. which may be electrically and optically inactive but which will remain in NTD Si and are unanneal-able below the melting temperature of Si (1412°C). We shall give be-low some comments on the impact of the unannealable defects on some devices.

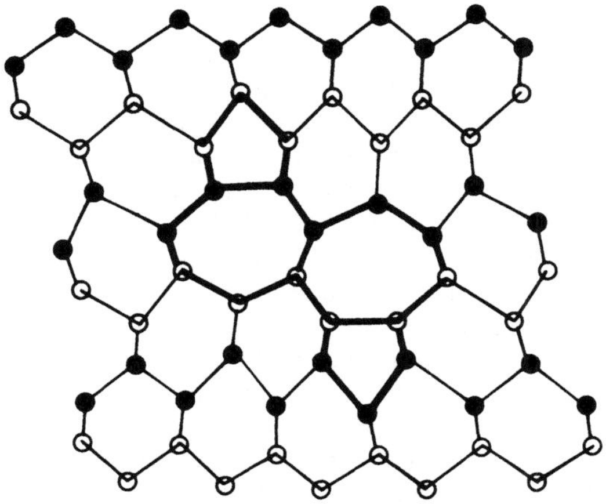

Fig. 12 The (110) plane projection of Fig. 11 resulting from stress in the $(\bar{1}\bar{1}1)$ plane applied in the <112> direction showing new bond arrangement

V. Impact of Defects on Devices

Assuming minor resistivity variations in NTD processed material then the major limiting factors in NTD Si for device use will be the unannealable physical imperfections discussed earlier. These imperfections include dislocations, stacking faults, voids, rod-like clusters of interstitials, interstitial loops, vacancy dislocation loops, and precipitates of impurities at dislocation loops or stacking faults. The association of these imperfections with trace diffusing metals can give rise to regions which can effectively shunt the junction space charge region and can affect a breakdown in silicon PN junctions. In addition, the 8-member ring tunnel shown in Figure 12 can act as a diffusion tunnel through which impurities can travel long distances in the crystal. One can think of these tunnels as dislocation pipes.

Large area high power silicon thyristors and rectifiers produced from NTD silicon requiring large values of carrier lifetime, material devoid of mechanical defects, and uniform resistivity will be limited by the degree of residual electrically inactive physical imperfections listed above which remain after heat treatment. For the case of rectifier devices made from large area (50 mm diameter) slices of silicon the residual mechanical imperfections of sizes ranging from say 10-20 μm may not be as important a limiting factor as for the case of very large scale integrated (VLSI) circuit systems where a large number of devices with 3 μm features are to be made on one slice of material and where the impact of residual unannealable physical defects of ∿10 μm size will be most important and be the limiting factor.

The impact of unannealable defects on NTD silicon used for infrared (IR) detectors (2-20μm) is minimized by annealing out the radiation-induced extrinsic photoconductivity by heating to ∼800°C, after which the remaining photoconductivity response would be due to the neutron transmutation doping impurity. The microstructures described above are not expected to have significant effects on the photoexcitation cross section in photoconductive IR detectors.

We wish to conclude this paper by emphasizing the fact that well defined microstructures remain in NTD silicon even after annealing to∼800°C, and while these microstructures may have little effect on the bulk resistivity they form the ultimate limits on what may be a futile attempt to achieve defect free NTD Si. Clearly, NTD Si with residual unannealable physical defects may not be useful in the production of VLSI circuits for which presently one uses 3μm dimensions for feature resolution with 1μm dimensions anticipated in the future. However, NTD Si should continue to give high quality large area uniformly doped silicon, and we suggest it might be profitable to irradiate at temperatures of $\gtrsim$500°C to produce NTD Si containing lower damage concentration than irradiations at T$\lesssim$190°C.

TABLE I

Time Response of Higher Order Bands to Photoexcitation De-Excitation, and Depletion (see text for sample details)

BAND	ABSORPTION TIME TO STEADY STATE	DECAY TIME TO STEADY STATE	EXCITATION	DEPLETION TIME TO STEADY STATE
9.54μm (1048 cm^{-1})	2 min 20 sec	25 sec	INITIAL	
9.54μm	29 sec	31 sec	SUBSEQUENT	
9.08μm (1101cm^{-1})	3 min 30 sec	1 min	INITIAL	
9.08μm	30 sec	20 sec	SUBSEQUENT	
9.54μm			DEPLETION	2 min 30sec
9.08μm			DEPLETION	4 min

Acknowledgements

The authors acknowledge the following former students in which we have used some of the results generated in their graduate research programs at RPI. Drs. Rosa Young, Akio Chen and Arne H. Kalma, also M. Rolle, R. Gruver and M. Mitchell. The financial support for the preparation of this paper was provided by the National Science Foundation Grant #EC779-08907. Finally the excellent work in typing the manuscript by Jean E. Mulson is affectionately acknowledged with thanks.

References

1. R.T. Young, J.W. Cleland, R.F. Wood and M.M. Abraham, J. Appl. Phys. 49(a), 4752 (1978).
2. J.M. Messe and P.J. Glairon, Neutron Transmutation Doping in Semiconductors, J.M. Meese Ed., Plenum Press (NY) p 109-128, p 291-305 (1979).
3. J.W. Cleland, P.H. Fleming, R.D. Westbrook, R.F. Wood and R.T. Young, ibid, p 261-279 (this paper contains references to earlier works).
4. B.C. Larson, R.T. Young, and J. Narayan, ibid p. 281-290.
5. B.J. Baliga and A.E. Evwaraye, ibid, p. 317-328.
6. H.Y. Fan and A.K. Ramdas, J. Appl. Phys. 30, 1127 (1959).
7. R.N. Thomas, T.T. Braggins, H.M. Hobgood and W.J. Takei, J. Appl. Phys. 49, 5, 2811 (1978).
8. H.M. Hobgood, T.T. Braggins, J.C. Swartz and R.N. Thomas, ibid (Reference 2 above) p. 65-90.
9. V.S. Vavilov, A.F. Plotnikov and V.D. Tkachev, Soviet Phys.-Solid State 4, 2522 (1963).
10. V.D. Tkachev, A.F. Plotnikov and V.S. Vavilov, Soviet Phys.-Solid State 5, 1322 (1964).
11. V.S. Vavilov, Soviet Phys.-Uspo 7, 797 (1965).
12. L.J. Cheng, Appl. Phys. Letters 24A, 729 (1967).
13. A.H. Kalma and J.C. Corelli, Phys. Rev. 173, 3, (734-745) (1968).
14. R.T. Young and J.C. Corelli, Phys. Rev. 5B, #4, 1455-1467 (1972).
15. R.F. Konopleva, S.R. Novikov and E.E. Rubinova, Soviet Phys.-Solid State 8 (2), 264 (1966).
16. R.F. Konopleva and S.R. Novikov, Soviet Physics-Semiconductors 2 (9), 1080 (1969).
17. V.A. Vorobeichik, R.F. Konopleva, S.R. Novikov and E.E. Rubinova, Soviet Physics Semicond. 5 (3), 397 (1971).
18. R.C. Newman and D.H.J. Totterdell, J. Phys. C. Solid State Phys. 8, 3944 (1979).
19. H.J. Stein, J. Electronic Mat. 4, 159 (1975).
20. J.C. Corelli, D. Mills, R. Gruver, D. Cuddeback, Y.H. Lee and J.W. Corbett, Radiation Damage and Defects in Semiconductors, Inst. Phys. Conf. Ser 31, 251 (1977).
21. Y.P. Koval, V.N. Mordkovich, E.M. Temper and V.A. Kharchenko, Sov. Phys. Semicond. 6, 1152 (1973).
22. C.S. Chen, R.V. Lowell and J.C. Corelli, Radiation Damage and Defects in Semiconductors, Inst. Phys. Conf. Ser. 16, 210 (1973).
23. V.N. Mordkovich, S.P. Solov'ev, E.M. Temper and V.A. Kharchenko, Sov. Phys.-Semicond. 8, 666.
24. Y.H. Lee, P.B. Brosious and J.W. Corbett, Radiat. Effects 22, 169 (1974).
25. Y.H. Lee, N.N. Gerasimenko and J.W. Corbett, Phys. Rev. B14, 4506 (1976).
26. D.F. Daly, J. Appl. Phys. 42, 864 (1971).
27. K.L. Brower, Phys. Rev. B14, 872 (1976).

28. L.J. Cheng, J.C. Corelli, J.W. Corbett and G.D. Watkins, Phys. Rev. 152, 555 (1966).
29. M.T. Mitchell, J.C. Corelli and J.W. Corbett, Radiation Damage and Defects in Semiconductors, Inst. Phys. Conf. Ser. 46, 317 (1979).
30. F. Carton-Merlet, B. Pajot and P. Vajda, ibid p 311 (Ref.#29)
31. M.T. Mitchell and J.C. Corelli, Radiat. Effects 41, 57 (1979).
32. W. Jung and G.S. Newell, Phys. Rev. 132, 648 (1963).
33. Y.H. Lee, private communication.
34. Y.H. Lee, Y.M. Kim and J.W. Corbett, Radiat. Effects, 15, 77 (1972).
35. D.F. Daly, J. Appl. Phys. 42, 864 (1971).
36. H.J. Stein, (ibid Ref. 2 above) p. 229-248.
37. T.Y. Tan, (to be published).
38. S.I. Romanov and L.S. Smirnov, Radiat Effects 37, 121 (1978).
39. W.K. Wu and J. Washburn, J. Appl. Phys. 48(10), 372 (1977).
40. W.K. Wu and J. Washburn, J. Appl. Phys. 48(9), 3747 (1977).
41. T.Y. Tan and J.W. Corbett (to be published).
42. I.G. Salisbury and M.H. Loretto, Phil. Mag A 39(3), 317 (1979).
43. E. Nes and J. Washburn, J. Appl. Phys. 42 (9), 3559 (1971).

ELECTRICAL PROPERTY STUDIES OF OXYGEN IN CZOCHRALSKI-GROWN NEUTRON-TRANSMUTATION-DOPED SILICON*

J.W. Cleland and N. Fukuoka**

Solid State Division
Oak Ridge National Laboratory
Oak Ridge, Tennessee 37830

ABSTRACT

Electrically active oxygen-related donors can be formed in Czochralski (Cz) Si either during crystal growth or during subsequent heat treatment; conventional n- or p-type dopant carrier concentrations are altered if these oxygen donors are present. Neutron transmutation doping (NTD) has been used to introduce a uniform concentration of ^{31}P in Si. However, oxygen donors can also be formed in NTD Cz Si during the process of annealing to remove NTD radiation damage. In the present experiments, the carrier concentration of Cz and NTD Cz Si samples was determined as a function of the initial dopant, oxygen, and ^{31}P concentration before and after isothermal or isochronal annealing. It is shown that low temperature (350-500°C) heat treatment can introduce a significant oxygen donor concentration in Cz Si and in NTD Cz Si that contains radiation-induced lattice defects. Intermediate temperature (550-750°C) heat treatment, which is intended to remove oxygen donors or lattice defects, can introduce other oxygen donors; annealing above 750°C is required to remove any of these oxygen donors. Extended (20 h) high-temperature (1000-1200°C) annealing can remove oxygen donors and lattice defects, but a significant concentration of oxygen donors can still be introduced by subsequent low temperature heat treatment. These results suggest that oxygen-related donor formation in NTD Cz Si at temperatures below 750°C may serve to mask any annealing study of lattice

*Research sponsored by the Division of Materials Sciences, U.S. Department of Energy under contract W-7405-eng-26 for the Union Carbide Corporation.

**Guest scientist from Osaka University, Osaka, Japan.

defects. It is concluded that annealing for 30 min at 750°C is sufficient to remove radiation damage in NTD Cz Si when the separate effects of oxygen donor formation are included.

INTRODUCTION

Neutron transmutation doping (NTD) has been used to introduce a precisely controlled and very uniform distribution of ^{31}P in Si.[1] In order to remove lattice damage introduced by recoiling atoms in the (n,γ) process and by fast neutron collisions with Si atoms it is necessary to anneal the Si after irradiation before the desired electrical properties of the P can be realized. The lattice damage associated with NTD Si is very complex. This damage consists of simple point defects, interstitial and vacancy complexes, larger defect clusters, and dislocation loops. The relative concentration of these defects depends on the irradiation temperature, the neutron energy spectrum and the neutron fluence. Annealing may result in the removal of some defects but also in the formation of other defect or defect-impurity complexes.

Electrical property studies have been used[2] to establish the annealing requirements necessary to obtain the anticipated carrier concentration and mobility for NTD float zone (FZ) single crystal Si of low ($\leq 3 \times 10^{16}$ cm^{-3}) oxygen content. Most of the simple recoil-induced defects were removed by annealing at 400°C. Annealing for 30 min at 750°C was sufficient to obtain the anticipated carrier concentration and mobility in NTD FZ Si irradiated up to a total fluence of $\geq 10^{20}$ n cm^{-2} in different reactor locales with thermal-to-fast ($t/\bar{f}$) neutron ratios from 1700 (low fluence) to 3 (high fluence). However, the minority carrier lifetime (MCL) was still degraded after annealing at 750°C, irrespective of neutron fluence or t/f ratio.[3] Recovery of the MCL required annealing at 1000°C; the extent of recovery was dependent on the irradiation temperature, neutron fluence, t/f ratio, time of annealing, and rate of cooling after annealing.

Infrared absorption (IR), x-ray diffuse scattering, and transmission electron microscopy (TEM) measurements have also indicated[4] that lattice defect complexes are introduced in NTD FZ Si as a consequence of irradiation or annealing. Dislocation loops of 50-100 Å diameter with a number density of 10^{14} cm^{-3} remained in NTD FZ Si irradiated with a fast ($\geq$ 1 MeV) neutron fluence of 10^{20} n cm^{-2} and annealed at 900°C; it was suggested that these loops may play a role in the trapping of minority carriers.[3]

The characterization of lattice damage introduced by neutron irradiation of Czochralski (Cz)-grown Si is more complex because Cz Si usually contains $\geq 5 \times 10^{17}$ oxygen cm^{-3}, and O or certain defect-O complexes may act as electrically active donors. The initial O concentration in Cz Si is a complex function of crystal growth conditions; in addition, the O content at the center of a 5 cm diameter ingot is usually larger by a factor of two or more than that near the edge. Oxygen donors are formed during cooldown after crystal growth; commercially available Cz Si ingots are usually post-annealed to reduce that O donor contribution. Capper and Wilkes[5] have described an empirical procedure used by certain Cz Si producers as consisting of post-annealing for 3 h at 650°C in N_2, followed by cooling under natural or forced convection. They examined 5 cm diameter dislocation-free Cz Si ingots and concluded that such annealing was not sufficient for close tolerance specifications, to the extent that reheating of ingot slices was necessary in order to further reduce the O donor concentration.

Kaiser[6] and Fuller[7] have suggested that O donors (possibly associated with vacancies) migrate when Si is heated at 350 to 500°C and that four O combine as SiO_4 to form a single donor. They did not observe any special interaction between O and n-type donor impurities as a consequence of heating doped samples, but they observed that O and p-type acceptor impurities also combined to form a product with donor-like properties. Kaiser[6] reported that the initial rate of O donor formation on heating was proportional to the fourth power of the O concentration and that the maximum donor concentration on extended annealing was proportional to the third power of the O concentration. The rate of donor formation was also a function of the annealing temperature and the time required to attain the maximum donor concentration decreased for material with an increased O concentration. Fuller[7] reported that a maximum donor concentration was observed after annealing for 70 h at 450°C and Kanamori[8] reported that a maximum donor concentration was observed after annealing for 64 h at 475°C. Both authors reported that these maxima were followed by a decrease in donor concentration with continued annealing at these temperatures.

Helmreich and Sirtl[9] have suggested that formation of SiO_4 is not required for thermal donor (TD) generation in Cz Si. They postulated that interaction between interstitial oxygen (O_i) and vacancies will result in the production of TD's if a single O_i becomes a substitutional O_s. They stated that such donors were stabilized against annealing by acceptor impurities. A rapid breakdown of TD complexes in favor of O_i reformation was observed in undoped Cz Si on annealing above 515°C, but some TD's remained in Cz Si that was doped with elements of a strong acceptor character even after annealing above 800°C. Kaiser[6] reported that those donors formed by SiO_4 or by O associated with B or Ga acceptors disappeared with first-order kinetics on annealing above 500°C, but that O-Al donor-like complexes required annealing at 1000°C.

Kanamori,[8] in contrast, has suggested that dislocations or an unknown impurity species X are required in Cz Si to remove a majority of the O donors at higher annealing temperatures after initial donor formation by extended annealing at 475°C. The ratio of remaining to initial donor concentration was as large as 0.5-0.7 in most of his samples of zero dislocation content n- or p-type Cz Si after initial donor formation at 475°C and reannealing for 20 h at 500°C. His experimental results were not adequate to determine the unknown species X in those samples of low ratio values, but he stated that the annealing behavior in those samples was similar to that reported by Fuller and Logan for Li-doped Cz Si.[10]

Kaiser[6] reported that heating Cz Si for 20 h at 1100°C was sufficient to "stabilize" the O such that no donors were introduced by subsequent annealing at 450°C. Baker[11] stated that heating Cz Si for 20 h at 1000°C was sufficient in that no resistivity change due to donor formation occurred with subsequent annealing at 450°C, even in Cz Si with a large O concentration. de Kock,[12] in contrast, observed that excess O was removed from solid solution by precipitation in Cz Si during high temperature (1000-1200°C) processing, but that some O remained in solution in dislocation-free Cz Si after such heat treatments, even at an O supersaturation ratio of 5. Previous results [6,7,11] have all indicated that a short (30 min) heat treatment at about 1350°C was sufficient to remove all of the O clusters or precipitates and restore most of the O to its normal position in the Cz Si lattice. Of course any subsequent low temperature (350-500°C) heat treatment may then introduce new O donors.[5-12]

Previous investigators[13-15] have reported that O-defect complexes that act as donors are also formed in NTD Cz Si by irradiation or by subsequent annealing to remove lattice damage, and that such donor complexes are stable against annealing up to 850°C,[13] 950°C[14], or 1000°C.[15] More recent experiments[16] indicated that annealing for 30 min at 750°C was sufficient to obtain the anticipated carrier concentration and mobility in either NTD FZ or Cz Si irradiated to introduce $\geq 10^{16}$ ^{31}P cm^{-3}, independent of initial O content or t/f neutron ratio. Any separate effect due to any O-defect complex was not identified in NTD Cz Si at those dopant concentrations.

This paper will discuss the results of annealing experiments using Cz Si samples of different initial dopant and O concentrations prior to irradiation as well as the results of annealing experiments using NTD Cz Si samples. Two reactor locales with different t/f ratios were used.[16] The thermal and fast ($\geq$ 1 MeV) neutron fluxes, as determined by dosimeters, were used to calculate the anticipated ratio of simple (n,γ) recoil events to fast neutron induced events, as previously indicated,[2] and to calculate the anticipated concentration of P introduced by transmutations. We determined the initial O_i content by conventional IR techniques. Carrier concentration and mobility values were obtained for thin 1 cm^2 van der Pauw (vdP) samples. Annealing was carried out in a quartz tube furnace in an inert atmosphere; the samples were quickly removed and air-cooled on a quartz plate.

EXPERIMENTAL RESULTS

(A) Annealing of Unirradiated Czochralski Silicon

Table 1 lists the initial donor (n) or acceptor (p) concentration, initial interstitial oxygen (O_i) concentration, initial rate of O donor formation on annealing at 450°C, and the maximum carrier concentration observed after extended (120-150 h) annealing at 450°C for samples from ten Cz Si ingots.[17]

Table 1. Ingot Characteristics of Silicon Samples Investigated

Ingot	Initial Donor (n) or Acceptor (p) Concentration (cm^{-3})	Interstitial Oxygen (O_i) Concentration (cm^{-3})	Rate of Donor Formation at 450°C (cm^{-3} hr^{-1})	Carrier Concentration after 120-150 hr at 450°C (cm^{-3})
A-1	$5.0x10^{13}$-n-	$6.1x10^{17}$	$4.0x10^{13}$	$2.0x10^{15}$-n-
A-3	$2.0x10^{13}$-n-	$7.5x10^{17}$	$1.1x10^{14}$	$5.5x10^{15}$-n-
A-4	$1.0x10^{14}$-n-	$2.0x10^{18}$	$3.4x10^{14}$	$1.6x10^{16}$-n-
B-1	$2.5x10^{14}$-n-	$2.4x10^{18}$	$1.6x10^{15}$	$2.0x10^{16}$-n-
M-85	$9.0x10^{13}$-n-	$8.9x10^{17}$	$4.0x10^{14}$	$8.5x10^{15}$-n-
M-86	$9.5x10^{13}$-n-	$8.5x10^{17}$	$4.3x10^{14}$	$1.3x10^{16}$-n-
M-65	$3.1x10^{14}$-n-	$1.8x10^{18}$	$4.0x10^{14}$	$9.9x10^{15}$-n-
M-73	$3.5x10^{14}$-n-	$1.3x10^{18}$	$2.7x10^{14}$	$5.0x10^{15}$-n-
W-1	$1.0x10^{14}$-p-	10^{18}	$2.0x10^{14}$	$3.6x10^{15}$-n-
10_B	$5.6x10^{16}$-p-	$8.0x10^{17}$	$2.0x10^{15}$	$5.0x10^{16}$-p-

Figure 1 is a graph of carrier concentration vs time of annealing at 450°C for samples from some of the ingots listed in Table 1. It is stated above that the initial O concentration in Cz Si is a complex function of different crystal growth conditions and that the O content may vary by a factor of two or more acoss a 5 cm diameter ingot slice. In the present experiments, the IR or vdP samples were not always taken from the same slice or slice locale. There was considerable variation in the initial rate of donor formation, and in the donor concentration after extended annealing for different samples from the same ingot or ingot slice, but the data of Table 1, plotted in Fig. 1, indicate that both the initial rate of donor formation and the donor concentration after extended annealing were dependent on the initial O_i concentration. Note that the apparent donor concentration after extended annealing at 450°C was almost always $\leq$ 1% of the initial O_i concentration.

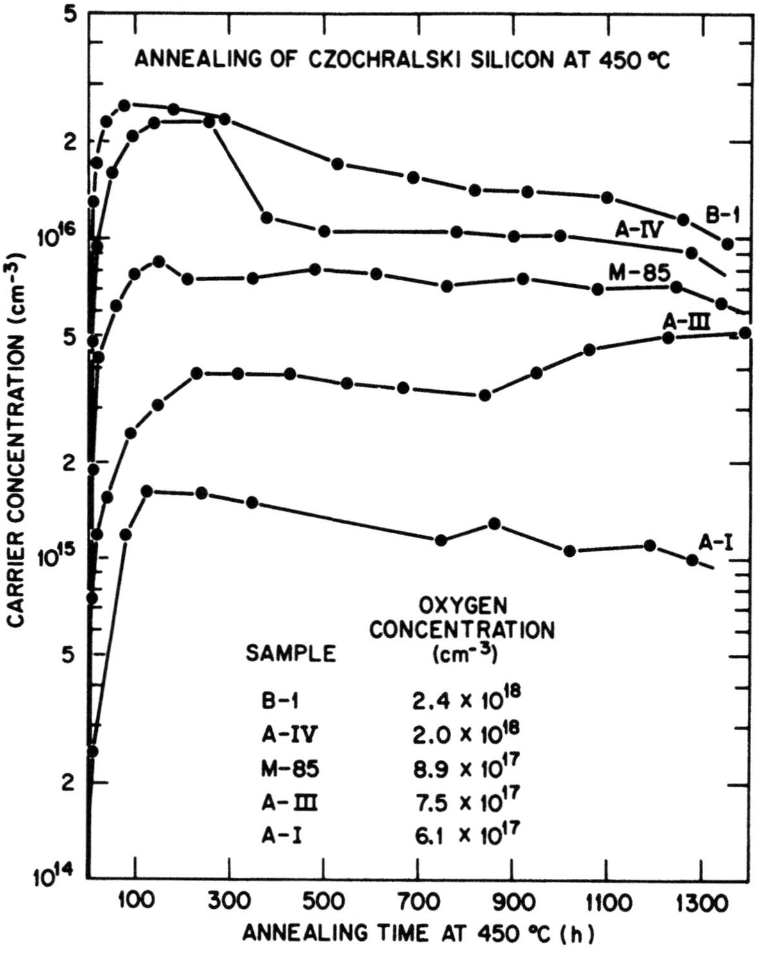

Fig. 1. Carrier concentration vs annealing time at 450°C for Cz Si samples.

Ingot B-1 was not post annealed after growth and the initial donor concentration in samples from that ingot was changed to a low apparent accceptor concentration by 30 min annealing at 750°C and rapid cooling. Ingots M-65 and M-73 were doped with P or B, respectively, and no additional O-dopant interaction was evident as a consequence of extended annealing in these samples at these low dopant concentrations. Ingot ^{10}B was doped with a stable isotope of B enriched to 94.5% abundance. The initial O_i concentration was somewhat less than that for the other ingots but the initial rate of donor formation was somewhat larger. Extended annealing reduced the apparent p-type acceptor concentration due to O-donor formation and the total change in concentration was larger than that observed in the high purity n-type A-1 and A-3 samples that contained more O_i initially. These data may indicate an enhanced O-acceptor interaction that forms additional donors in p-type Cz Si on low temperature (350-500°C) heat treatment, as has also been reported by others.[6,7,9]

Fuller[7] and Kanamori[8] reported a maximum O-donor concentration after annealing Cz Si at 450 or 475°C for about 70 h. The data shown in Fig. 1 indicates that the time required to reach a maximum carrier concentration was not a very sensitive function of the initial O_i concentration and the carrier concentration decreased on extended (150-1300 h) annealing for some of the samples investigated. Samples B-1 and A-IV of Fig. 1 were reannealed for 30 min at 750°C in order to remove most of the O donors and were then reannealed at 450°C. Both the initial rate of O-donor formation and the maximum carrier concentration attained were reduced by about a factor of two by this treatment; the carrier concentration again decreased on extended annealing. Other samples were selected from the center or near the edge of ingot slices, and from different sections of the various ingots. No effect of radial or axial position was observed upon extended annealing at 450°C, other than that attributed to the initial O_i concentration. The data shown in Fig. 1 and that obtained on the other samples do indicate that several competing processes may occur under extended annealing conditions, but that the initial O_i concentration is the primary factor governing O-donor formation in Cz Si via low-temperature annealing.

Previous investigators[6,7,9] have indicated that O donors are introduced in Cz Si by annealing at 350-500°C, but not at higher temperatures. We selected additional samples representative of Cz Si of high, medium, and low initial O_i content from among the ingots listed in Table 1. Figure 2 is a graph of carrier concentration vs extended annealing time at 550°C, followed by 450°C. These data indicate that significant O-donor formation occurs on annealing at 550°C in these samples, irrespective of initial O_i content. The carrier concentration did not decrease on extended annealing at 550°C and no evidence was obtained for additional O-donor formation at 450°C after a near saturation concentration of O donors was

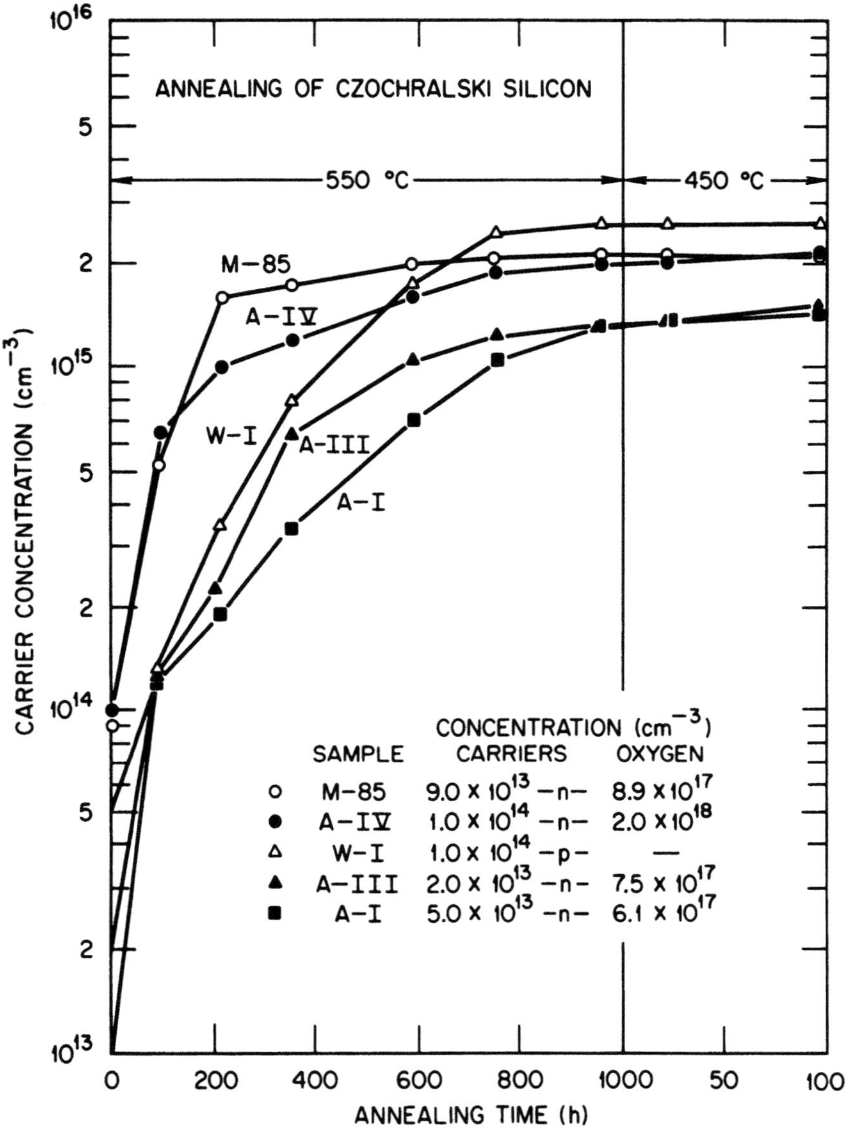

Fig. 2. Carrier concentration vs annealing time at 550 and 450°C for Si samples.

introduced at 550°C. Other samples of Cz Si of high, medium, and low initial O_i content were selected for extended annealing at five different temperatures. Figure 3 is a graph of the resulting carrier concentrations vs time. These data indicate that significant concentrations of O donors were introduced on extended annealing at either 550 or 650°C, and that some O-donor formation still occurs at 750°C (possibly even 850°C). Note that the apparent rate of donor formation increases with an increase in anneal-temperature from 550 to 750°C in these samples.

Kaiser[6] and Fuller[7] have indicated that annealing above 500°C is sufficient to remove O donors in Cz Si unless they are associated with impurities of a strong acceptor character and Kanamori[8] has indicated that annealing above 550°C is required for total O donor removal unless dislocations or an unknown impurity are present. Accordingly, samples from five of the ingots listed in Table 1 were heat treated at 450°C to introduce a large O-donor concentration and were then reannealed at 550°C as shown in Fig. 4. The maximum O-donor concentration was initially reduced by more than 90% in each sample, after about 25 to 50 h at 550°C; however, formation of other O donors then occurred on extended annealing at 550°C. Note that the saturation concentration of O donors after extended annealing is comparable to that shown for the samples in Fig. 3, which did not contain O donors initially. Similar data was obtained for other samples, after an initial heat treatment at 450°C to introduce a large O-donor concentration, followed by extended reannealing at 650°C. The results were almost identical to those shown in Figs. 3 and 4, as regards the saturation concentration of O donors, but the time required for removal of the initial O donors and for formation of the new O donors decreased for the samples reannealed at 650°C.

Since more than 90% of the O donors introduced at 450°C were removed by annealing at higher temperatures, before new O donors were formed, it was decided to examine the higher temperature stability of O donors that were initially introduced in Cz Si by extended annealing at 550°C or higher. Figure 5 shows the carrier concentration vs annealing time for samples of high, medium, and low initial O_i concentration, after 300 h at 550°C, and after reannealing for 20 h at 100°C temperature intervals from 650 to 1150°C. There is little, if any, evidence for removal of O donors formed at 550°C below a reannealing temperature of 750°C; extended (20 h) reannealing above 850°C is required to remove most of the O donors. Very similar results to those shown in Fig. 5 were obtained for O donors that were formed in similar samples after extended annealing at 650°C. These data indicate that (1) a significant concentration of O donors can be formed in Cz Si on annealing at 550°C or higher; (2) the maximum O-donor concentration is almost independent of the initial O_i concentration; and that extended high temperature annealing is required to remove these O donors.

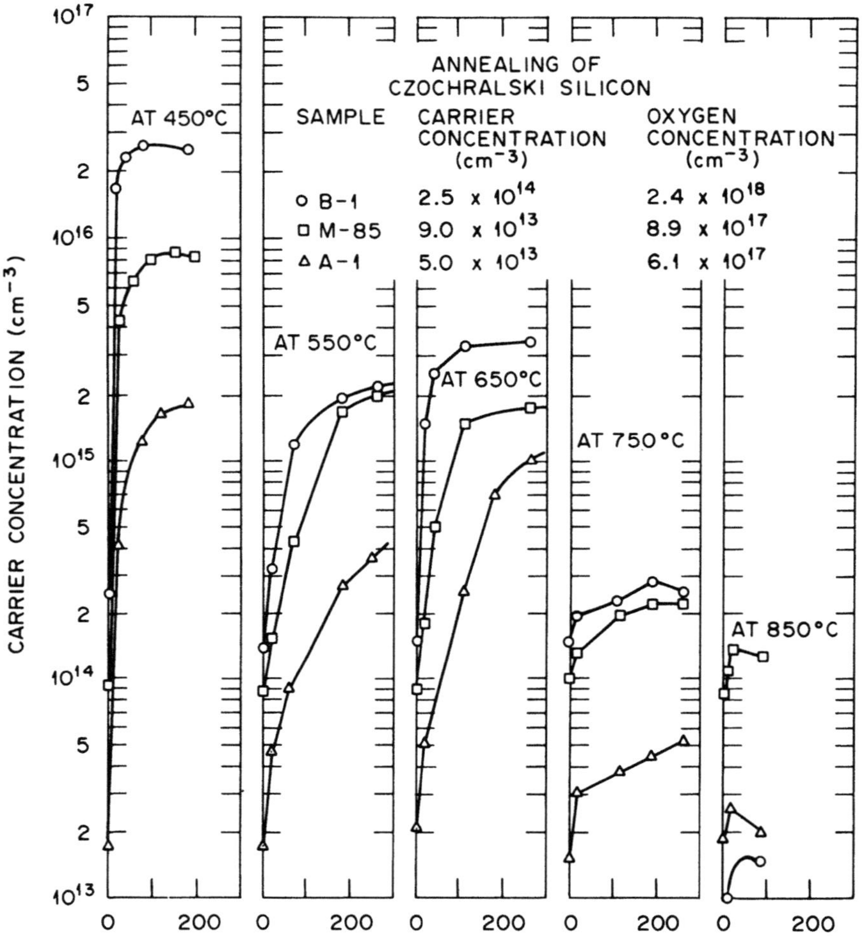

Fig. 3. Carrier concentration vs annealing time at 450 to 850°C for Cz Si samples.

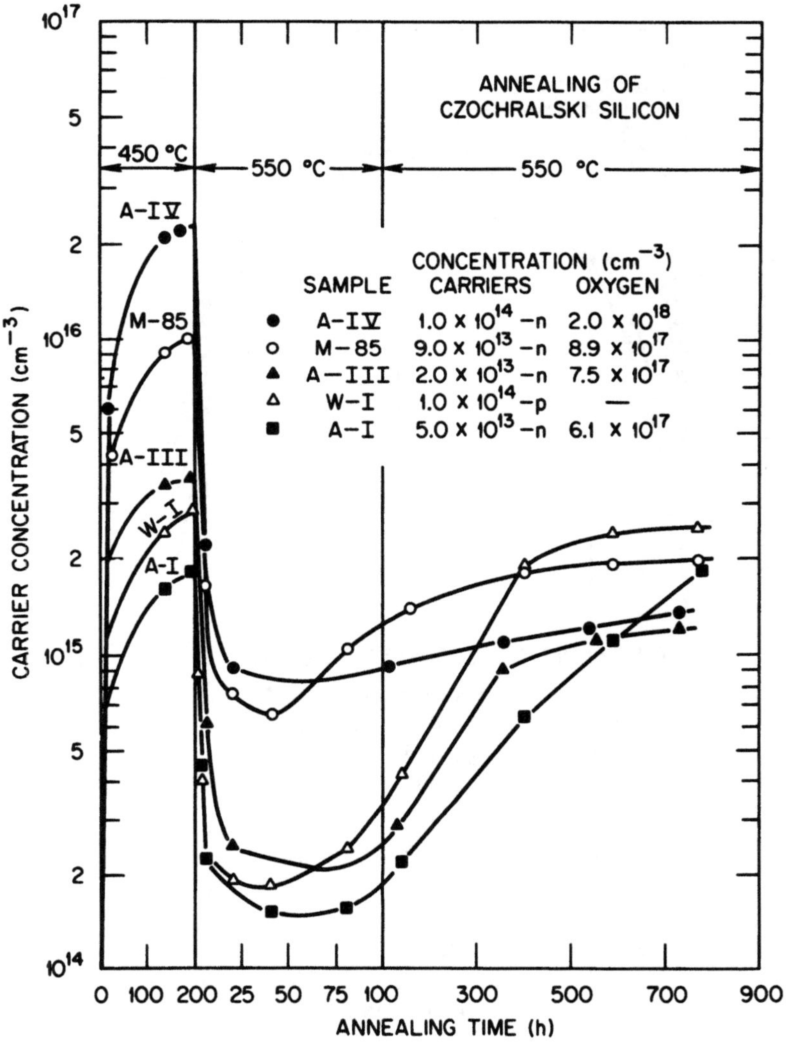

Fig. 4. Carrier concentration vs annealing time at 450 and 550°C for Cz Si samples.

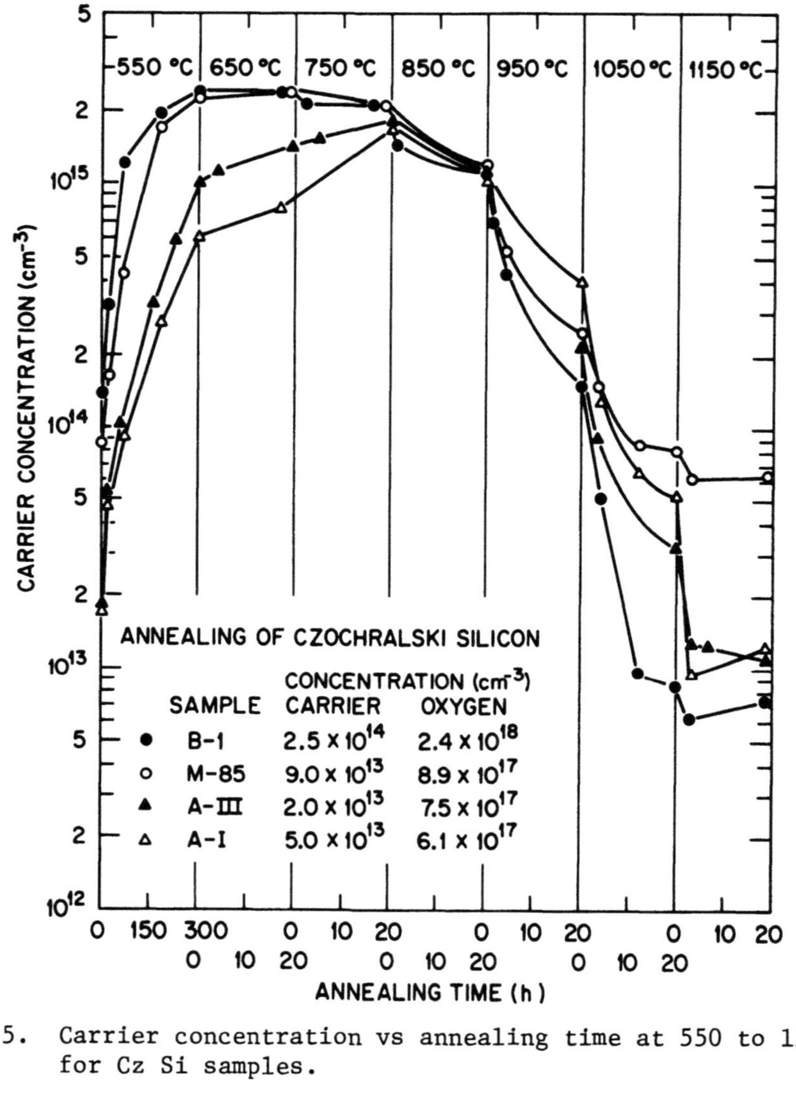

Fig. 5. Carrier concentration vs annealing time at 550 to 1150°C for Cz Si samples.

Previous investigators[6,11] have reported that heat treatment for 20 h at 1000°C or higher is sufficient to eliminate any subsequent low temperature O-donor formation in Cz Si of large initial O_i concentration, although de Kock[12] observed that some O_i remained in dislocation-free Cz Si after high temperature heat treatment. In the present experiments, samples from the ingots listed in Table 1 were annealed for 20 h at 1000, 1100, or 1200°C and air cooled. The carrier concentration after that heat treatment was almost identical to the initial value for each sample so that no evidence of any defect or impurity introduction or removal due to this heat treatment was observed. Table 2 lists the initial (n) or acceptor (p) concentration for each sample and the carrier concentration after extended (120-150 h) annealing at 450°C, for each set of samples after each initial heat treatment. Figure 6 indicates the increase in carrier concentration at 450°C for representative samples from the ingots of high (B-1), medium (M-85), and low (A-1) initial O_i content. All of the results listed in Table 2 and shown in Fig. 6 were such as to indicate a significant increase in O-donor concentration on annealing at 450°C after high temperature heat treatment (with the sole exception of the sample from ingot M-85 after heat treatment at 1000°C only). The samples were reannealed a second time for 20 h at 1000, 1100, or 1200°C, and again no evidence of any defect or impurity introduction or removal was observed other than removal of the O donors formed at 450°C. The results of extended annealing at 450°C on these samples were almost identical to those shown in Fig. 6. All of these data indicate that prior heat treatment for 20 or 40 h at 1000 to 1200°C is not sufficient to stabilize some of the O_i in Cz Si against subsequent O-donor formation by extended low temperature heat treatment.

(B) Irradiation Conditions

Table 3 lists the thermal and fast ($\geq$ 1 MeV) neutron fluence, t/f ratio, and the anticipated concentration of ^{31}P introduced in Si per hour of irradiation in the D-4 and D-6 locales of the D_2O tank in the Bulk Shielding Reactor (BSR) at the Oak Ridge National Laboratory (ORNL). These data were used to estimate the apparent concentration of (n,γ) recoil and of fast neutron-induced events as previously indicated[2] for NTD FZ Si. If there is a correspondence between the rate of lattice defect introduction by neutron irradiation in FZ and Cz Si, as would be anticipated, the data of Table 3 can be used to estimate that the total lattice defect concentration remaining in NTD Cz Si after ambient temperature irradiation is about 10^{16} cm^{-3} or 10^{14} cm^{-3} per h of irradiation in the D-4 and D-6 locales, respectively.

(C) Isothermal Annealing of NTD Cz Si at 450°C

Samples from ingots B-1, A-3, and A-1 listed in Table 1 were irradiated in the D-6 locale for times of 1 to 100 h and then an-

Table 2. Carrier Concentration of Cz Si Samples Before and After Heat Treatment.

Ingot	Initial Donor (n) or Acceptor (p) Concentration (cm^{-3})	Carrier Concentration after 20 h at 1000°C and 120-150 h at 450°C (cm^{-3})	Carrier Concentration after 20 h at 1100°C and 120-150 h at 450°C (cm^{-3})	Carrier Concentration after 20 h at 1200°C and 120-150 h at 450°C (cm^{-3})
A-1	5.0×10^{13}-n-	4.4×10^{14}-n-	9.0×10^{14}-n-	9.2×10^{14}-n-
A-3	2.0×10^{13}-n-	7.2×10^{15}-n-	6.6×10^{15}-n-	7.0×10^{15}-n-
A-4	1.0×10^{14}-n-	1.5×10^{16}-n-	1.5×10^{16}-n-	1.7×10^{16}-n-
B-1	2.5×10^{14}-n-	1.5×10^{16}-n-	1.4×10^{16}-n-	9.5×10^{15}-n-
M-85	9.0×10^{13}-n-	8.3×10^{13}-n-	5.6×10^{14}-n-	7.1×10^{15}-n-
M-86	9.5×10^{13}-n-	1.3×10^{16}-n-	1.2×10^{16}-n-	1.3×10^{16}-n-
M-65	3.1×10^{14}-n-	5.9×10^{15}-n-	1.0×10^{16}-n-	8.9×10^{15}-n-
M-73	3.5×10^{14}-p-	3.9×10^{15}-n-	4.9×10^{15}-n-	7.9×10^{15}-n-
10_B	5.6×10^{16}-p-	5.0×10^{16}-p-	5.1×10^{16}-p-	5.1×10^{16}-p-

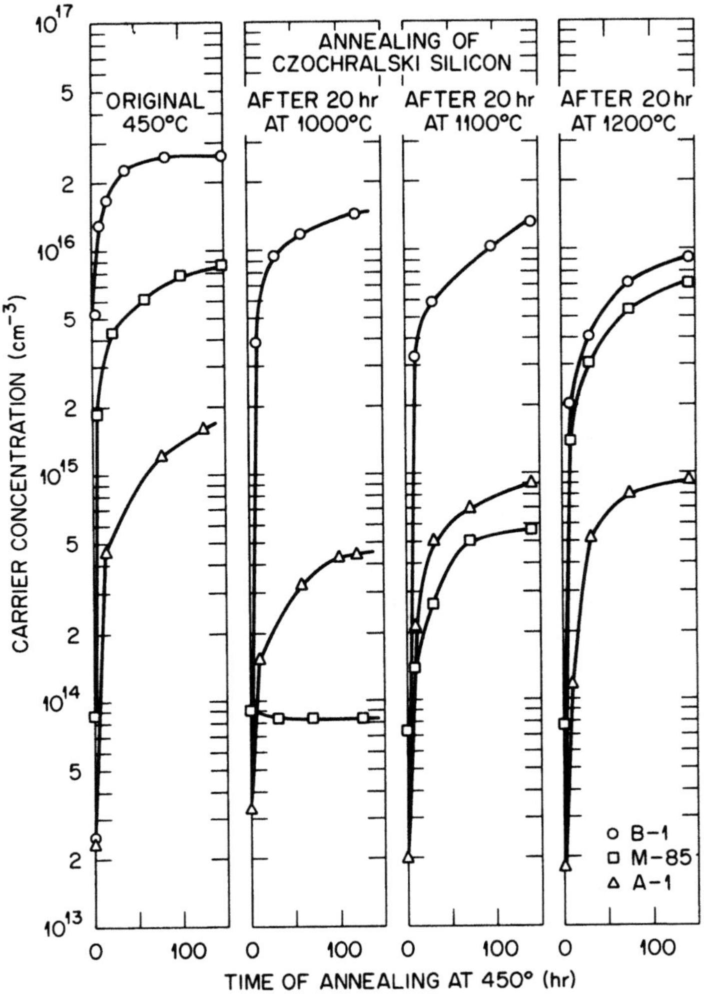

Fig. 6. Carrier concentration vs annealing time at 450°C for Cz Si samples after high temperature heat treatment.

Table 3. Neutron fluence in the D_2O tank, Bulk Shielding Reactor, ORNL

Reactor Locale	Irradiation Time (h)	Thermal Neutron Fluence (cm^{-2})	Fast Neutron Fluence (cm^{-2})	Thermal/Fast Neutron Ratio (t/f)	Phosphorus Concentration (cm^{-3})
D-4	1	3.2×10^{16}	4.7×10^{15}	7	6.5×10^{12}
D-6	1	2.9×10^{15}	3.6×10^{13}	80	5.8×10^{11}

nealed for $\geq$ 100 h at 450°C. The initial rate of O-donor formation and the O-donor concentration after extended annealing were almost identical to the values obtained on unirradiated samples (as listed in Table 1) for these samples irradiated with a fast neutron fluence of up to 3.6×10^{15} cm^{-3}. Most of the (n,γ) recoils may have annealed at 450°C, but these results indicate that the simultaneous presence of $\leq 10^{16}$ fast neutron-induced defects (D_f) cm^{-3} had little if any apparent effect on the rate of formation or final concentration of O donors as a consequence of extended annealing at 450°C.

Other samples from the same ingots were irradiated in the D-4 locale for times of 1 to 800 h. Figure 7 is a graph of the carrier concentration vs annealing time at 450°C for some of the B-1 samples. These data indicate that the rate of O-donor formation and the final O-donor concentration after extended annealing at 450°C were reduced as a function of increasing D_f concentration. Note that the carrier concentration was much larger than the anticipated ^{31}P concentration in samples B-1 to B-3 in Fig. 7, following irradiation and annealing at 450°C, and that a considerable concentration of O donors was still introduced in sample B-3 despite an estimated 10^{18} D_f cm^{-3}. A small amount of O-donor formation may have occurred in sample B-4 on annealing at 450°C, but the presence of almost 10^{19} D_f cm^{-3} evidently masked any observation of extensive O-donor formation or the presence of 5.4×10^{15} P cm^{-3}. Results very similar to those described above were also obtained on the A-3 and A-1 samples that contained less oxygen.

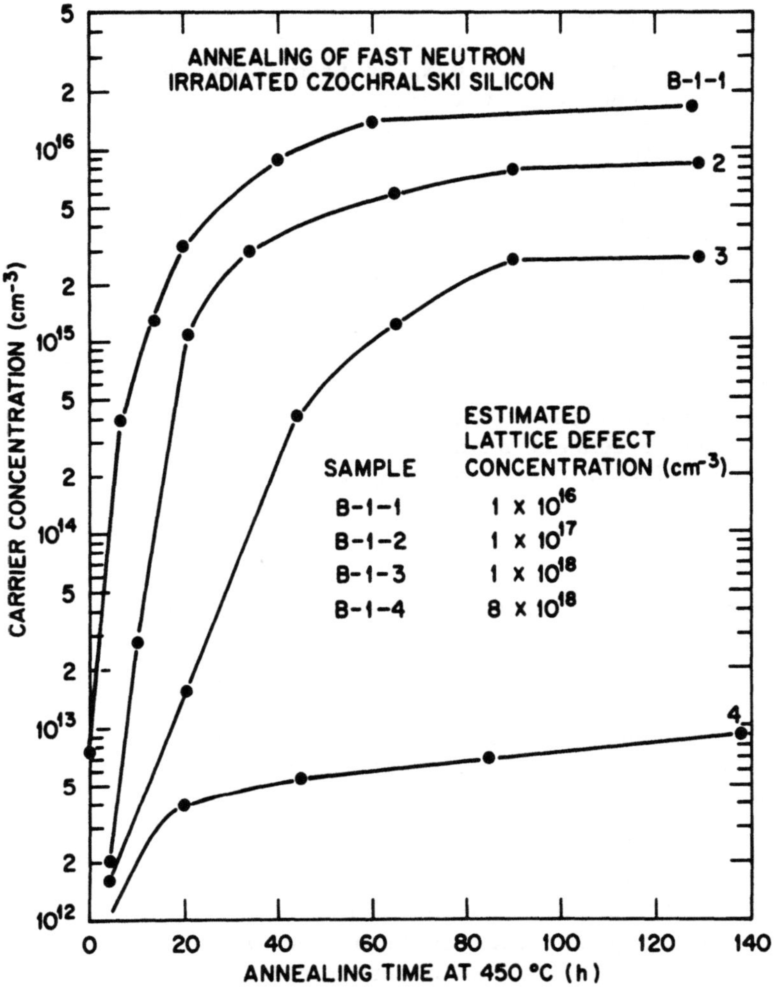

Fig. 7. Carrier concentration vs annealing time at 450°C for NTD Cz Si samples.

(D) Isothermal Annealing of Cz and NTD Cz Si at 750°C

Since previous results[2,16] have indicated that 30 min annealing at 750°C is sufficient to remove fast neutron-induced lattice defects in NTD FZ Si, selected samples from some of the ingots listed in Table 1 were given a similar heat treatment before extended annealing at 450°C to form O donors. Other samples were given a similar heat treatment after O-donor formation at 450°C, after neutron irradiation to introduce lattice defects, or after irradiation and O-donor formation at 450°C. Samples from those ingots that were probably post-annealed after growth to remove O donors did not indicate any significant change in carrier concentration after heat treatment for 30 min at 750°C. Samples that contained O donors initially, that contained O donors after extended annealing at 450°C, or contained O donors and D_f, all indicated a residual excess donor concentration $\geq 10^{14}$ cm^{-3} after the 30 min at 750°C heat treatment. There was a small variation in the final carrier concentration in different samples from the same ingot, that may have been due to the initial O_i content, the number of O donors introduced, or the D_f content after irradiation. However, every sample investigated had an additional donor concentration $\geq 10^{14}$ cm^{-3} except for the post-annealed, unirradiated Cz Si samples. The ratio of remaining-to-initial O-donor concentration was ≤ 0.1 in almost every sample, but these data indicate that annealing for 30 min at 750°C is sufficient to remove D_f, but is not sufficient to remove all of the O donors in either Cz or NTD Cz Si that contains such donors initially.

(E) Isothermal Annealing of NTD Si at 500 and 600°C

A sample (M-1) of n-type, high purity (2.23×10^{12} cm^{-3}) FZ Si of low ($\leq 3 \times 10^{16}$ cm^{-3}) O_i content was irradiated to introduce an estimated 6×10^{16} lattice defects cm^{-3} and 3.8×10^{13} P cm^{-3}. Figure 8, a graph of carrier concentration vs annealing time at 500°C, indicates that the apparent net donor concentration in sample M-1 was less than the anticipated P concentration after extended annealing at 500°C. Subsequent results (not shown) gave a donor concentration of 3.1×10^{13} cm^{-3} after annealing for 125 h at 600°C and 3.9×10^{13} cm^{-3} after 30 min at 750°C; the last value is in good agreement with the anticipated concentration of P. Equivalent data is shown in Fig. 8 for samples from ingot A-1 of Table 1 that contained 8.5×10^{17} O_i cm^{-3} as unirradiated (A-1-2), or as irradiated to introduce 10^{16} (A-1-3) or 10^{17} (A-1-4) D_f cm^{-3}. These results indicate that the introduction of O donors due to annealing at 500°C was delayed by the presence of D_f. Note that the apparent donor concentration of A-1-4, that contained 10^{17} D_f cm^{-3}, remained at a low 10^{12} cm^{-3} value during an annealing time of about 30 h before any O-donor formation was observed.

Similar experiments were carried out on samples from ingot B-1 of Table 1 that contained 2.4×10^{18} O_i cm^{-3} as unirradiated (B-5),

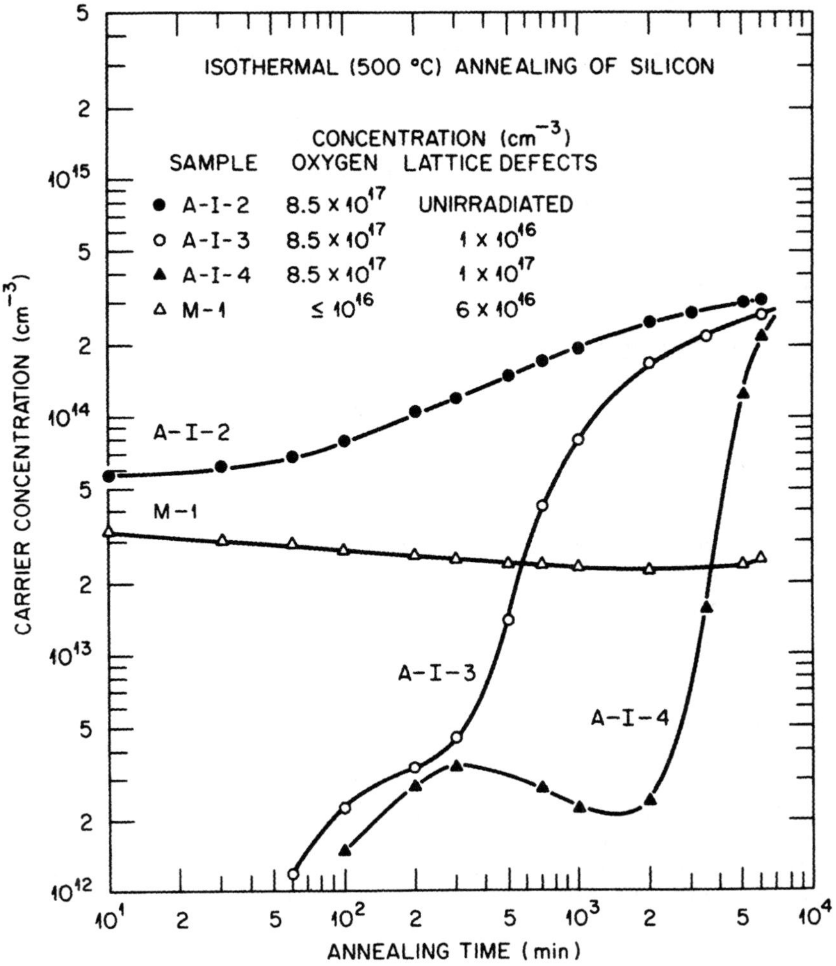

Fig. 8. Carrier concentration vs annealing time at 500°C for FZ and Cz Si samples.

or as irradiated to introduce 10^{16} (B-6) or 10^{17} (B-7) D_f cm^{-3}. Figures 9 and 10 show that O donors were introduced at 500°C and removed at 600°C. The formation of donors is delayed by the presence of D_f (Fig. 9). The apparent donor concentrations after extended annealing at 500°C were about 2.3, 1.8, and 0.9 x 10^{16} cm^{-3} for the three samples that contained very few, 10^{16}, or 10^{17} D_f cm^{-3}. The data of Fig. 10, indicates that 90-95% of the O donors were removed by annealing for about 150 h at 600°C, irrespective of the initial D_f concentration. These results demonstrate that the presence of D_f may impede the formation of O donors in NTD Cz Si by annealing at 500°C, but they do not indicate the presence of any significant concentration of any special type of lattice defect-O donor complex that is stable against annealing at temperatures above 500°C. The relatively small increase in donor concentration as a consequence of continued annealing of 600°C (Fig. 10) is not due to removal of D_f, since it occurred in all samples, but may be due to the formation of some more complex type of oxygen donor at this annealing temperature.

(F) Isothermal Annealing of NTD Si at 750 and 950°C

Previous experiments[2] have demonstrated that annealing for 30 min at 750°C is sufficient to obtain the anticipated carrier concentration and mobility in NTD FZ Si, irrespective of neutron fluence or t/f ratio. In more recent experiments,[16] samples of FZ and Cz Si were irradiated simultaneously and annealed for 30 min at 750°C. The donor concentration measured in the NTD FZ Si samples was always nearly identical to the P concentration, as calculated over a wide range of neutron fluence. The donor concentration as measured in the NTD Cz Si samples, however, was larger than the calculated P concentration, particularly for low fluence irradiations. No change was observed in the donor concentration in any of the NTD FZ Si samples as a consequence of extended annealing at 750°C, but the donor concentration increased in each NTD Cz Si sample by about 10^{15} cm^{-3}, irrespective of an initial D_f concentration that was increased from about 10^{14} to 10^{18} cm^{-3} in different samples. This additional donor concentration in the NTD Cz Si samples was decreased substantially by subsequent annealing at 950°C, but some fraction ($\leq$ 10%) remained even after extended annealing at 950°C.

This type of experiment has been continued. Figure 11 is a graph of donor concentration vs time of annealing at 750 and 950°C, for selected samples from the A-111 and B-1 ingots of Table 1, after irradiation in the D-6 locale for 1 to 800 h to introduce 5.8 x 10^{11} to 4.6 x 10^{14} P cm^{-3}. The apparent donor concentration in the A-111 samples was nearly identical to the anticipated P concentration after annealing for 30 min at 750°C. It was stated in Section (A) that ingot B-1 was not post-annealed, and that the initial donor concentration in B-1 samples was changed to an apparent low acceptor

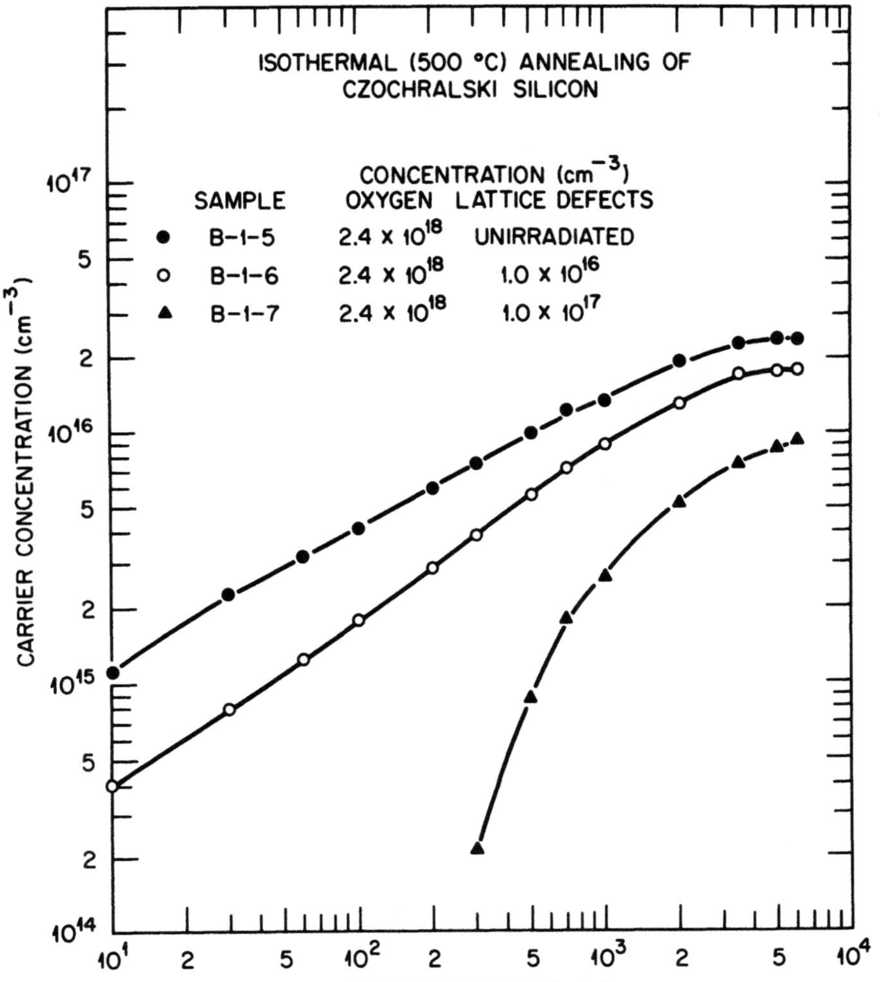

Fig. 9. Carrier concentıation vs annealing time at 500°C for Cz Si samples.

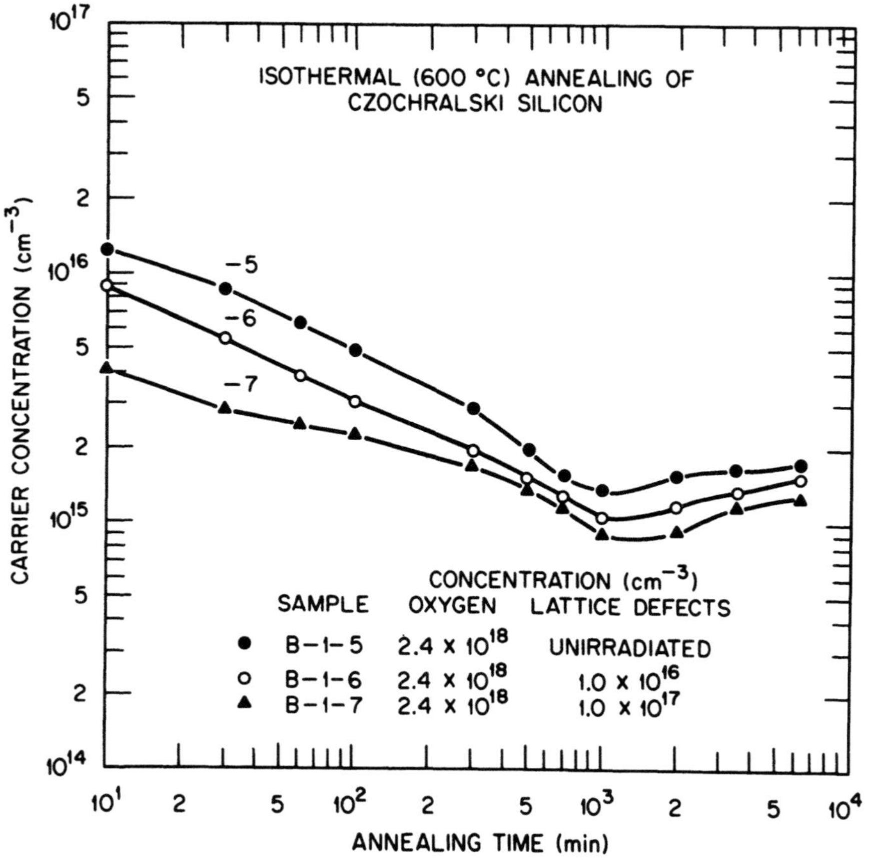

Fig. 10. Carrier concentration vs annealing time at 600°C for Cz Si samples.

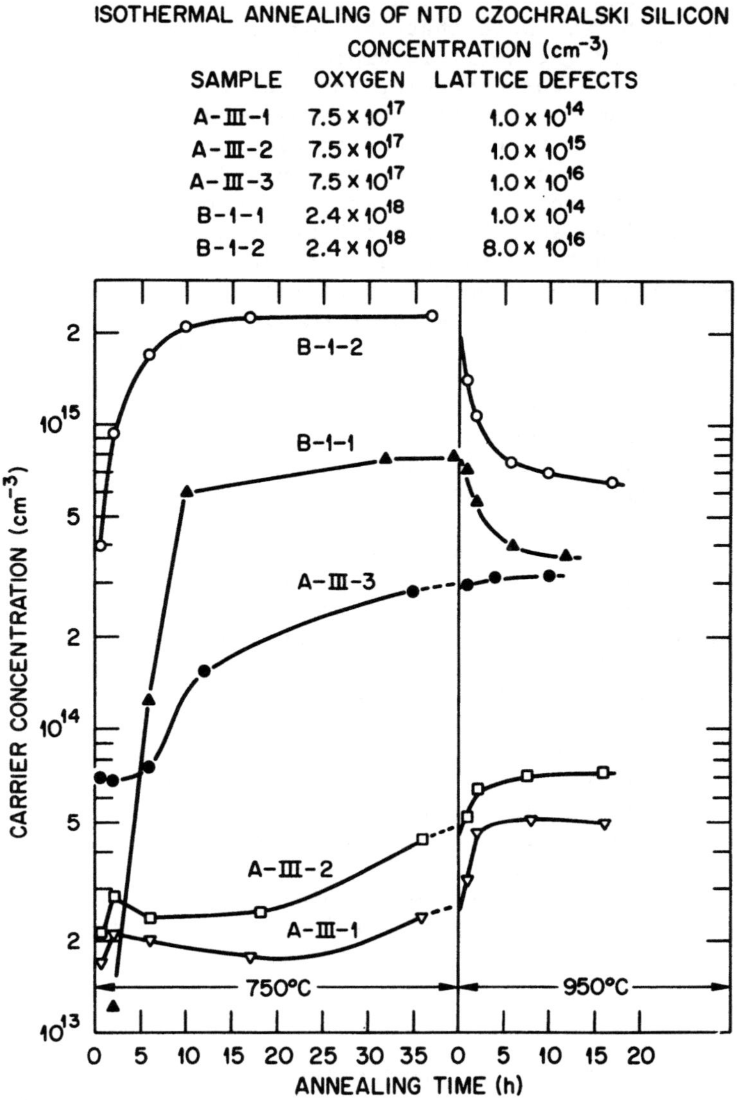

Fig. 11. Carrier concentration vs annealing time at 750 and 950°C for NTD Cz Si samples.

concentration after annealing for 30 min at 750°C. The apparent donor concentration in the B-1 samples shown in Fig. 11 was much larger after irradiation and annealing for 30 min at 750°C than would be anticipated if the initial O-donor concentration had been removed by the first annealing treatment. The donor concentration increased in both the A-111 and B-1 samples upon extended (30-40 h) annealing at 750°C. Note that the apparent increase was only about 4×10^{12} cm^{-3} for the A-111-1 sample that contained about 10^{14} D_f cm^{-3}; whereas, the corresponding increase was about 5×10^{14} cm^{-3} for the B-1-1 sample that also contained about 10^{14} D_f cm^{-3}. The initial D_f concentration in these samples was presumably identical, and the initial O_i concentration was only different by about a factor of 2, but the apparent increase in donor concentration on extended annealing at 750°C was different by a factor of more than 100. The fact that the increase in donor concentration in the B-1-1 sample was larger than the D_f concentration, by a factor of 5, would also appear to argue against the formation of a special type of lattice defect O complex as a consequence of irradiation or extended annealing at 750°C.

It is evident from the data of Fig. 11 that the apparent donor concentration increased to some extent in the A-111 and B-1 samples as a consequence of extended annealing at 750°C, and that the increase was larger for those samples that contained more D_f initially. However, there is no evident correlation between the initial D_f concentration and the increase in donor concentration on extended annealing. Note also that the apparent donor concentration continued to increase in the A-111 samples, that contained less O_i, as a consequence of annealing at 950°C; whereas, the corresponding data for the B-1 samples indicates a substantial reduction in donor concentration. Some fraction of an excess donor concentration remained, over that anticipated, in every sample investigated, but the excess amount was $1\text{-}3 \times 10^{15}$ cm^{-3} in the B-1 samples and $0.3\text{-}2 \times 10^{14}$ cm^{-3} in the A-111 samples. The fact that the initial increase in donor concentration on extended annealing was not a linear function of the initial D_f concentration, and the fact that donors were introduced at 950°C in the A-111 samples, but removed in the B-1 samples, also suggests that donor addition or removal at the different annealing temperatures was a function of initial O_i content but was not a sensitive function of the D_f concentration.

DISCUSSION

The primary result of these experiments is a better understanding of the electrical properties of O donors in Cz Si and of the annealing requirements to remove such donors and thus obtain the anticipated carrier concentration and mobility in Cz or NTD Cz Si. Commercially available Cz Si is usually post-annealed to remove O donors, but some such donors may remain. Any subsequent heat treatment above about 300°C can form O donors at an initial rate $\geq 10^{15}$ cm^{-3} hr^{-1} in Cz Si that contains $\geq 10^{18}$ O_i cm^{-3}. The maximum O donor concentration on extended ($\geq$ 100 h) annealing can exceed 1% of the initial O_i concentration. There is no evident interaction between O and donor impurities in n-type Cz Si,but previous investigators[6,7,9] have reported that O and acceptor impurities form some type of donor in p-type Cz Si. They stated that annealing above about 500°C can reduce the O-donor concentration in n-type Cz Si, or in p-type Cz Si doped with B or Ga, but that O-Al donors may have been formed that require annealing above 1000°C.

In the present experiments, it was determined that a small fraction ($\sim$ 1%) of any O donors introduced in Cz Si by heat treatment at 350-500°C remained after subsequent heat treatment to about 1000°C. Most of the O donors formed at 350-500°C were removed by annealing for 20-50 h at 550°C, 5-10 h at 650°C, or 30 min at 750°C, but new O donors were then formed by extended annealing at these temperatures. An O-donor concentration in excess of 10^{15} cm^{-3} was introduced in samples of Cz Si of different initial O_i concentration by annealing for 900 h at 550°C or for 100 h at 650°C; these O donors were stable against higher temperature annealing to at least 850°C.

High temperature (1000-1200°C) heat treatment for 20 h has been reported[6,11] to be sufficient to eliminate any subsequent low temperature (350-500°C) O-donor formation in Cz Si, although de Kock[12] observed that some O_i remained after 1200°C heat treatments. In the present experiments, it was demonstrated that prior heat treatment for 20-40 h at 1000-1200°C was not sufficient to prohibit some O-donor formation in samples of Cz Si of different initial O_i concentration by subsequent low temperature heat treatment.

The initial rate of O-donor formation and the final concentration of O donors introduced in NTD Cz Si by annealing at 450°C was not altered by the simultaneous presence of up to $\leq 10^{16}$ cm^{-3} fast neutron-induced lattice defects (D_f). Both the initial rate and the final concentration were reduced after comparable annealing at 450°C in more heavily irradiated samples, but a donor concentration in excess of 10^{15} cm^{-3} was formed in a NTD Cz Si sample that contained 10^{18} D_f cm^{-3}.

Previous studies[1,2] have indicated that 30 min annealing at 750°C was sufficient to obtain the anticipated carrier concentration and mobility in NTD FZ Si irradiated up to a large ($\geq 10^{20}$ cm^{-2}) neutron fluence. In the present experiments, it was determined that comparable annealing reduced the O-donor concentration in NTD Cz Si, but a residual concentration of $\geq 10^{14}$ cm^{-3} remained irrespective of the D_f concentration. These results indicate that formation of O donors in NTD Cz Si, by heat treatment at 450°C, and any subsequent removal of O donors by 30 min annealing at 750°C, was not altered significantly by the simultaneous presence of lattice defects. No evidence was obtained for the formation of an O-lattice defect complex that reputedly acts as a donor and is stable against high temperature (850-1000°C) annealing.[13-15] A residual donor concentration of $\geq 10^{14}$ cm^{-3} was observed in every sample, irrespective of initial $\overline{O}_i$ and D_f concentration, after annealing, but that was probably due to incomplete removal of all O donors.

The primary result of iochronal annealing studies on NTD FZ or Cz Si samples that contained 10^{16} or 10^{17} D_f cm^{-3} was that the presence of D_f impeded the formation of O donors in Cz Si samples at 500°C, but 90-95% of the O donors were removed on annealing at 600°C, irrespective of D_f concentration. These data may indicate some type of interaction between O and D_f as a consequence of irradiation or partial annealing, but they do not indicate the presence of a special type of O-defect complex that is stable against higher temperature annealing. It was demonstrated that O donors are formed in unirradiated Cz Si by isothermal annealing at 500 or 600°C and that such donors are stable against annealing to about 850°C, at which temperature the introduction and removal of that type of O donor in Cz Si was totally independent of any initial D_f concentration.

Some type of O donor was also formed in NTD Cz Si samples by extended (30-40 h) annealing at 750°C. The apparent increase in donor concentration was a factor of 5 larger than the D_f concentration for samples with a large initial O_i concentration, but the total increase in donor concentration was not a linear function of the initial D_f concentration. The apparent donor concentration continued to increase in samples of low initial O_i content, on extended annealing at 950°C, but decreased in samples of large O_i content under similar annealing conditions. The fact that the initial increase in donor concentration on extended annealing at 750°C was not a linear function of the D_f concentration, together with the fact that donors were added at 950°C in samples of low O_i content, but removed in samples of large O_i content, would also indicate that donor addition or removal was a function of the O_i content but was not a function of the D_f concentration.

SUMMARY

It has been demonstrated that one can introduce O donors in Cz or in NTD Cz Si by low temperature (300-500°C) annealing. The donor concentration on extended annealing is a sensitive function of the O_i content, but is not a function of the lattice defect content up to about 10^{16} D_f cm^{-3}. Some fraction of the O-donor concentration remains in Cz or NTD Cz Si after isochronal or isothermal annealing up to about 1000°C. That fraction is also a sensitive function of the O_i content, but is not a function of the D_f content. The results of isochronal annealing studies are such as to indicate that the presence of D_f impedes the formation of O donors on annealing at 500°C, but 90-95% of the O donors are removed on annealing at 600°C. It was also demonstrated that one can introduce O donors in unirradiated

Cz Si by annealing at 500 or 600°C, and that such donors are stable against annealing to about 850°C. Introduction and removal of that type of O donor is totally independent of any initial D_f concentration. Some type of O donor is also formed in NTD Cz Si by extended (30-40 h) annealing at 750°C, but the increase in donor concentration in different samples is not a linear function of the D_f concentration. All of these results serve to indicate that annealing for 30 min at 750°C is sufficient to remove virtually all of the D_f in NTD FZ or Cz Si and virtually all of the O donors in NTD Cz Si, irrespective of neutron fluence or t/f ratio. Previous observations of an O-defect-induced donor, that required high temperature annealing, may have been due to the presence of O donors prior to irradiation, or to the introduction of additional O donors during annealing.

ACKNOWLEDGMENTS

The authors wish to thank L.E. Katz, Bell Telephone Laboratories, Allentown, PA, and H.M. Hobgood, Westinghouse R and D Center, Pittsburgh, PA, for providing the B-1, ^{10}B, and W-1 ingots, respectively, P.H. Fleming for experimental assistance, and G.E. Jellison, Jr., D.H. Lowndes, and F.W. Young, Jr. for many useful discussions.

REFERENCES

1. "Proceedings of the Second International Conference on Neutron Transmutation Doping in Semiconductors," Jon M. Meese, ed., Plenum Press, New York, (1979).
2. R. T. Young, J. W. Cleland, R. F. Wood and M. M. Abraham, J. Appl. Phys. 49: 4752 (1978).
3. R. D. Westbrook and J. W. Cleland, Extended Abstracts, J. Electrochem. Soc. 125: No. 3, 144C (1978).
4. B. C. Larson, R. T. Young, and J. Narayan, p. 281 in "Proceedings of the Second International Conference on Neutron Transmutation Doping in Semiconductos," Jon M. Meese, ed., Plenum Press, New York, (1979).
5. P. Capper and J. G. Wilkes, Appl. Phys. Lett. 32:187 (1978).
6. W. Kaiser, Phys. Rev. 105:1751 (1957): Phys. Rev. 112:1546 (1978).
7. C. S. Fuller, Chemical Reviews 59: No. 1, Feb. 1959.
8. A. Kanamori, Appl. Phys. Lett. 34:287 (1979).
9. D. Helmreich and E. Sirtl, Semiconductor Silicon/1977, J. Electrochem. Soc. PV-77-2, p. 626 (1977).
10. C. S. Fuller and R. A. Logan, J. Appl. Phys. 28:1427 (1957).
11. J. A. Baker, "Oxygen in Silicon," ASTM Review Article-E-521 (available on request from Dow Corning, Inc., Hemlock, Michigan).
12. A. J. R. deKock, p. 103 in "Physics of Semiconductors," B. L. H. Wilson, ed., Conference Series No. 43, Institute of Physics, Bristol and London.

13. R. N. Thomas, J. Appl. Phys. 49(5):2811 (1978).
14. H. M. Liaw and C. J. Varker, Semiconductor Silicon/1977, J. Electrochem. Soc. PV-77-2, p. 46 (1977).
15. V. N. Mordkovich, S. P. Solov'ev, E. M. Temper and V. A. Kharchenko, Sov. Phys. Semicond. 8 No. 1: July 1974.
16. J. W. Cleland, P. H. Fleming, R. D. Westbrook, R. F. Wood and R. T. Young, p. 1 in "Proceedings of the Second International Conference on Neutron Transmutation Doping in Semiconductors," Jon M. Meese, ed., Plenum Press, New York (1979).
17. The authors wish to thank L. E. Katz, Bell Telephone Laboratories, Allentown, PA, and H. M. Hobgood, Westinghouse R and D center, Pittsburgh, PA, for providing the B-1, ^{10}B and W-1 ingots, respectively.

IMPURITY INTERACTIONS WITH STRUCTURAL DEFECTS IN IRRADIATED SILICON

R.C. Newman

J.J. Thomson Physical Laboratory

University of Reading Reading RG6 2AF

ABSTRACT

Silicon irradiated in a reactor will suffer damage as a result of the γ-flux, β-recoils and fast neutron collisions. Damage of the first type should resemble that produced by fast electron irradiation. Direct neutron collisions are most easily investigated in material that has been subjected to high fluences in reactors which may be less than ideal for the production of NTD silicon. Damage produced in these two ways will be discussed and special emphasis will be given to the information obtained from infrared absorption measurements, especially those relating to boron, carbon and oxygen complexes. It will be shown that irradiation and annealing procedures can lead to non-equilibrium defects involving these impurities. We shall also report on recent small angle neutron scattering SANS measurements which allow the sizes of clusters and the absolute numbers of defects involved to be estimated.

1. INTRODUCTION

Much early work was concerned with the damage produced by electron irradiations carried out with silicon samples held at temperatures below 300K. It became clear that intrinsic defects are mobile below 300K and small concentrations of impurities play an important role because they act as traps for the primary defects.[1] At low damage concentrations they also control the position of the Fermi level which determines the charge state and hence the mobility of various complexes.[2] Some impurities are normal dopants, but others may be present only as a result of unavoidable or inadvertant contamination. The importance of oxygen was recognised at an early stage[3-5] but it was much later that the presence of carbon was demonstrated[6] and even later that its deleterious effects on device performance became apparent.[7]

The research on damage moved on with investigations of the effects of higher energy electrons, fast neutrons and heavier ions, and with the samples irradiated at room temperature.[8] The interpretation of measurements, now related to clustered defects has become progressively more difficult. However, it could be conjectured that if samples were annealed to a sufficiently high temperature under clean conditions, the defects should anneal out and we should again be left with a "good" crystal. A primary purpose of this paper is to question the proposition that annealed NTD silicon contains the defects that we would normally expect to find in melt grown crystals. Some measurements that we shall discuss indicate that this is not always so. Firstly, we shall describe briefly the types of damage produced by reactor irradiations. Most emphasis will then be given to a review of impurity pairing and clustering; other discussion will be devoted to the severe damage produced by primary collisions of fast neutrons with silicon lattice atoms.

2. THE NATURE OF THE DAMAGE

Silicon irradiated in a nuclear reactor will suffer damage because of (a) the incident flux of γ-rays, (b) recoils produced by γ and charged particle emission, and (c) direct collisions with fast neutrons. Typically, we should consider material that has received a thermal neutron fluence of the order of 10^{18} cm^{-2}, which means that there will have been an associated fast neutron fluence in the range 10^{15} to 10^{17} cm^{-2}. Calculations indicate that this latter radiation is the most important in producing displacements of lattice atoms.[9] For a spectrum averaged neutron cross section of 2·9 barns[10], a fluence of 10^{17} will produce about 10^{16} primary collisions per cm^3. In the region of a direct collision we expect about 10^3 displacements, and silicon interstitials thus generated will move outwards leaving behind a vacancy rich zone.[11] Vacancies are unstable at room temperature and coalesce to form divacancies[12] or larger aggregates, while others recombine with silicon interstitials so that the number of effective displacements is less than the calculated number. For a fast neutron dose of 10^{17} cm^{-2}, these heavily damaged regions are not expected to overlap significantly and we must take account of the migration of vacancies and interstitials into the surrounding regions of relatively good crystal where they can be captured by impurities. These diffusing intrinsic defects are of course no different from those generated more homogeneously by the other types of irradiation. Annealing processes will have to be considered for both the relatively lightly damaged matrix as well as for the heavily damaged regions near primary collisions.

3. DIAGNOSTIC TECHNIQUES

As we shall be concerned with infra-red and n^o-scattering measurements we shall make a few comments about these techniques. Infra-red absorption arising from defects in a silicon crystal may arise as a result either of electronic transitions or vibrational transitions[13] and either type of absorption may be in the form of broad features or lines (fig. 1). "Strain" in the crystal arising from structural defects, internal electric fields, etc., will cause a smearing out of the sharp edges in the density of electronic states at the valence and conduction bands.[14] Associated electronic absorption below the band gap is then observed and has been called near edge absorption. Likewise, the continuum of phonon modes up to the Raman frequency ω_R at 523 cm^{-1} becomes infra-red active. In addition, there may be specific electronic bands due to particular defects and vibrational bands which may be either below or above ω_R, when the modes become localized spatially (LVM) and show strong isotope effects. Lines become broadened when strain is present and we shall discuss this effect in relation to the 9μm oxygen band. Unfortunately, a measurement of the integrated absorption of an optical feature does not allow the concentration of the defect to be determined, unless a prior calibration of the dipole moment has been effected.[15]

Neutron scattering measurements[16-18] require a high flux of well collimated monochromatic incident particles with a wavelength λ sufficiently great to prevent Bragg reflections from occurring. For the work on silicon discussed below a wavelength of $9{\cdot}0 \pm 0{\cdot}8$Å was used and this allowed a range of scattering vectors $|k| = 4\pi \sin\theta/\lambda$ of $0{\cdot}01$ to $0{\cdot}10$Å^{-1} to be investigated; the scattering angle is 2θ. The physical principle involved in the measurements is similar to the optical scattering produced by small inhomogeneities in a medium and which can be commonly observed as haloes around a bright light. There are advantages compared with X-ray and TEM investigations.[19] Dynamical scattering is negligible because the scattering factor for neutrons is small and specially thin samples are not required as for TEM studies. Thus a series of measurements can be carried out on a bulk sample following an irradiation and subsequent anneals. The technique is quantitative because the cross sections are known and it is possible to calculate absolute numbers of defects that are present as well as obtaining information about their sizes. The main disadvantage is that of limited sensitivity and fractional concentrations of defects of the order of 10^{-3} are commonly required.

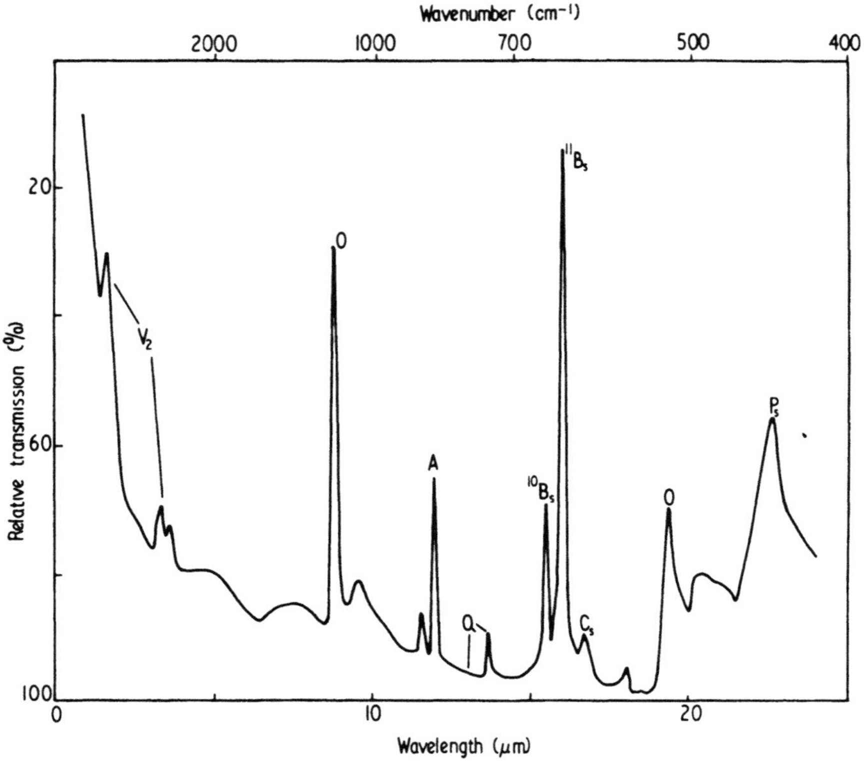

Fig. 1. IR spectrum of CZ silicon doped with boron and phosphorus to 10^{19} cm^{-3} and irradiated with 2MeV electrons to eliminate free carriers. Absorption features are near edge absorption 1-5μm : V_2, electronic bands : O, oxygen LVM : A, [O-V] LVM : Q, boron interstitial centre : B_s, substitutional boron lines : C_s, carbon LVM : P_s, phosphorus resonant mode : band mode continuum below 523 cm^{-1}.

4. DAMAGE IN THE MATRIX

It is generally true that all the simpler defects[20] anneal by 600°C and that strain in the crystal is removed after anneals to a somewhat higher temperature of 750°C. However, previous work does pose unanswered questions. In particular the mechanism of the anneals and the resulting new arrangements of atoms are not usually determined. For example, it is well known that divacancies anneal around 450K in neutron irradiated material,[21-22] but the fate of the vacancies lost in this stage is uncertain. It is no simple matter to keep track of vacancies or self-interstitials when they cluster, and so it would seem more profitable to ask similar questions about defects involving impurity atoms and where we might hope to track the impurity. For example, it is known that after a low temperature irradiation, substitutional aluminium atoms $A\ell_s$ capture mobile silicon interstitials and exchange sites. On annealing to about 150°C the $A\ell_i$ atoms become mobile and are trapped by further $A\ell_s$ atoms to form $[A\ell_s^- - A\ell_i^{2+}]$ deep donor complexes.[1] Thus the irradiation has brought together two impurity atoms that were originally immobile and well separated in the crystal. This process occurs because of the high diffusion coefficient of the interstitial species. The diffusion rates of defects with an adjacent trapped vacancy, such as $[P_s - V]$, are also high.[23] We can therefore ask whether we obtain pairing of phosphorus with other impurities during an anneal when the E-centre becomes mobile. The next question to ask is whether annealing to even higher temperatures causes such pairs of impurities to dissociate, or do they react with a second intrinsic defect to form a pair of adjacent substitutional impurities. It would then take an anneal at a much higher temperature to dissociate this type of complex. Answers to these questions do not appear to be available in general, but we do have some data for boron, phosphorus, carbon and oxygen.

It is pertinent to start with the work of Angress et al. who were interested in the infrared vibrational absorption produced in silicon by the presence of boron, phosphorus and arsenic impurities.[24-25] Their samples were doubly doped with boron and donor atoms to achieve a large degree of electrical compensation during growth and the final compensation was effected by a relatively small dose of 2MeV electron irradiation at 300K. The spectra (figs. 1 and 2) showed localized modes (LVM) from substitutional boron and a continuum of absorption below ω_R, including a resonant mode at 441 cm^{-1} due to substitutional phosphorus (366 cm^{-1} for As_s). On further irradiation of the samples, all the sharp features showed a reduction in strength which we can now interpret in a simplistic way : the boron atoms capture silicon interstitials while the phosphorus atoms capture vacancies. Subsequently, similar samples were investigated with the emphasis on the effects of the radiation[26] : the removal of boron from substitutional sites was quantified : lines due to boron paired with substitutional donors were identified [B-P], [B-As] and [B-Sb][27-28] : lines due to an interstitial boron defect (Q-centre) were found : lines from an unrelated centre (P-centre) involving two adjacent boron atoms[29] were analysed (fig.2) and it was verified that both the Q and P-centres were produced by the irradiation. The structure of the Q-centre which is produced in FZ material with a low carbon content has not been determined but it is doubtful if it is an isolated boron interstitial. The latter defect anneals in p-type silicon at room temperature according to EPR data and the problem of the interpretation has been fully discussed by Watkins.[30] Annealing these samples led to changes in

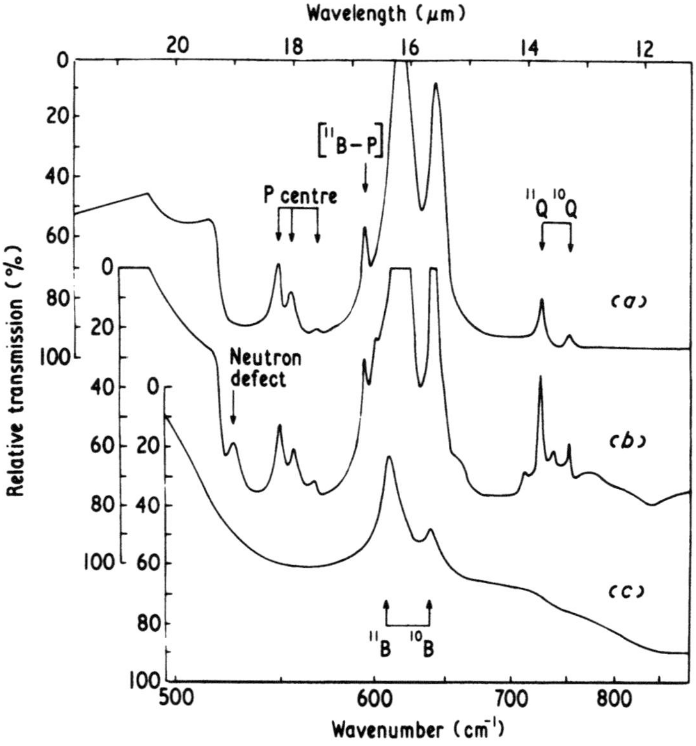

Fig. 2. IR spectrum of FZ silicon doped with boron and phosphorus to 5×10^{19} cm^{-3} showing boron LVM, the Q-centre (Bi), the P-centre (B-B pairs) and band mode absorption. (a) after 5×10^{18} e$^-$ cm^{-2} (b) 10^{18} n^o cm^{-2} and (c) 10^{20} n^o cm^{-2}.

the absorption. In particular, at about 200°C the B_i defects (Q-centres) were removed with a corresponding increase in the strength of the lines from [B_s-P_s] pairs. It was concluded that either an E-centre migrates to an interstitial boron atom, or a B_i is liberated and migrates to and combines with the E-centre to form a very stable defect that would not show electrical or spin resonance activity. Such samples could be heat treated at say 1000°C, re-irradiated at room temperature and re-examined to find whether the enhanced concentration of [B-P] complexes remained. Such a complicated procedure is necessary because an anneal at a temperature above about 350°C causes the return of free carrier absorption, thus necessitating the second irradiation. So far only one such experiment has been performed.[26] A particular sample similar to the one used to obtain the spectrum in fig.1, initially showed no detectable absorption from [B-P] pairs after growth and e^-irradiation (2MeV) to a dose of 5×10^{18} cm^{-2}. The sample was annealed up to 800°C and then further irradiated, where upon the absorption line due to [B-P] pairs was large and the estimated concentration of the pair defect was greater than 10^{18} cm^{-3}.

Similar effects have been observed[31] in material subjected to relatively low doses of fast neutrons up to 10^{18} cm^{-3}, indicating that the nature of the damage in the matrix is similar to that produced by 2MeV electrons. It is not until the n^o-dose is greater than about 3×10^{18} cm^{-3} that significant broadening of absorption lines occurs as shown in fig.2. Very heavy n^o-irradiations of 10^{19} - 10^{20} cm^{-2} cause nearly all the boron and phosphorus atoms to be removed from their original substitutional sites and there is a considerable build up of strain as evidenced by very strong near edge absorption and band mode one-phonon absorption characteristic of that found in pure silicon after n^o-irradiation.[22,32] It might have been conjectured that the concentration of the Q-centre would simply increase with increasing n^o dose. For low doses this is so, but then saturation effects occur and for higher doses there is a reduction in strength as the concentration of B_s falls (fig.2). This behaviour cannot be attributed entirely to the fact that an increasing volume of the crystal is becoming highly damaged corresponding to primary n^o collisions, because similar behaviour is observed in the electron irradiated samples. It is our opinion that the interstitial boron defects act as nucleation centres for the growth of larger clusters of self interstitials. The presence of such clusters could account for a broad absorption band observed in the spectral range from 900 cm^{-1} to 500 cm^{-1} in samples irradiated to very high n^o-doses; this band has never been observed in any sample that did not contain boron. On annealing, the broad feature is reduced in strength and there is continued growth of [B-P] centres, even after the Q-bands have annealed (fig.3). Free carriers start to return at about 300°C and by 400°C the resistivity drops so that optical measurements cannot be continued. The conclusion is that irradiation followed by annealing leads to appreciable pairing of substitutional donors and acceptors in concentrations higher than those found in as-grown crystals.

Another feature of interest is that the electrical recovery of different samples containing 2×10^{19} boron cm^{-3} (no compensating donors) was initially lower for a sample given a n^o dose of 5×10^{19} cm^{-2}, compared with another with a dose of 10^{19} cm^{-2}. However, at about 500°C, the curves crossed and "complete recovery" occurred more readily in the samples given the higher dose.[33]

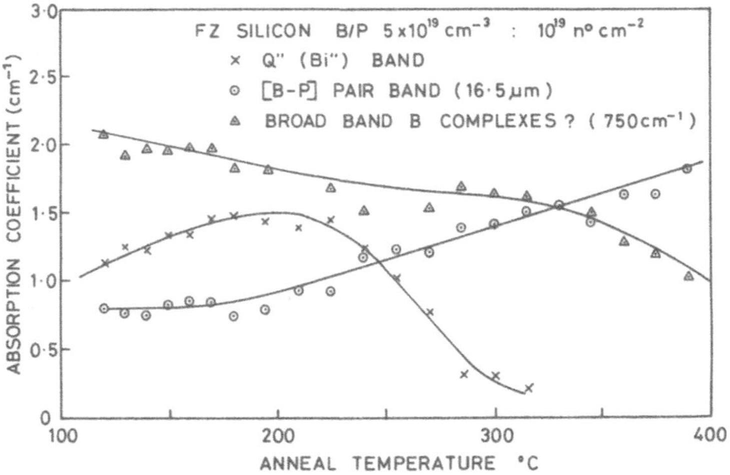

Fig. 3. Annealing behaviour of FZ silicon containing boron and phosphorus after a fast n^{o}-irradiation of 10^{19} cm^{-2}. Note steady increase in band due to [B-P] nearest neighbour pairs, even when the Q-band (Boron interstitial) has annealed.

The behaviour of neutral carbon is considered next. Self interstitials produced in an irradiation below room temperature are captured by C_s to produce bonded carbon interstitials which give rise to characteristic LVM absorption lines[34-35] : in p-type material these defects are also paramagnetic.[36] Interstitial carbon atoms migrate at room temperature and are then captured by other impurities. A well known example is the formation of $[O_i - C_i]$ pair in CZ crystals that gives the LVM line at 11·6µm (865 cm^{-1})[37] and previously designated as C(3) by us.[38] The C(3) defect is interesting because it behaves in a similar way with respect to radiation dose as the boron Q-centre. Thus the concentration of C(3) centres initially increases with dose, but then decreases as the concentration of C_s falls[39] in both neutron and electron irradiated material.[40] We have performed irradiations on carbon doped CZ material and found that eventually <u>all</u> of the carbon can be lost from those sites that lead to LVM absorption (fig.4). An unanswered question is, where is the carbon? Is it the same question as where was the boron? The carbon is not present as precipitated silicon carbide as this gives rise to characteristic absorption near 12µm.[41] It has been suggested that graphitic inclusions might form, or the C(3) centres nucleate aggregates of self interstitials.

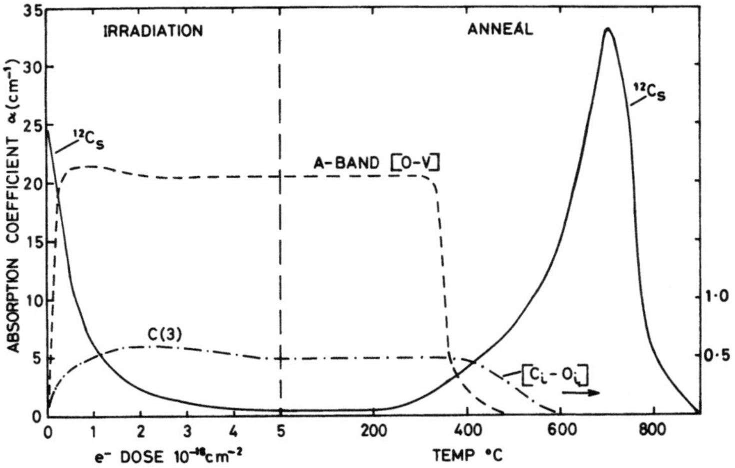

Fig. 4. Effect of 2MeV electron irradiation and annealing on a CZ crystal doped with carbon. Note removal of carbon from substitutional sites and its regeneration. SiC precipitates form after 800°C anneal.

Mobile C_i atoms are also trapped by remaining C_s impurities to form a pair structure in which the two atoms occupy inequivalent sites. This centre is paramagnetic[42] and it is thought to be responsible for LVM lines near 18μm but it has not been possible to interpret any clear isotope effects in samples containing enriched ^{13}C. This defect may be analogous to the boron P-centre (fig.2).

As mobile carbon is trapped by carbon and mobile boron is trapped by boron there is clearly the possibility of a hybrid structure being formed. Recently,[43] we have demonstrated that this is so. Samples of silicon containing about 10^{16} cm^{-3} of boron together with supersaturated carbon at a concentration of 10^{18} cm^{-3} showed the normal hole photo-ionization (electronic) absorption spectrum at a nominal 4°K appropriate to boron. There was no evidence for the presence of the [boron-X] spectrum, corresponding to defects with trigonal symmetry and with a somewhat smaller ionization energy than the isolated acceptor.[44] Other workers have provided strong evidence indicating that the latter centre is a boron-carbon nearest neighbour pair.[44,45] Our samples were then electron irradiated to doses of about 10^{18} cm^{-2} which removed the p-type conductivity, after which they were annealed at 600°C. Following these treatments we found that [B-X] centres were present in a concentration of about 20% of the unpaired boron acceptors. Heat treatment alone produced only a much smaller concentration of 1% of [B-X]. These observations are important because of the application of NTD to compensating unwanted boron acceptors in indium doped material to be used in the manufacture of extrinsic silicon infrared detectors.[46] We surmise that there could be simultaneous generation of unwanted [In-C] pair centres if the crystal contained carbon impurities and in addition [B-P] or [In-P] pairs might be produced.

It is of interest to speculate about the possibility of carbon forming pairs or more complicated defects with other impurities that may be present. Indeed we have generated [$A\ell_s$-X] pairs analogously to [B-X]. The former centre may then trap a mobile aluminium interstitial to give the defect [$A\ell_i^{2+}$ - $A\ell_s^-$ - C_s^o] with trigonal symmetry. We have previously proposed that this centre may account for the EPR spectrum designated as (SiG 20) by Watkins[1], leaving the spectrum (SiG 19) to be ascribed to the simpler [$A\ell_i^{2+}$ - $A\ell_s^-$] pair defect.[15,47]

Two types of annealing behaviour can be identified for heavily e^--irradiated silicon containing carbon. The first is the usual reduction in the concentration of defects such as [O_i-C_i] which occurs within a temperature range 300-500°C and there is generation of other [C-O] complexes which anneal at higher temperatures. Of more interest perhaps is the return of the lost carbon. It is seen from fig.4 that this is a gradual process starting near 200°C when the C(3) centre starts to anneal, but it continues up to 700°C, which is well above the temperature for the removal of C(3) centres. After an anneal at 800°C there was precipitation of carbon to form SiC particles.

In fast n^o-irradiated material the behaviour appears to be generally similar. The amount of substitutional carbon lost in the irradiation increases with doses up to 10^{19} cm^{-2}. Irradiation to higher doses "reduces" the amount of carbon lost from substitutional sites suggesting that there is then appreciable recombination of C_i with vacancies, to regenerate C_s. This is illustrated in Fig.5 for an irradiated phosphorus doped sample which also contained carbon, (similar results were obtained for high resistivity material). On annealing, it is seen that the sample given a dose of 5×10^{19} cm^{-2} recovers more rapidly than the others. Optical measurements cannot be continued to high temperature because of the return of free carrier absorption. Parallel resistivity measurements[33] show that the high temperature recovery of the donor activity of the phosphorus is also the greater in the sample given the higher dose of irradiation with a cross over in the region of 600°C.

5. DAMAGE DUE TO FAST NEUTRON COLLISIONS

In each primary collision a silicon atom will acquire kinetic energy and the effects locally will be akin to those produced by the

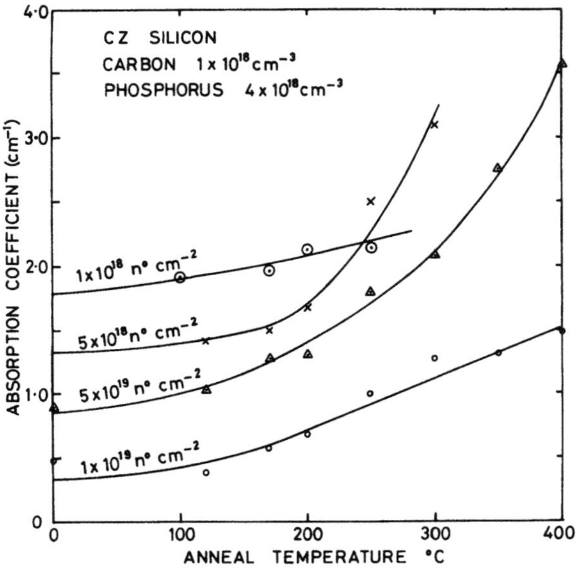

Fig. 5. Recovery of substitutional carbon in n°-irradiated phosphorus doped silicon after annealing.

implantation of silicon by a silicon atom with an energy of the order of 100 keV, but at a very low flux. The early work of Gossick[11] and Bertolotti et al[48] appears to verify this model. As the neutron dose is increased, the highly damaged zones will start to overlap, and there will be a build up of strain. The presence of strain can be detected either from subtle changes in ESR spectra or IR absorption features, or more directly from the scattering of X-rays, electrons in the electron microscope or by slow neutrons. For NTD silicon, where the fast n°-dose is low, information has to be derived primarily from the spectroscopic techniques because of the limited sensitivity of the diffraction methods where doses of up to 10^{20} cm^{-2} often have to be used. At this high dose there will be about 10^{22} cm^{-3} individual displacements, representing some 20% of the total number of lattice atoms. Nevertheless the silicon remains crystalline, implying that much of the damage is annealed during the irradiation. However, we would expect this material to be rather homogeneous in character, albeit different from the starting crystal.

The divacancy is a very important defect in relation to primary collisions. In either n or p-type silicon the defect shows an EPR spectrum, and various electronic absorption bands in the infrared at 1·7, 3·3 and 3·9μm[49] have been assigned quite unambiguously. Fortunately the 1·7μm IR band is also observed when neutral non-paramagnetic divacancies are present in high resistivity material. Measurements have shown that the activation energy for the anneal of divacancies is the same as that for the geometrical reorientation from say a [111] axis to a [$1\bar{1}1$] axis, indicating that the anneal results from the migration of the defect as an entity through the crystal. Annealing has been studied in n^o-irradiated crystals when we should expect to observe optical absorption from divacancies in the matrix and the regions near primary collisions, but EPR spectra would only be expected for charged defects in the former regions. These expectations have been realised by Daly and Noffke[50] who observed a difference in the annealing in the two regions and concluded that there was a very high concentration of divacancies near a primary collision. Thus the lower anneal temperature found in n^o-irradiated material compared with e^--irradiated crystals is explained, but the fate of the divacancies is not. The outstanding question is whether recombination with self-interstitials or the growth of larger vacancy aggregates is the predominant process. The reduction in the near edge absorption and band mode absorption would indicate a significant reduction in the crystal disorder, down to about 10% of that originally present.[21-22] Other measurements however, indicate only a much smaller change in the strain at this stage (fig.6).

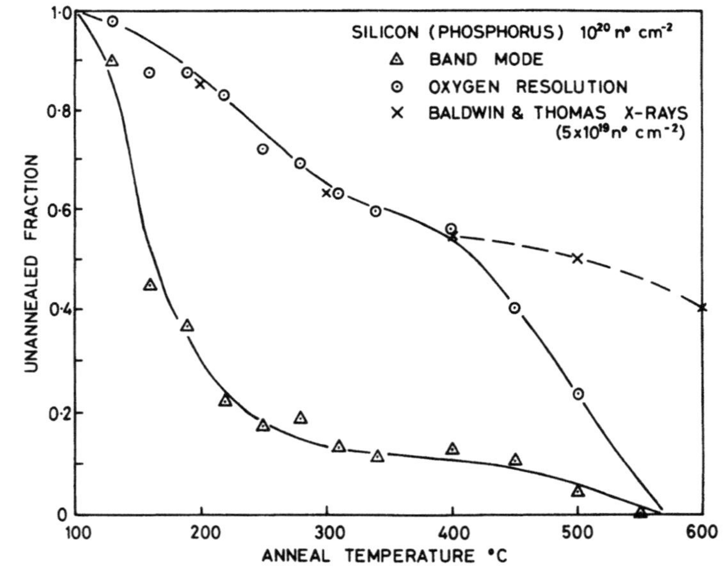

Fig. 6. Annealing of CZ silicon after 10^{20} n^o cm^{-2}. Note, resolution of oxygen 9µm band closely follows X-ray data for residual strain.

The 9μm absorption band due to interstitial oxygen shows fine structure which is easily resolved at 77K. The structure is related to "rotations" of the oxygen about the [111] axis between the two neighbouring silicon atoms to which it is bonded.[51] If the crystal is strained inhomogeneously this motion will be modified and a loss in resolution of the 9μm band is expected, as is observed in crystals doped with a high concentration of germanium.[52] Measurements of samples irradiated to doses in the range 10^{16} to 10^{20} n° cm^{-2} showed that the loss in resolution could be correlated with the growth in the band mode absorption, the near edge absorption and the concentration of divacancies.[53] However, the annealing of the strain did not correlate with the anneal of these various absorption features (fig.6), although there was much better correlation with the anneal of two localized mode lines at 530 and 530 cm^{-1} labelled "neutron defect" in (fig.2) which are induced by n°-irradiation[54] but not e^--irradiation; the defects giving rise to these bands do not involve impurities but their structure has not yet been determined. The annealing is also similar to that found from anomalous X-ray transmission data,[55] particularly if corrections are made to the high temperature measurements as indicated in the paper quoted.

Recently Beddoe et al.[17] found that silicon irradiated to a dose of 10^{20} n° cm^{-2} at 40°C showed small angle n° scattering which is reproduced in fig.7 for a sample in its as-irradiated state, after an anneal at 110°C and following subsequent anneals up to 700°C. The irradiation defects were not uniformly distributed and there were between 2-5 x 10^{20} cm^{-3} displaced atoms (vacancies plus interstitials) in the as-irradiated sample. The concentration of divacancies in a similar piece of material was estimated to be 3 x 10^{19} cm^{-3} from IR measurements, giving a total number of displaced atoms of 1·2 x 10^{20} cm^{-3} (vacancies plus interstitials). The two concentrations are very comparable, showing that **50% of the vacancies** present were divacancies. It was not surprising therefore, that there was a very large anneal in the SANS at about 100°C. This anneal was mainly of the scattering at small wavevectors corresponding to spatial inhomogeneities with a radius of 150Å and a defect density of about 10%. Also present were smaller regions with a radius of about 50Å and a defect concentration of up to 50%. The data indicated a volume density of these defects of about 10^{16} cm^{-3} which annealed slowly and were still detected up to temperatures of 700°C. Two models were considered in which the two types of damaged regions were concentric, or were spatially separated, but it was not possible to distinguish between these possibilities. What was clear however, was that the first annealing stage either led to the formation of very large defects about 1000Å in diameter, or vacancies recombined with self interstitials.

Some of our specimens have been examined by Dr. G.R. Booker using weak beam dark field TEM but he failed to detect contrast in

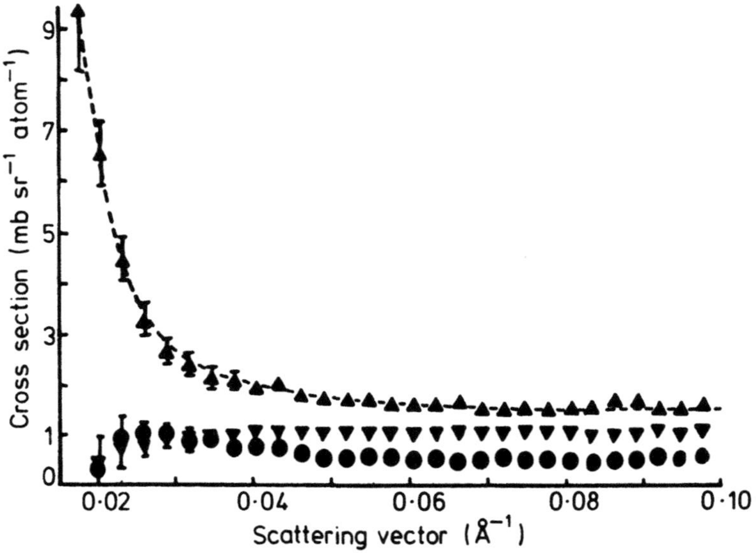

Fig. 7. Neutron scattering data for silicon after a fast n^o dose of 10^{20} cm^{-2} ▲; after an anneal at 110°C ▼; after further anneals up to 700°C ●.

any samples whether they were annealed or not and it was concluded that large loops were not produced. On the other hand, Larson et al.[56] did observe small defects by TEM, for one sample given a dose of 10^{20} n^o cm^{-3}, and they estimated a number density of 2×10^{16} cm^{-3}, very similar to the SANS data. The TEM data that have been reported are very variable in the sense that loops are sometimes seen but not always, and so it is necessary to enquire about the conditions of the irradiations. All our samples have been irradiated in the AWRE Herald reactor with water cooling at 40°C so that divacancies do not anneal. Other reactors operate at much higher temperatures and annealing during the irradiation is expected and would lead to significant differences.

6. SUMMARY AND CONCLUSIONS

We have shown that irradiation of silicon followed by annealing can lead to aggregation of impurities which would be well separated spatially in as-grown material. So far, we have no evidence for complexes such as [P-C]. The temperature of the samples during irradiation is an important parameter, as discussed previously,[57] but little attention seems to have been paid to variations in damage related to this factor. There is a problem of interpreting a number density of defects of only 10^{16} cm^{-3} in material where the number density of fast n^o collisions is 10^{19} cm^{-3}. The observations might reflect statistical fluctuations or special events such as points where (n^o, p) or (n^o, α) reactions occur. Alternatively, the highly damaged regions might be associated with interstitial clusters, possibly nucleated by residual impurities, including carbon and oxygen. Evidence for this type of process has been obtained and the model could provide a simple answer to the question of where do the interstitials go. Preliminary data suggest that material that has received a fast neutron dose of greater than 5×10^{19} cm^{-2} anneals at a lower temperature than less heavily irradiated samples.

ACKNOWLEDGMENTS

Thanks are due to Dr. G.R. Booker of Oxford University for TEM studies, to Dr. R.J. Stewart for helpful discussions and the SRC for their financial assistance.

REFERENCES

1) G.D. Watkins, Radiation Damage in Semiconductors, (Dunod, Paris, 1965), p 97.
2) L.C. Kimerling and C.P. Carnes, J. Appl. Phys. 42, 3548 (1971)
3) W. Kaiser, P.H. Keck and C.F. Lange, Phys. Rev. 101, 1264 (1956).
4) G.D. Watkins, J.W. Corbett and R.M. Walker, J. Appl. Phys. 30, 1198 (1959).
5) G. Bemski, J. Appl. Phys. 30, 1195 (1959).
6) R.C. Newman and J.B. Willis, J. Phys. Chem. Solids 26, 373 (1965).
7) N. Akiyama, Y. Yatsurugi, Y. Endo, Z. Imayoshi and T. Nozaki, Appl. Phys. Lett. 22, 630 (1973).
8) H.J. Stein, Neutron Transmutation Doping in Semiconductors (Plenum, New York, 1979) Ed. J.M. Meese, p 229.
9) J.M. Meese, Neutron Transmutation Doping in Semiconductors (Plenum, New York, 1979) Ed. J.M. Meese, p 1.
10) M.A. Kirk and L.R. Greenwood, Neutron Transmutation Doping in Semiconductors (Plenum, New York, 1979) Ed. J.M. Meese, p 143.
11) B.R. Gossick, J. Appl. Phys. 30, 1214 (1959).
12) G.D. Watkins, J. Phys. Soc. Japan 18, Suppl. II, 22 (1963).
13) R.C. Newman, Defects and their structure in non-metallic solids (Plenum, New York, 1975), Ed. B. Henderson and A.E. Hughes, p 221.
14) J.E. Fischer, Phys. Rev. 181, 1368 (1969).
15) R.C. Newman, Infrared Studies of Crystal Defects, (Taylor and Francis, London, 1973).

16) S. Gupta, E.W.J. Mitchell, R.J. Stewart and G. Kostorz, Phil. Mag. A37, 227 (1978).
17) R.E. Beddoe, S. Messoloras, E.W.J. Mitchell and R.J. Stewart, Defects and Radiation Effects in Semiconductors 1978 (I.o.P. London, 1979) Ed. by J.H. Albany, Conf. Ser. No. 46 p 258.
18) R.E. Beddoe, G. Kostorz, S, Messoloras, E.W.J. Mitchell and R.J. Stewart, submitted to Phys. Rev. Lett. (1980).
19) H.J. Stein, Radiation Effects in Semiconductors (Gordon and Breach, London, 1971), Ed. J.W. Corbett and G.D. Watkins, p 125.
20) J.W. Corbett and J.C. Bourgoin, Point Defects in Solids. Vol.2 Semiconductors and Molecular Crystals (Plenum, New York, 1975) Ed. J.H. Crawford, Jr. and L.M. Slifkin, p 1.
21) L.J. Cheng and J. Lori, Phys. Rev. 171, 856 (1968).
22) R.C. Newman and D.H.J. Totterdell, J. Phys. C : Solid St. Phys. 8, 3944 (1975).
23) E.L. Elkin and G.D. Watkins, Phys. Rev. 174, 881 (1968).
24) J.F. Angress, A.R. Goodwin and S.D. Smith, Proc. R. Soc. A287, 64 (1965).
25) J.F. Angress, A.R. Goodwin and S.D. Smith, Proc. R. Soc. A308, 111 (1968).
26) A.R. Bean, S.R. Morrison, R.C. Newman and R.S. Smith, J. Phys. C : Solid St. Phys. 5, 379 (1972).
27) V. Tsvetov, W. Allred and W.G. Spitzer, Localized Excitations in Solids (Plenum, New York, 1968) Ed. R.F. Wallis p.185.
28) R.C. Newman and R.S. Smith, Localized Excitations in Solids (Plenum, New York, 1968), Ed. R.F. Wallis, p 177.
29) R.C. Newman and R.S. Smith, Phys. Lett. A24, 671 (1967).
30) G.D. Watkins, Phys. Rev. B12, 5824 (1975).
31) K. Laithwaite, R.C. Newman and D.H.J. Totterdell, J. Phys. C : Solid St. Phys. 8, 236 (1975).
32) M. Balkanski and W. Nazarewicz, J. Phys. Chem. Solids 23, 573 (1962).
33) K. Laithwaite and R.C. Newman, unpublished results.
34) A.R. Bean and R.C. Newman, Solid St. Commun. 8, 175 (1970).
35) A. Brelot and J. Charlemagne, Radiation Effects in Semiconductors (Gordon and Breach, London, 1971), Ed. by J.W. Corbett and G.D. Watkins, p 161.
36) G.D. Watkins and K.L. Brower, Phys. Rev. Lett. 36, 1329 (1976).
37) R.E. Whan, J. Appl. Phys. 37, 3378 (1966).
38) R.C. Newman and A.R. Bean, Radiation Effects in Semiconductors, (Gordon and Breach, London, 1971), Ed. by J.W. Corbett and G.D. Watkins, p 155.
39) L.J. Cheng and P. Vajda, J. Appl. Phys. 40, 4679 (1969).
40) M.R. Brozel, R.C. Newman and D.H.J. Totterdell, J. Phys. C : Solid St. Phys. 8, 243 (1975).
41) A.R. Bean and R.C. Newman, J. Phys. Chem. Solids 32, 1211 (1971).
42) K.L. Brower, Phys. Rev. B9, 2607 (1974).
43) J. Woodhead and R.C. Newman, to be published.
44) C.E. Jones, D. Schafer, W. Scott, R.J. Hager, Bull. Am. Phys. Soc. 24, 276 (1979).
45) R. Baron, J.P. Baukus, S.D. Allen, T.C. McGill, M.H. Young,

H. Kimura, H.V. Winston and O.J. Marsh, Appl.Phys.Lett.34 257 (1979).

46) H.M. Hobgood, T.T. Braggins, J.C. Swartz and R.N. Thomas, Neutron Transmutation Doping in Semiconductors (Plenum, New York, 1979) Ed. J.M. Meese, p 65.

47) S.D. Devine and R.C. Newman, J. Phys. Chem. Solids, 31, 685 (1970).

48) M. Bertolotti, D. Sette and G. Vitali, J. Appl. Phys. 38, 2645 (1967).

49) L.J. Cheng, J.C. Corelli, J.W. Corbett and G.D. Watkins, Phys. Rev. 152, 761 (1966).

50) D.F. Daly and H.E. Noffke, Radiation Effects 10, 191 (1971).

51) D.R. Bosomworth, W. Hayes, A.R.L. Spray and G.D. Watkins, Proc. Roy. Soc. A317, 133 (1970).

52) H.J. Hrostowski and B.J. Alder, J. Chem. Phys. 33, 980 (1960).

53) D.H.J. Totterdell and R.C. Newman, J. Phys. C : Solid St. Phys. 8, 589 (1975).

54) M.T. Lappo and V.D. Tkachev, Sov. Phys. Semicond. 4, 1882 (1971).

55) T.O. Baldwin and J.E. Thomas, J. Appl. Phys. 39, 4391 (1968).

56) B.C. Larson, R.T. Young and J. Narayan, Neutron Transmutation Doping in Semiconductors, (Plenum, New York, 1979), Ed. J.M. Meese, p 281.

57) L. Katz and E.B. Hale, Neutron Transmutation Doping in Semiconductors, (Plenum, New York, 1979), Ed. J.M. Meese, p 307.

DEFECT PRODUCTION DURING NEUTRON DOPING OF Si

J. M. Meese[†*!], M. Chandrasekhar[†*], D. L. Cowan[*], S. L. Chang[†*], H. Yousif[*], H. R. Chandrasekhar[*] and P. McGrail[†#]

University of Missouri

Columbia, Missouri

ABSTRACT

It is of interest to observe and classify the various defects which exist immediately after irradiation since it is at this point that differences in the neutron energy spectra of different reactors are most apparent. We will discuss methods of estimating the displacement rate at which Si atoms are dislodged from their normal lattice sites during a neutron irradiation induced displacement cascade. We also will present the results of a variety of measurements using EPR, optical absorption, Raman scattering and DLTS which have been used to observe a number of defects which have been produced in Si by room temperature neutron irradiation. The numbers of displacements observed by these techniques will be compared to a simple Kinchin-Pease cascade model.[1-3]

INTRODUCTION

The neutron transmutation process has a number of merits for a variety of semiconductor applications, however, the process is not without certain drawbacks.[4] Perhaps the most serious problem is the radiation damage induced by fast neutron knock-ons and recoil displacements which accompany gamma and beta emission.[4,5] This radiation damage must be annealed by heat treatment and the minority carrier lifetime is degraded during this process.[6] The primary unanswered question of NTD research is whether certain classes of defects remain unannealed or whether impurity contamination during annealing is the cause of this lifetime degradation.[6]

At the last NTD conference held at the University of Missouri, papers by both Malmros[6] and Cleland, et al.[7] suggested that the minority carrier lifetime was slightly higher after annealing for silicon irradiated in light water reactors with a high fast neutron component than for silicon irradiated in the highly thermalized neutron spectrum of heavy water reactors. Can additional radiation damage be used effectively to reduce the number of recombination centers, or are the results of these papers indicative of historical accidents in which lifetimes improved as annealing experience increased or experimental procedures improved? Are there significant differences in the types of defects produced in different reactor spectra or are the defects the same but the relative concentrations of various defects altered in some significant way?

This paper is directed toward these and related questions. At present, we cannot explain why different reactors yield differing minority carrier lifetimes in NTD-Si. The concentrations of recombination centers in annealed NTD-Si are far too low to be investigated by most of the conventional microscopic defect identification tools available today. We can describe, however, the types and concentrations of defects which we have observed immediately following neutron irradiation in the hope that this information will serve as a clue to further studies.

SIMPLE CASCADE MODELS AND REACTOR FLUX SPECTRA

The distribution in energy of neutrons at a particular reactor irradiation position is described by the differential neutron flux energy spectrum $d\psi/dE_n$ which is a function of the neutron energy, E_n. This quantity describes the number of neutrons/cm^2/sec per unit energy range as a function of neutron energy. The total flux is given by

$$\psi_T = \int_0^\infty \frac{d\psi}{dE_n} \, dE_n \ .$$

Accurate measurement of this differential flux spectrum is very difficult. The flux spectrum is usually obtained from activation data on a variety of threshold reactions and by iterative calculation.[8] Given a particular differential flux spectrum, cascade theory allows an estimate of the rate, dN_D/dt, at which atoms are displaced in this flux. If N_T is the concentration of target atoms (5×10^{22} Si atoms/cm^3) and N_D the concentration of displaced atoms, the displacement rate is given by[2,3]

$$dN_D/dt = N_T \int_{E_d/\lambda}^{\infty} \sigma_d(E_n) \frac{d\psi}{dE_n} dE_n \tag{1}$$

where the displacement cross section, $\sigma_d(E_n)$, and the lowest incident neutron energy which can cause displacements, E_d/λ, are characteristic of the material being irradiated, while the differential flux, $d\psi/dE_n$, is characteristic of the reactor in which the irradiation is performed.

The displacement threshold energy, E_d, is the minimum lattice atom recoil energy following a neutron collision which must be achieved in order to break crystal bonding to produce a displacement. For elastic scattering of the neutron through a center of mass angle θ, the recoil energy of the struck Si atom is given by[2,3]

$$E = \lambda E_n \sin^2 \frac{\theta}{2} \tag{2}$$

where $\lambda = 4M_n M_{Si}/(M_n + M_{Si})^2 = 0.133$ depends on the neutron and Si atom masses. The minimum neutron energy which can cause a displacement, $E_n^{(min)} = E_d/\lambda$, is obtained from Eq. (2) for $E = E_d$ and $\sin^2 \theta/2 = 1$ ($\theta = \pi$) corresponding to a direct hit with a recoil energy at the displacement threshold. Although the displacement energy varies considerably for different orientations, we shall assume an average value of $E_d = 25$ eV in the calculations to follow although displacement energies of half of this have been observed for particular orientations.

The displacement cross section must be calculated from cascade theory. For a given incident neutron energy, $E_n > E_n^{(min)}$, all recoil energies from λE_n down to E_d are possible [See Eq. (2).]. The total displacement cross section is given by an integration over all possible recoil energies, E, i.e.,

$$\sigma_d(E_n) = \int_{E_d}^{\lambda E_n} \frac{d\sigma}{dE} \nu(E) dE \tag{3}$$

where $d\sigma/dE$ is the differential neutron scattering cross section and $\nu(E)$ represents the total number of atoms which are displaced by the primary knock-on atom as it loses energy in traveling through the lattice by making collisions with other lattice atoms (displacement energy loss) and with electrons (ionization energy loss).

Although a number of cascade theories of considerable refinement exist, it is sufficient to use a simple Kinchin-Pease[1] model with ionization correction to obtain order of magnitude correct displacement rates which are useful in estimating the amount of damage different reactor spectra will produce.[2,3] We shall also neglect inelastic neutron scattering and assume that the elastic scattering is isotropic. Under these assumptions[2,3]

$$\frac{d\sigma}{dE} = \frac{\sigma_e}{\lambda E_n} \tag{4}$$

where σ_e is the total elastic scattering cross section and

$$\nu(E) \cong \begin{cases} E/2E_d \ , & E_d \leq E \leq E_i \\ E_i/2E_d \ , & E > E_i \end{cases} \tag{5}$$

The quantity, E_i, is called the ionization threshold and is given by[2,3]

$$E_i \cong M_{Si}(\text{keV}) = 28 \text{ keV} \ .$$

The use of this ionization threshold in Eq. (5) approximates the more precise Lindhard energy loss model[10] for $\nu(E)$ and is based on the fact that low energy recoils lose most of their energy by displacements while high energy recoils lose energy primarily by ionization.

Substituting Eqs. (4) and (5) into (3), we obtain

$$\sigma_d(E_n) = \begin{cases} \dfrac{\sigma_e \ \lambda E_n}{4E_d} \ , & E_n \leq E_i/\lambda \\ \dfrac{\sigma_e \ E_i}{4E_d} \left(2 - \dfrac{E_i}{\lambda E_n}\right) \ , & E_n > E_i/\lambda \end{cases} \tag{6}$$

where $E_i/\lambda \cong 0.2$ MeV for Si and where we have neglected E_d compared to λE_n. If we neglect the ionization losses at high recoil energies, then $\nu(E) = E/2E_d$ and $\sigma_d(E_n) = \sigma_e \ \lambda E_n/4E_d$. A plot of these approximate displacement cross sections is shown in Fig. 1. For very large neutron energies, the cross section with ionization correaction approaches $2\sigma_e E_i/4E_d$. The cross section is within 10% of this saturation value at $5E_i/\lambda \cong 1.0$ MeV. Since the flux above 1.0 MeV falls rapidly with increasing neutron energy in most reactors, it follows that most of the displacements are caused by the lower energy neutrons below 1.0 MeV. We shall make use of this fact when we approximate the differential flux in a typical thermal reactor.

In order to estimate dN_D/dt using Eqs. (1) and (6), we must know $d\psi/dE_n$ as a function of neutron energy E_n. The differential flux of thermal reactors consists of a Maxwellian distribution at very low thermal energies below the displacement threshold, a degraded fission spectrum at neutron energies above 1.0 MeV and a down scattered spectrum proportional roughly to $1/E_n$ between the

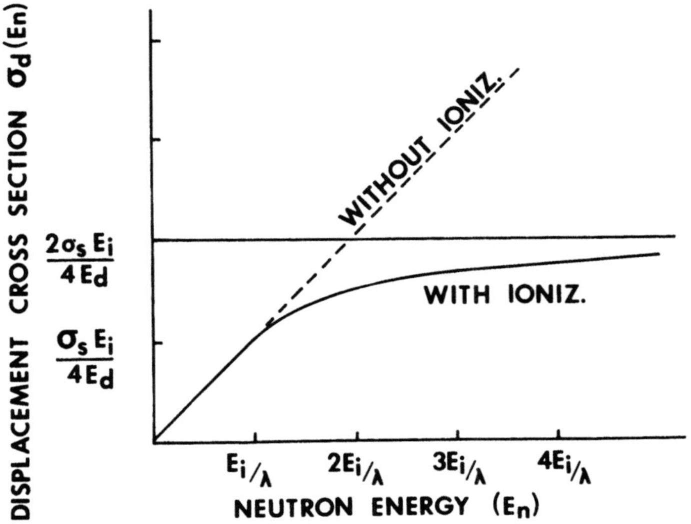

Fig. 1. Displacement cross section vs. neutron energy for simple Kinchin-Pease model with and without the ionization threshold, E_i.

thermal and fission spectra.[2] For Si, the ionization energy loss of high energy recoils implies that, for those neutrons above 1.0 MeV, only the total number of neutrons is important in determining the displacement rate, not their distribution in energy. For a second row reflector position at MURR, a light water reactor, the spectrum above E_d/λ can be approximated by the expression

$$\frac{d\psi}{dE_n} = \begin{cases} K/E_n \ , & E_d/\lambda \leq E_n \leq E_f \\ 0 \ , & E_n > E_f \end{cases} \tag{7}$$

where $K = 7.8 \times 10^{11}$ n/cm^2/sec and where $E_f = 3.27$ MeV is selected as an upper cutoff limit to the fission spectrum such that the total number of neutrons above 1.0 MeV is equal to

$$\psi_T(> 1.0 \text{ MeV}) = \int_{1.0 \text{ MeV}}^{E_f} (d\psi/dE_n)dE_n \ .$$

The above approximate spectrum derived from the reactor design data yields a calculated Cd-ratio on ^{59}Co of 26.6 compared to experimental values of 25-30 where the Cd-ratio is defined by the expression[11]

$$R_{Cd} = 1 + \frac{\psi_{th}\sigma_o}{K(R.I.)}$$

where ψ_{th} is the thermal flux, σ_o = 37.5 barns[12] the 2200 m/sec ^{59}Co activation cross section, and K the source strength used in Eq. (7). The resonance integral defined by[11]

$$R.I. = \int_{0.4\ eV}^{\infty} \sigma_a(E_n)\,\frac{dE_n}{E_n} = 75 \text{ barns}$$

is the integral of the ^{59}Co activation cross section for a $1/E_n$ spectrum.[12]

For order of magnitude calculations of dN_D/dt, the neutron spectrum can always be approximated by a series of power laws. For example, the intense pulsed neutron source (IPNS) being built at Argonne National Laboratory will have an approximate differential flux spectrum given by[13]

$$\frac{d\psi}{dE_n} = \begin{cases} 430\ K\ E_n^{\ 1.38} , & E_n \leq 0.1 \text{ MeV} \\ 1.8\ K\ E_n^{\ -1} , & 0.1 < E_n \leq 0.5 \text{ MeV} \\ K\ E_n^{\ -1.84} , & 0.5 < E_n \leq 10 \text{ MeV} \end{cases}$$

where K = 4.13 x 10^{11} $n/cm^2/sec/MeV^x$.

Using Eqs. (1), (6) and the spectrum in Eq. (7), we find that

$$\frac{dN_D}{dt} = N_T\ \sigma_e\ K\ (\frac{\lambda E_f}{4E_d})\ f(\beta) \tag{8}$$

where we have neglected E_d compared to E_i and where $\beta = E_i/\lambda E_f$ = 6.23 x 10^{-2} and

$$f(\beta) = \beta^2 + 2\beta\ \ell n(1/\beta) = 0.357 \tag{9}$$

is the ionization loss correction factor. If we ignore ionization loss, the displacement rate is given by Eq. (8) with $f(\beta) = 1$. Using σ_e = 3 barns, N_T = 5 x 10^{22} cm^{-3}, K = 7.8 x 10^{11} $n/cm^2/sec$, E_f = 3.27 MeV, λ = 0.1332, E_i = 28 keV and E_d = 25 eV; the displacement rate is found to be

$$\frac{dN_D}{dt} = \begin{cases} 5.1 \times 10^{14} \text{ disp/cm}^3\text{/sec, (no ioniz)} \\ 1.8 \times 10^{14} \text{ disp/cm}^3\text{/sec, (with ioniz)} \end{cases}$$

A similar calculation for IPNS yields 1.2×10^{14} disp/cm^3/sec and 3.5×10^{14} disp/cm^3/sec with and without ionization energy loss corrections. A calculation of the displacement rate by energy group for IPNS is instructive. We find that the displacement rate ratios are

$$2.12 \times 10^{-3}:2.86:3.55, \text{ without ionization}$$

and

$$2.12 \times 10^{-3}:2.48:0.776, \text{ with ionization.}$$

It is clear that the highest energy group (E_n > 0.5 MeV) suffers the greatest reduction in the displacement rate when ionization corrections are made. Although the flux at MURR above E_d/λ is almost seven times that of IPNS, the MURR displacement rate is only 1.7 times as high illustrating the importance of the energy spectrum of the neutron source. The results of these displacement rate calculations are shown in Table 1.

Two other quantities of interest in cascade theory are the mean recoil energy and the mean number of displacements per cascade. We can obtain the mean neutron energy of the MURR spectrum from the expression

Table 1. Comparison of calculated displacement rates for two very different neutron energy spectra (MURR-row 2 and IPNS).

SOURCE	ψ(> 0.1 MeV) (n/cm^2/sec)	DISPLACEMENT RATE (disp/cm^3/sec)	
		with ionization	without ionization
MURR	1.35×10^{13}	2×10^{14}	5.1×10^{14}
IPNS	2×10^{12}	1.2×10^{14}	3.5×10^{14}

$$\overline{E}_n = \frac{1}{\psi_T}\int_{E_d/\lambda}^{E_f} E_n \frac{d\psi}{dE_n}\, dE_n = \frac{E_f}{\ell n \frac{\lambda E_f}{E_d}} = 0.335 \text{ MeV}$$

where $\psi_T = \int_{E_d/\lambda}^{E_f} \frac{d\psi}{dE_n}\, dE_n$.

The mean recoil energy is obtained by realizing that for an incident neutron of energy E_n all recoil energies between E_d and λE_n are equally likely. Therefore, the spectral average of the mean of these values yields the mean recoil energy, $\overline{E}$, as

$$\overline{E} = \frac{1}{\psi_T}\int_{E_d/\lambda}^{E_f} \frac{\lambda E_n + E_d}{2} \frac{d\psi}{dE_n}\, dE_n$$

$$\cong \frac{\lambda E_f}{2\, \ell n \frac{\lambda E_f}{E_d}} = 22.3 \text{ keV} \qquad (10)$$

where we have neglected E_d compared to λE_f. Notice that $\overline{E} = \lambda \overline{E}_n/2$.

The mean number of displacements per cascade can be obtained from the definition

$$\frac{dN_D}{dt} = N_T\, \sigma_e\, \psi_T\, \overline{\nu} \; .$$

Comparing this to Eq. (8), we find that

$$\overline{\nu} = \frac{\lambda E_f\, f(\beta)}{4E_d\, \ell n \frac{\lambda E_f}{E_d}} \qquad (11)$$

$$= 159 \text{ displacements/cascade.}$$

It should also be noted that

$$\overline{\nu} = \frac{\overline{E}}{2E_d}\, f(\beta) \; . \qquad (12)$$

The mean recoil energy, $\overline{E}$, and mean number of displacements per cascade, $\overline{\nu}$, will be used to analyze some of the experimental data to follow.

For the MURR reactor position described above, the thermal flux is 4×10^{13} n/cm^2/sec. Since the mean recoil energy imparted to a Si atom after neutron capture and gamma emission is about 474 eV,[14] $\bar{\nu} = \bar{E}/2E_d = 9.5$ displacements are created on the average for each thermal capture. The thermal neutron displacement rate is given by $N_T\ \sigma_c\ \psi_{th}\ \bar{\nu} = 3 \times 10^{12}$ disp/cm^3/sec where $\sigma_c = 0.16$ barns. The phosphorus production rate, $N_T\ \sigma_a\ \psi_{th} = 6.7 \times 10^9$ P/cm^3/sec. Therefore, the ratio of phosphorus atoms to gamma recoil displacements to fast neutron displacements is 1:448:30000 for this reactor position.

In a heavy water reactor the Cd-ratio on gold might be above 500. The corresponding Cd-ratio on gold for the above MURR reactor position is 4.27. Therefore, if the character of the down scattered spectrum is similar in the two reactors and the thermal fluxes are equal, the fast neutron displacement rate should be about (4.27/500) lower than for the light water reactor. The ratios above then become 1:448:256. We see that there are about 2 gamma recoils for each fast neutron displacement and about 700 total displacements for each phosphorus atom produced in the heavy water reactor.

We shall describe in the sections to follow the types of defects which we have observed by a number of different experimental techniques.

OPTICAL ABSORPTION AND EPR MEASUREMENTS

At the previous NTD conference we presented a survey of the annealing observed by resistivity measurement for samples irradiated to different phosphorus concentrations in a light water reactor.[15] These experiments are summarized in Fig. 2 which shows 15 min. isochronal anneals for three samples irradiated to produce final resistivities of approximately 1 Ω-cm, 100 Ω-cm and intrinsic. The samples were initially 5250 Ω-cm p-type. After irradiation the material remained p-type but type converted to n-type at different temperatures dependent on the concentration of phosphorus added. We have also observed that n-type material is p-type following irradiation suggesting that more defect energy levels occur in the lower half of the band gap than the upper half. There is obviously a wealth of annealing stages in Fig. 2. Above 600°C, the recovery is in part dominated by phosphorus electrical activation, however, additional defect annealing stages are also observed.

Optical absorption and EPR measurements have been performed on both float zone and Czochralski in order to identify microscopically the types and concentrations of defects which exist following transmutation doping. Both the FZ and CZ samples had an initial resistivity of 250 Ω-cm and were n-type. These samples were irradiated to a thermal fluence of 1.34×10^{19} n_{th}/cm^2 corresponding to a dis-

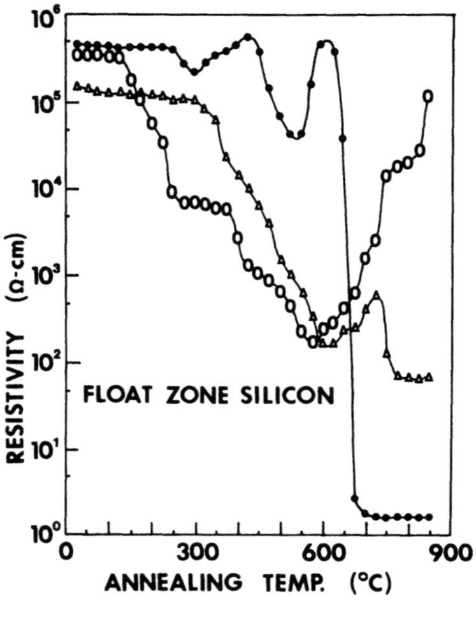

Fig. 2. Dose dependence of isochronal annealing of detector grade NTD float zone silicon.

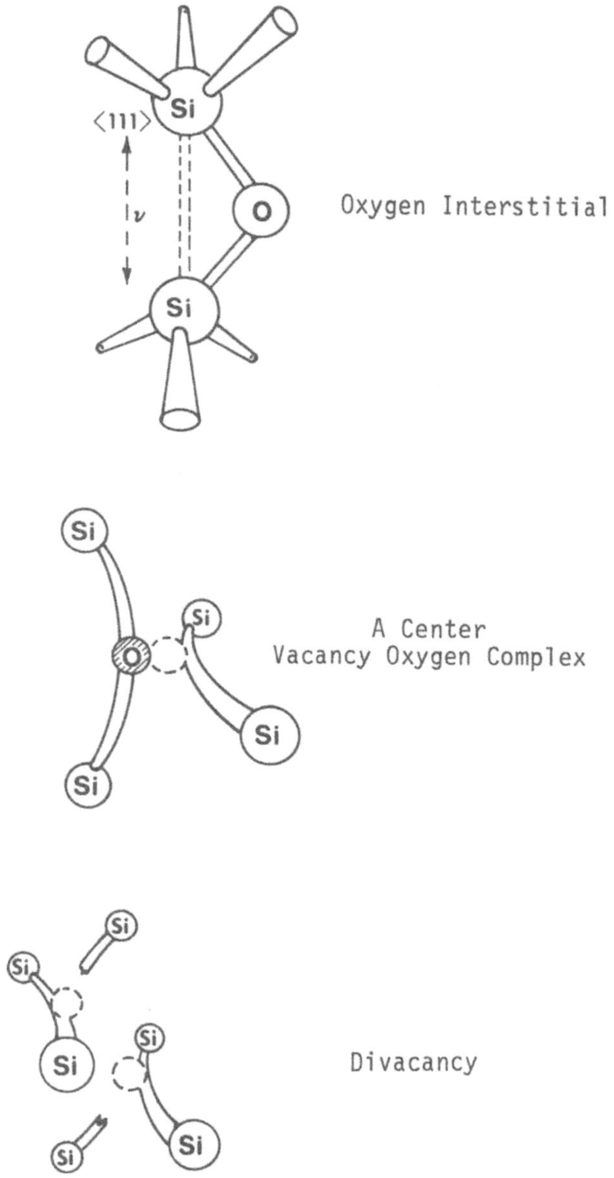

Fig. 3. Atomic configurations of defects found by optical absorption.

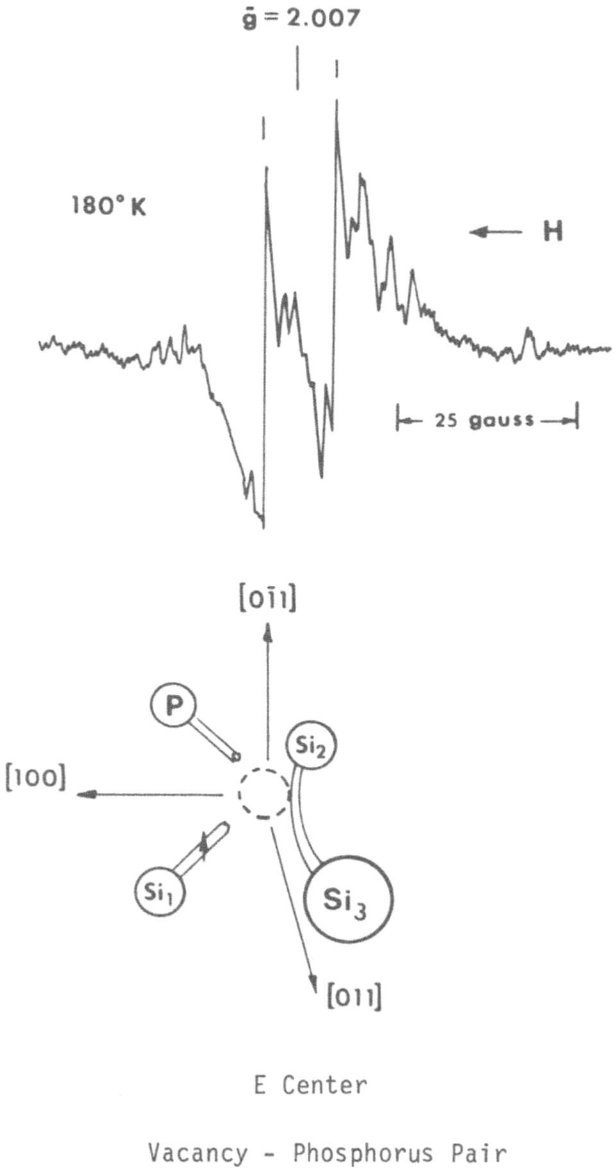

Fig. 4. EPR spectrum and atomic configuration of the E-center. Spectrum at 9.10 GHz.

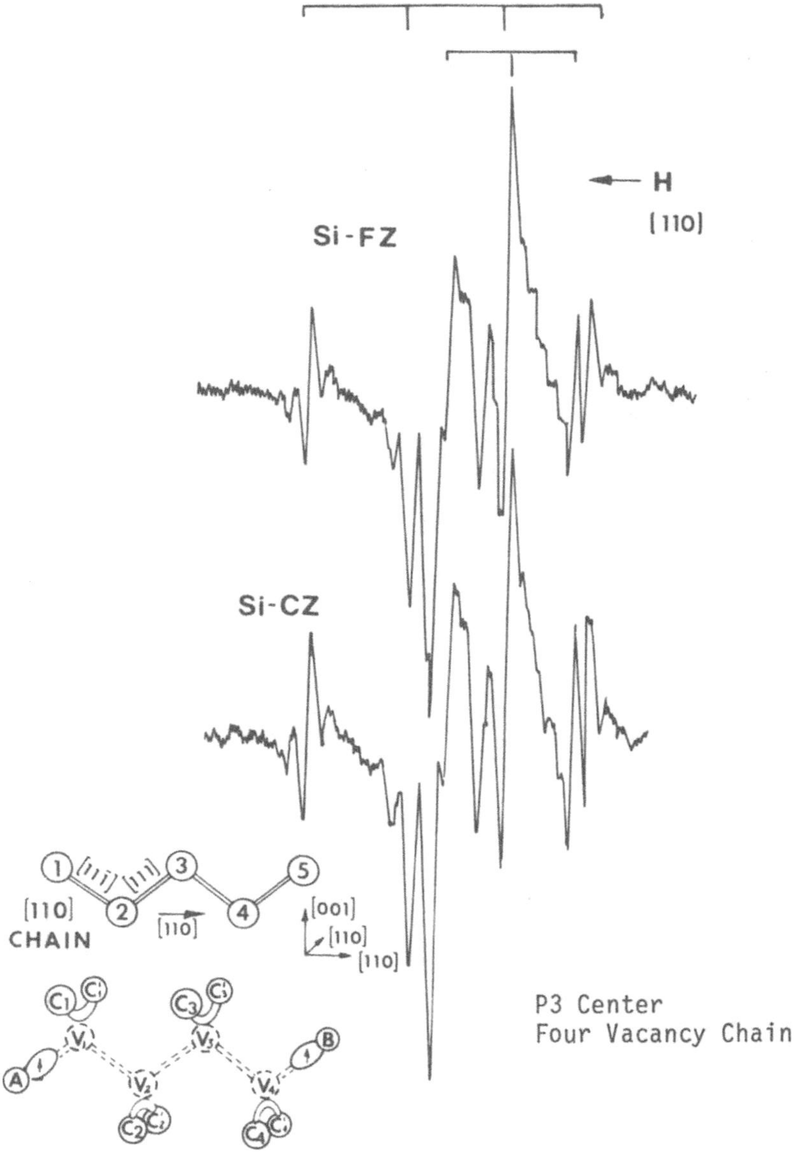

Fig. 5. EPR spectrum of P-3 four vacancy chain, the dominant defect concentration in room temperature NTD silicon irradiated to ϕ_{th} = 1.34 x 10^{19} n/cm^2 in both CZ and FZ. Spectra taken at room temperature at 9.10 GHz.

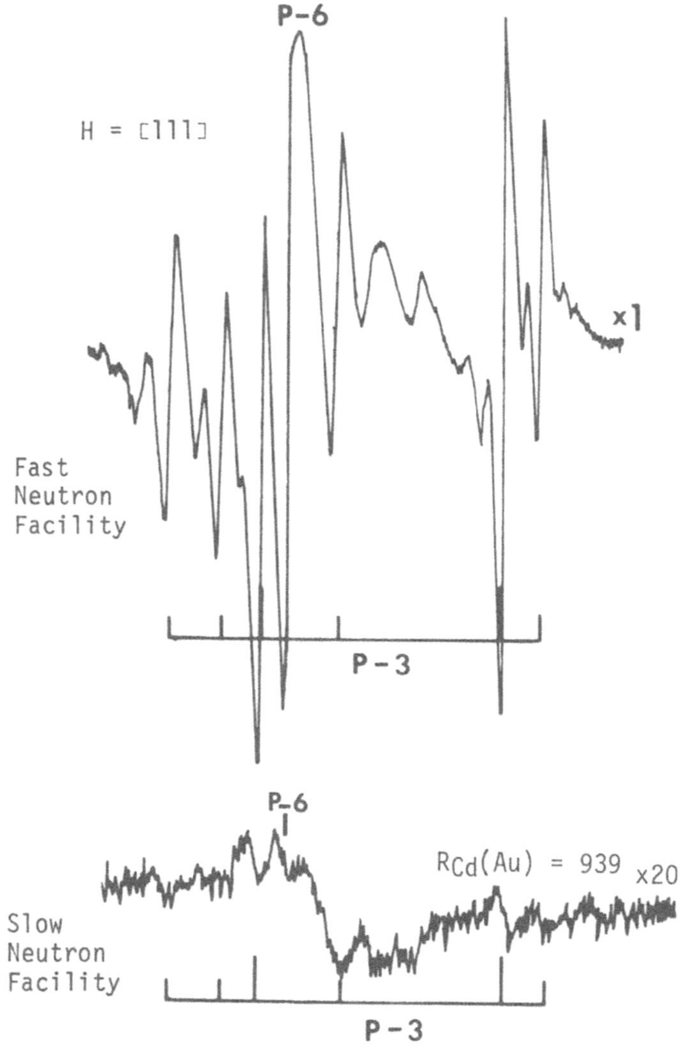

Fig. 6. Room temperature (9.1 GHz) EPR spectra for samples irradiated a) in Argonne CP-5 fast facility (top trace) and b) in Cd-ratio of 939 on Au. The P-3 and P-6 centers are indicated.

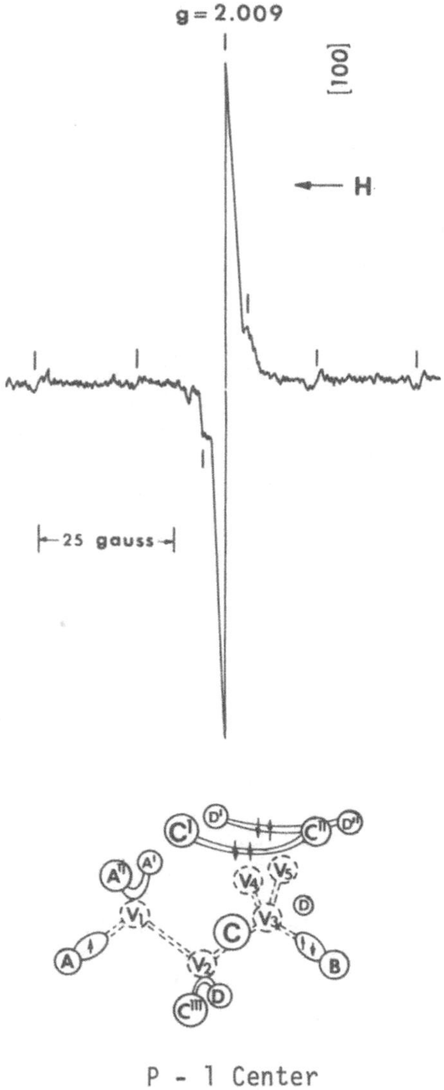

Fig. 7. EPR spectrum and atomic configuration of the P-1 center. Room temperature data at 9.10 GHz.

placement concentration of the order of 4 to 6 x 10^{19} disp/cm^3 according to the simple Kinchin-Pease model described by Eq. (8).

Figure 3 shows the atomic configurations of those defects observed by optical absorption while Figs. 4-7 show defects observed by EPR. The E-center,[16] a vacancy-phosphorus close pair is somewhat difficult to observe in NTD-Si but is observed occasionally at low temperatures. The four-vacancy P-3 center,[17] the divacancy center[18] and the P-6(IX) center[19] dominate what can be seen immediately after irradiation. The A-center[20] vacancy oxygen pair and oxygen interstitial[21] occur only in Czochralski material. The oxygen interstitial, which is observed before irradiation, decreases in intensity as the A-center concentration increases.

The P-1 (five vacancy) center[19] can usually be observed after room temperature irradiation, however, its concentration is lower than 10% of the P-3 center concentration. The P-1 center does not become the dominant defect structure until the divacancy anneals. It apppears that this defect can be formed directly during irradiation or from P-3 centers which have attracted additional vacancies during vacancy defect annealing. Because P-3 centers are converted to P-1 centers during divacancy annealing, it is not entirely clear at what temperature the P-3 center would anneal if the divacancies were not present.

Figure 5 illustrates another interesting aspect of the P-3 centers. The top EPR trace is for float zone Si while the bottom trace is for Czochralski Si. Both samples were irradiated simultaneously to the same fluence. If these spectra are normalized to sample mass, it is found that the concentration of P-3 centers in both samples are equal within 1%. This is indeed a surprise since the A-center concentration in the Czochralski sample was 3 x 10^{17} cm^{-3} while the P-3 concentrations in both samples was about 3.5 x 10^{17} cm^{-3} (1.4 x 10^{18} vacancies/cm^3), i.e., the vacancy concentration in A-centers is 20% of the vacancy concentration in P-3 centers. One would expect A-centers to deplete the concentration of P-3 centers if the P-3 centers had been formed by a condensation of isolated vacancies. It appears that the P-3 centers are remnants of the original displacement cascades.

The concentration of P-6(IX) centers is roughly equal to that of P-3 centers. The P-6 center is thought to be a di-interstitial defect in a positive change state.[22] We have observed that the ratio of P-3 to P-6(IX) concentration is approximately constant for these irradiations, for irradiations at Argonne's CP-5 reactor in both their fast flux facility[8] and in a Cd-ratio on Au of 939 (see Fig. 6.). Since the concentration of P-6(IX) is fixed with the concentration of P-3 centers, we must also conclude that P-6 is also a cascade remnant.

Table 2. Concentration of observed vacancy defects relative to calculated concentration of displacements (light water reactor).

DEFECT BOOKKEEPING

Calculated displacements = 4 to 6 x 10^{19} cm^{-3}

(1.34 x 10^{19} n_{th}/cm^2, 7.6 x 10^{12} n (> E_d/λ)/cm^2/sec.)

DEFECT		CONCENTRATION OBSERVED (cm^{-3})		VACANCY CONCENTRATION (cm^{-3})
Divacancy (V-2)		4.4 x 10^{17}		0.88 x 10^{16}
4-vacancy (P-3)		3.5 x 10^{17}		1.4 x 10^{18}
5-vacancy (P-1)		< 3 x 10^{16}		< 0.14 x 10^{18}
Vacancy-Phos. (E)		< 2 x 10^{15}		< 0.002 x 10^{18}
Vacancy-Oxy. (A)		3 x 10^{17}		0.3 x 10^{18}
			TOTAL	2.8 x 10^{18}
	observed/calculated ≅	0.05 to 0.07 (5 to 7%)		

For bookkeeping purposes, we have summarized the concentration of vacancy containing defects in Table 2. The total vacancy concentration of these defects is 2.8 x 10^{18} cm^{-3} while the total concentration of displacements is 4 x 10^{19} cm^{-3}. Therefore, only 7% of the calculated displacements have been observed.

It is instructive to compare the defect concentrations above with the total concentration of displacement cascades. For the above experiment, the cascade concentration, using N_D = 6 x 10^{19} cm^{-3} and $\overline{\nu}$ = 159 from the previous section, is given as

$$[\text{cascade concentration}] = N_D/\overline{\nu} = 3.77 \times 10^{17}\ cm^{-3} . \qquad (13)$$

Since the mean recoil energy of the secondary making this cascade is ∿ 22 keV, then the range of this recoil is on the average about 125 Å.[5] Assuming that the radius of this cascade track is no larger than the range, it is easy to show that the overlap of cascades is not likely for this fluence. It should also be noted that the concentration of cascades is almost identical to the concentration of P-3 cascade remnants. In other words, there is about one four-vacancy P-3 center created in each cascade.

To understand this, we must consider the energy loss of the primary knock-on (PKA) ion along the displacement track. As the ion moves along the track displacements are made producing simple Frenkal pairs. As the recoil secondary energy is lost by this process, the ion's mean free path between collisions drops to a value below the separation distance between atoms along the track.[2,3] Once this happens, every atom along the track must be displaced creating a Brinkman displacement spike.[23] An estimate of the energy below which a spike is created, E_s, is given by[2]

$$E_s = \frac{(\pi a)^4 \; N_T^{4/3} \; E_a}{16 E_d} = 937 \text{ eV} \qquad (14)$$

where $E_a \cong 20$ eV $(Z^2)^{7/6} = 9448$ eV and $a = a_0/Z^{1/3} = 2.2 \times 10^{-9}$ cm where $a_o = 0.53$ Å is the Bohr radius. This energy should be compared with the mean energy per displacement $= \overline{E}/\overline{\nu} = 140$ eV and with the energy required to make a four-vacancy center which can be obtained roughly from Eq. (12) with $\overline{\nu} = 4$ and $f(\beta) = 1$ as

$$\overline{E}(\text{P-3}) = 2E_d\overline{\nu} = (50 \text{ eV})\ \overline{\nu} = 200 \text{ eV} . \qquad (15)$$

Considering the approximations of the above models, and the possibility of some defect recombination, the agreement between Eqs. (14) and (15) is reasonable. We conclude, therefore, that our 22 keV primary ion loses energy along the track by creating ionization and single displacements at a rate of about 140 eV per Frenkel pair.

Once the spike energy is reached, a track of displacements results in the formation of a P-3 four-vacancy center. Since the P-6(IX) interstitial defect complex occurs in equal numbers with P-3 regardless of the reactor spectrum, it most likely is formed from the atoms ejected from the displacement spike also. In fact, the very low concentration of P-3 and P-6 centers observed in the high Cd-ratio samples can be accounted for quantitatively by the contamination of the thermal flux by a small fast neutron component, indicating that a knock-on Si, with an energy less than or of the order of a typical gamma recoil knock-on ($\overline{E} \cong 474$ eV), has a negligible probability for producing a large cluster center like the P-3. This lattice disorder is not, however, isolated as recognizable isolated point defects. Only the fast neutrons accomplish this by way of the spike mechanism.

It should also be noted from Table 2 that there is about one divacancy per cascade. It is not known at this time if the divacancy is also a result of the displacement spike, however, the energy required to form the divacancy is from Eq. (15) about 100 eV. The sum of the divacancy and four-vacancy energies is an appreciable fraction of the total spike energy.

RAMAN SCATTERING

In Table 2 of the previous section, only 5 to 7% of the calculated displacements are observed as isolated vacancy defects. Where are the rest of the displacements? A broad EPR resonance observed for the sample irradiated in a high Cd-ratio suggests that considerable disorder exists in the lattice which is not resolvable into point defects. Raman scattering has also provided an observation of highly disordered material which appears in many ways (but not all) to be amorphous.

It is well known that high dose ion implants will drive Si amorphous. The extent to which the material becomes amorphous, however, depends on the density of displacement energy deposited. A damage energy density of $\sim 2 \times 10^{23}$ eV/cm^3 will produce an amorphous layer by ion implantation.[24] For higher fluences than the previous section, radiation damage has been observed by Raman Scattering.

The samples were irradiated at MURR to a fluence of between 3.5 to 7 x 10^{19} n/cm^2 corresponding to a fast neutron fluence ($E_n > E_d/\lambda$) of 0.7 to 1.3 x 10^{19} n/cm^2. For the highest fluence, the concentration of displacements is calculated to be 3.6 x 10^{20} cm^{-3} corresponding to a cascade density of about 2 x 10^{18} cm^{-3}. This is about an order of magnitude higher than in the previous section. Consequently, 0.7% of the crystal volume is damaged (3.6 x 10^{20}/5 x 10^{22}).

Stein[5] has estimated that the displacement energy density for a 22 keV recoil is $\sim 10^{21}$ eV/cm^3. The ratio of this energy density to that required to produce totally amorphous material[24] is 0.5% which is in good agreement with the damage fraction of crystal volume we obtain for our reactor spectrum.

We have observed features in the Raman spectra due to first order processes that arise from the destruction of translational invariance in the lattice and with Raman peaks corresponding to maxima in the vibrational density of states (VDOS) averaged over the Brillouin Zone (BZ). These new peaks are very similar to those seen in the Raman spectra of amorphous Si.[25]

Figure 8 shows the Raman spectrum of CZ Si before and after irradiation to a thermal fluence of 7 x 10^{19} n/cm^2. Before irradiation, the spectrum shows the allowed first order process, the zone center optic phonon at a frequency of 520 cm^{-1}. The weaker modes are overtone processes such as the second order transverse acoustic mode, 2TA(X) at 300 cm^{-1}.

Upon irradiation, two new peaks appear at ∿ 95 and 490 cm^{-1} superimposed on an enhanced background scattering below the optic phonon frequency. The new features appear in a Γ_1 scattering geometry and are first order. The peak intensities of these new features scale with fluence.

In the insert of Fig. 8 we show schematically the origin of these new peaks. In fully crystalline Si, which has two atoms per unit cell, the BZ would have acoustic (LA and TA) and optic (LO and TO) branches as in panel (a) of Fig. 8. The zone center phonon mode (520 cm^{-1}) would be the only allowed first order process ($\vec{k}$ = 0) which is Raman active. If a defect were introduced on the average in every alternate cell so that the new super unit cell is double the original cell, the BZ would be halved and is equivalent to a folding of the BZ back into the zone center once (panel (b) in Fig. 8). By extending this argument to fully amorphous Si, where translational invariance is completely lost, the entire BZ is folded into the zone center. The spectrum obtained consists of broad peaks representative of the vibrational density of states averaged over the entire BZ.

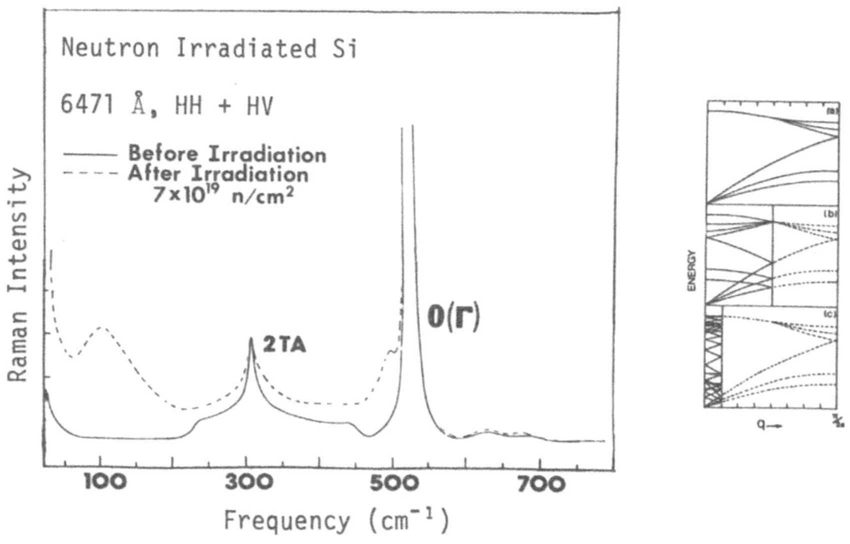

Fig. 8. Stokes Raman spectra of Si before and after neutron irradiation. The insert shows a schematic of the BZ for a crystal with 2, 4 and 16 atoms per unit cell.

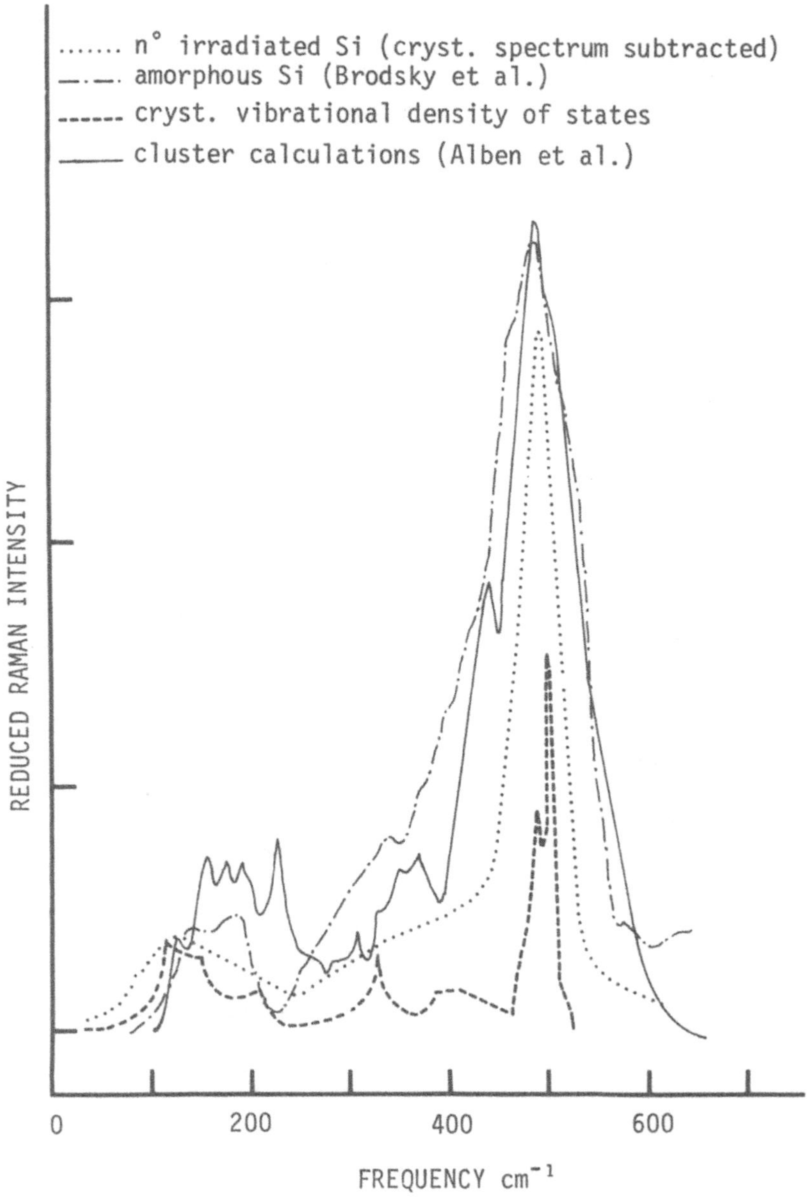

Fig. 9. The reduced Raman spectrum of neutron irradiated and amorphous Si and the crystalline VDOS.

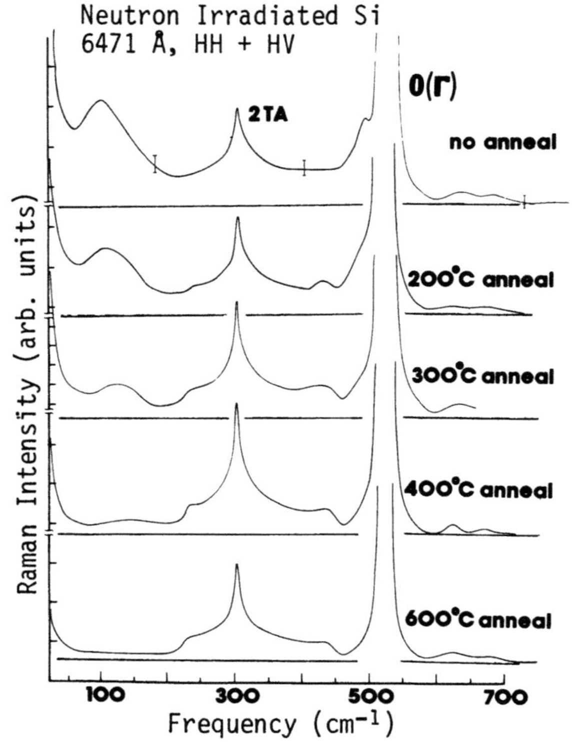

Fig. 10. The Raman spectrum of sample CZ 2.1c after successive isochronal anneals.

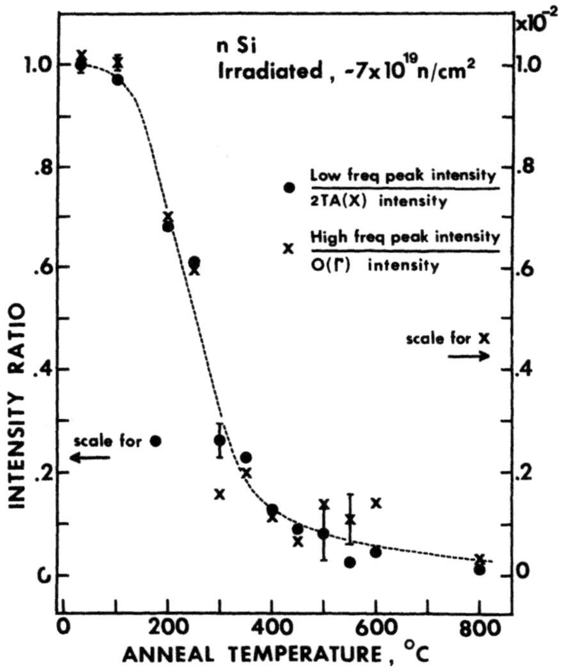

Fig. 11. The intensity of the amorphous-like TA and optic peaks produced by neutron irradiation as a function of anneal temperature.

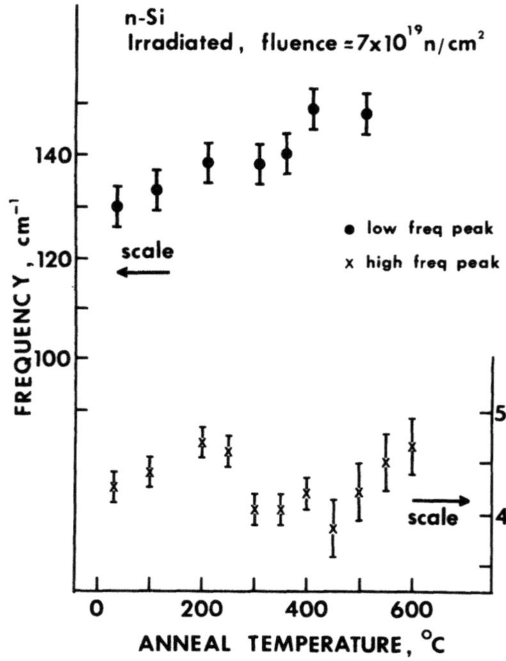

Fig. 12. The frequency shifts of the high and low frequency amorphous-like Raman peaks as a function of anneal temperature.

The comparison between the irradiated sample spectrum and amorphous Si[26] is shown in Fig. 9 where the reduced spectra, $I_R(\omega)$, are obtained by[27]

$$I_R(\omega) = \omega \frac{(1 - e^{\hbar\omega/kT})}{(\omega_L - \omega)^4} I_M(\omega)$$

from the measured Raman intensities $I_M(\omega)$. The resulting spectrum $I_R(\omega)$ contains only effects due to transition matrix elements and the vibrational density of states (VDOS).

Also shown on Fig. 9 are cluster calculations by Alben, et al.[28] and the crystalline VDOS obtained from a shell model fit to neutron scattering data. The broad neutron damage peaks do indeed correspond in frequency to peaks in the VDOS. We therefore attribute the low frequency peak to an average over the TA modes, the central background to an average over LA modes, and the high frequency peak to optic modes LO and TO.

These amorphous-like zones anneal in a way suggestive of crystal regrowth of amorphous Si as shown in Figs. 10 and 11. In addition to a decrease in the irradiation induced peak intensity, the low frequency peak shifts to higher energies as we anneal as shown in Fig. 12. The peak which is at 125 cm^{-1} in the reduced Raman spectrum prior to annealing, shifts up to 150 cm^{-1} before it disappears. We attribute this to the fact that as we approach crystallinity, contributions at the zone center arise mainly from the edges of the BZ, corresponding to a few folds of the BZ (see Fig. 8 insert). This would shift the peak to higher energies. The single phonon mode due to a folded TA(X) branch would yield a frequency of 150 cm^{-1} in good agreement with our data. The high frequency peak (optical branch) is not expected to shift in any systematic way in agreement with data of Fig. 12.

The Raman experiment seems to be telling us that highly disordered material, probably amorphous, is present after irradiation and that this disorder can only be formed if most of the calculated displacements are involved in this disorder. EPR on this sample has failed to find the dangling bond signal characteristic of amorphous Si, however. We must conclude that the radiation damage observed by Raman scattering and by the broad EPR background resonance, if not amorphous Si, at least suggests a highly disordered material of closely spaced defects in which most of the displacements have participated.

DLTS SPECTRA AND ANNEALING

To obtain energy levels in NTD-Si we have used deep level transient spectroscopy (DLTS).[29] Previous DLTS data has been presented by Guldberg[30] for n-type material annealed above 425°C and by Baliga and Evwaraye[31] for both p- and n-type material annealed above 425°C. We have purposely limited the fluence and radiation damage to low values in order to preserve the diodes so that spectra could be obtained from room temperature up to anneal temperatures of 500°C. Diodes were generally p-n junctions. Most data for this preliminary survey were taken on collector-to-base junctions of commercial transistors for convenience, however, spectra were also obtained on both Schottky and p-n junctions fabricated on boron and phosphorus doped material. No significant differences were observed between the commercial diodes and those fabricated. Two limitations must be pointed out however. The defect levels in these samples is far below the levels observed in the previous experiments. Also, the sample material is Czochralski.

Table 3. Capture cross sections and energy levels of electron and hole traps after irradiation. Data of C. E. Barnes is taken from Ref. 32.

Trap	Energy Level (eV)	σ (cm^2)
MURR DATA:		
E_1	0.16	6.7×10^{-15}
E_2	0.24	4.6×10^{-15}
E_3	0.42	2.2×10^{-15}
H_1	0.15	1.8×10^{-15}
H_2	0.19	5.2×10^{-16}
H_3	0.37	2.9×10^{-15}
H_4	0.40	1.5×10^{-15}
C. E. BARNES:		
E_1	0.17	---
E_2	0.23	---
E_3	0.44	---
H_2	0.20	---
H_3	0.38	---

It was generally observed that the defect production rates were about a factor of ten lower in p-type material than n-type. Therefore, n-type samples were irradiated at MURR to a thermal fluence of about 1.8 x 10^{15} n/cm^2 while p-type samples were irradiated to a thermal fluence of 1.8 x 10^{16} n/cm^2. The total fast fluence above E_d/λ was $\sim$ 2 x 10^{14} and 2 x 10^{15} n/cm^2 respectively for n-type and p-type material. Only majority carrier traps were investigated. Signals were analyzed using a broad band lock-in amplifier with a sample and hold input amplifier to gate off the bias pulse signal.

Barnes[32] has also reported similar data which is unpublished except in abstract form. His data was taken using a boxcar detection system. Table 3 shows a comparison of our data with those of Barnes. The agreement on ionization energies is generally within 0.01 eV with our ionization energies being consistently lower. The carrier capture cross sections we report in this section are obtained from thermal data and may be significantly different than those cross sections obtained by a pulse width technique.

Isochronal annealing of the DLTS spectra in both n-type and p-type material has been completed up to 500°C. Diodes failed after this temperature. Figure 13 shows the electron trap spectra after each 15 min. isochronal anneal.

A total of seven electron traps were observed. The absolute concentrations of these defects is shown in Fig. 14 as a function

Table 4. Energy levels (E_C - E_T) and carrier capture cross sections for electron traps in n-type material.

Level	(E_C - E_T) (eV)	σ (cm^2)
E_1	0.16	4.8 x 10^{-15}
E_2	0.24	4.6 x 10^{-15}
E_3	0.42	3.7 x 10^{-15}
E_4	0.34	2.5 x 10^{-15}
E_5	0.27	7.5 x 10^{-12}
E_6	0.21	1.0 x 10^{-14}
E_7	0.52	5.8 x 10^{-14}

of annealing temperature. The defect energy levels and capture cross sections are shown in Table 4.

The position of the E_3 peak shifts slowly to the right as the annealing temperature is increased. Each shift is accompanied by a drop in defect concentration (Fig. 14). We have designated the shifted peaks E_3(I), E_3(II) and E_3(III). No energy level data was obtained for E_3(I), however, E_3(II) = 0.42 eV, 2.5 x 10^{-15} cm^2 which is within experimental error of the values for E_3 = 0.42 eV, 3.7 x 10^{-15} cm^2. The shifts in E_3(III) are caused mainly by changes in the carrier capture cross section as shown in Table 5.

The ionization energy for E_3(III) = (0.45 ± 0.02)eV appears to be slightly higher than that of E_3 = (0.42 ± 0.02)eV. This suggests that peak E_3 is composed of at least two very closely spaced energy levels which anneal at different temperatures.

New peaks which are produced by annealing are the E_4, E_5, E_6 and E_7 levels. The E_4 peak has been observed by Barnes.[32] This peak grows rapidly from 250 to 350°C and then decays in concentration. The E_5 peak appears as the E_4 peak anneals suggesting a relationship between these two defects. There is also some suggestion from the shape of the peak that E_5 might also be a double level. At still higher temperatures (500-550°C) the E_6 and E_7 levels appear.

We have also performed transient current[33] measurements on the n-type material after irradiation but before annealing as shown in Fig. 15. The different spectra of this figure correspond to different bias pulse voltages for a fixed D.C. reverse bias. As the

Table 5. Ionization energies and capture cross sections for the E_3(III) level.

Anneal Temperature (°C)	Energy of Level (eV)	Capture Cross Section (cm^2)
300	0.45	7.4 x 10^{-15}
350	0.45	5.1 x 10^{-15}
400	0.43	1.8 x 10^{-15}
450	0.47	4.2 x 10^{-14}

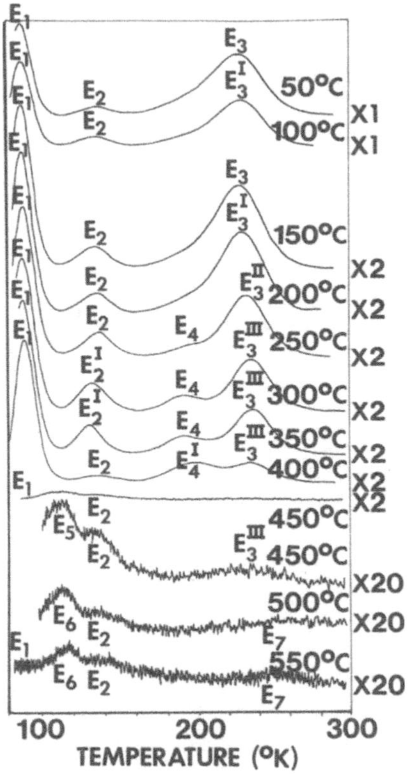

Fig. 13. DLTS electron traps in n-type neutron irradiated Si as a function of isochronal annealing (100 Hz, $\phi_{th} = 1.8 \times 10^{15}$ n/cm^2).

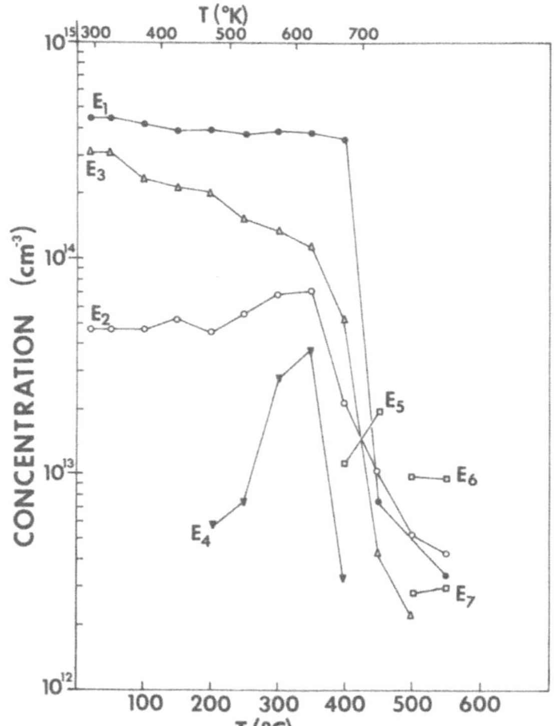

Fig. 14. Concentration of electron trap levels vs. isochronal annealing temperature (data from Fig. 13).

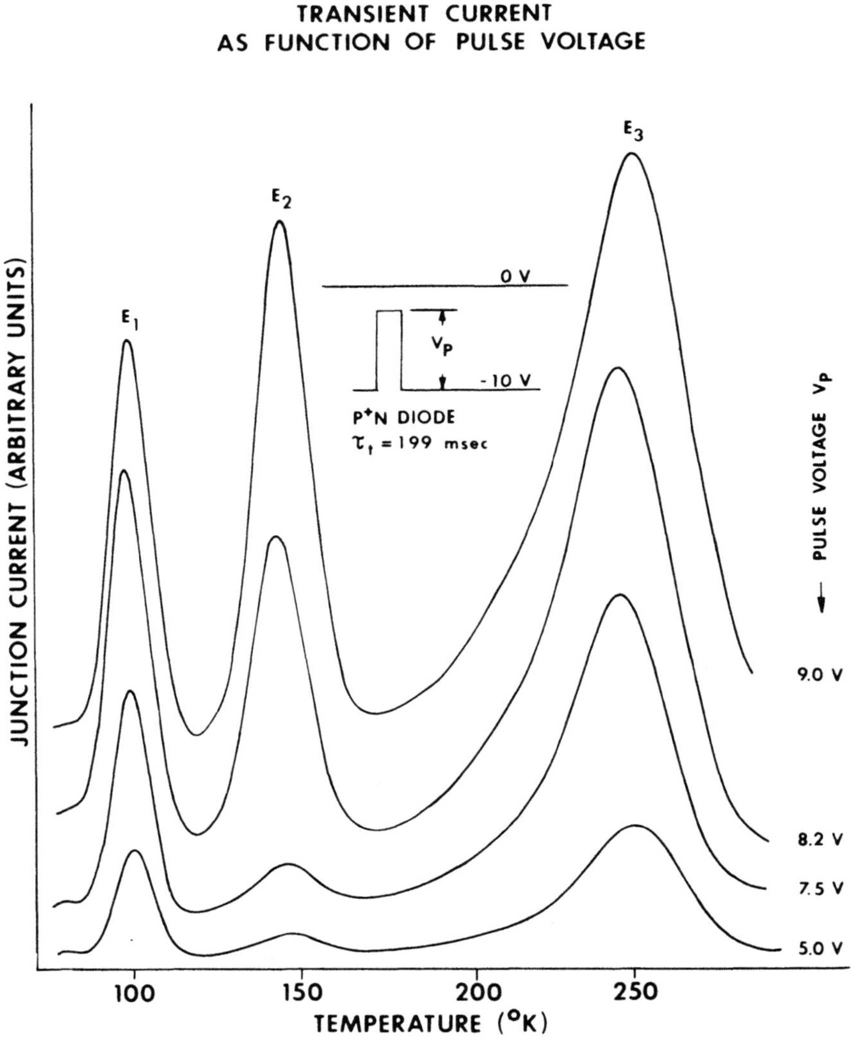

Fig. 15. TCS spectra of a neutron irradiated p^+n diode for various pulse voltages.

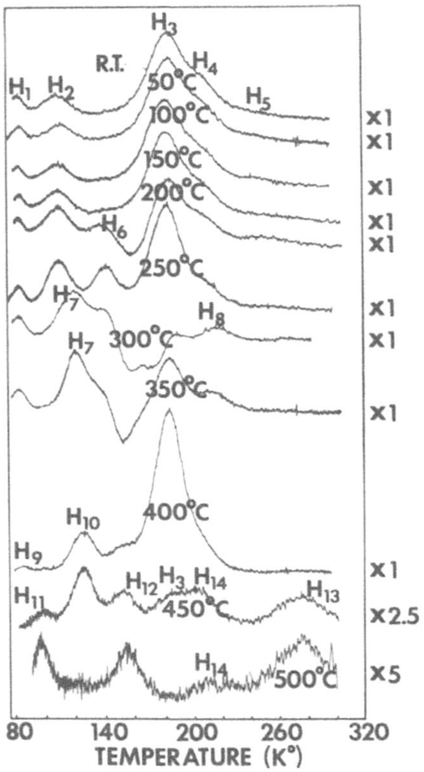

Fig. 16. 10 Hz DLTS hole traps in p-type neutron irradiated Si as a function of isochronal anneals (ϕ_{th} = 1.8 x 10^{16} n/cm^2).

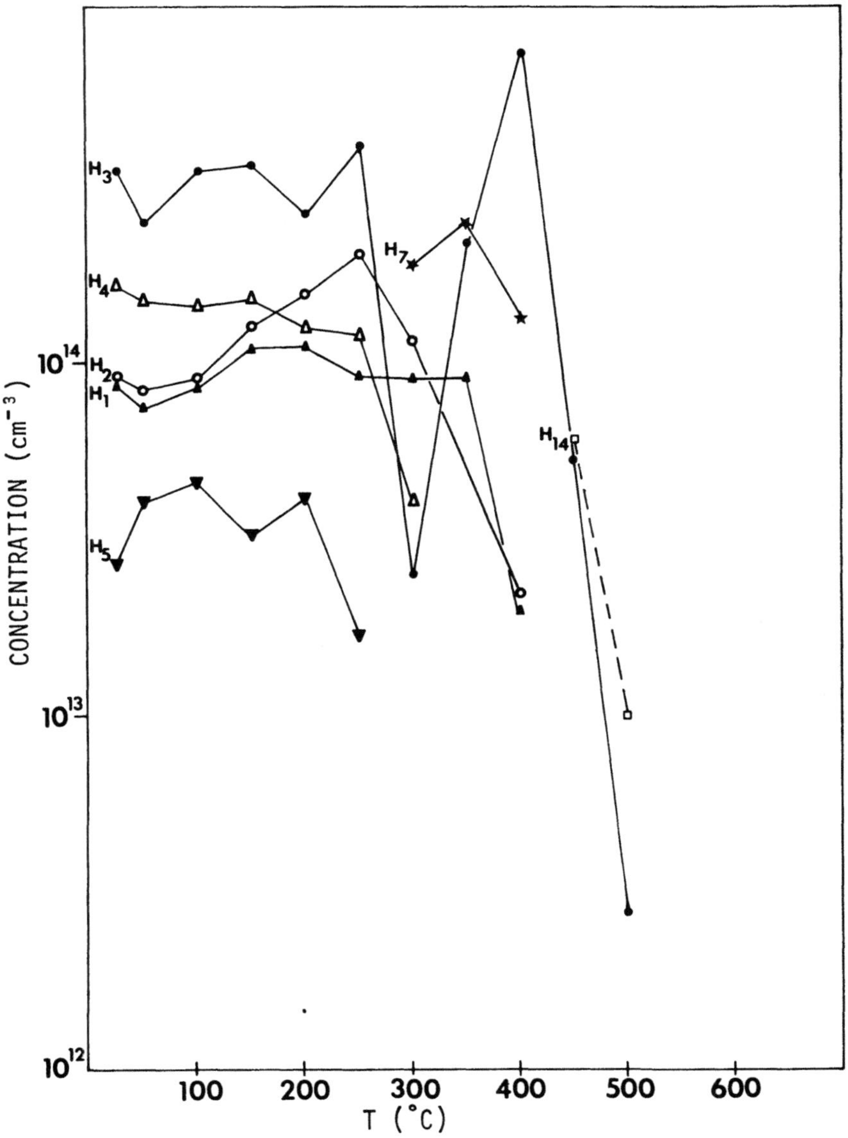

Fig. 17. Concentration of hole traps H_1-H_5, H_7 and H_{14} vs. annealing temperature (data from Fig. 16).

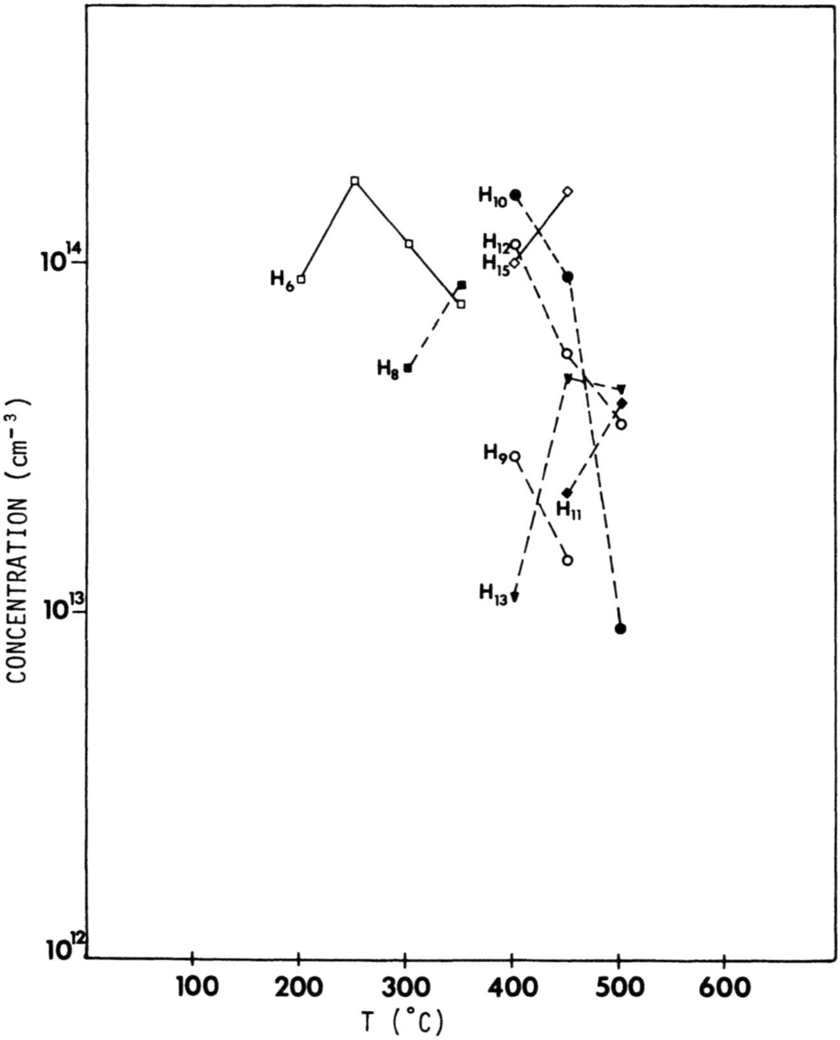

Fig. 18. Concentration of hole traps H_6 and H_8-H_{13} vs. annealing temperature (data from Fig. 16).

bias pulse voltage approaches zero, additional traps are sampled which are physically located very close to the junction. It is clear from Fig. 15 that the E_2 trap has a higher concentration near the junction than the other traps since its amplitude increases relative to the other two traps. The same experiment measuring junction capacitance does not show this effect. Lang[34] has shown, however, that the sensitivity of a spatial variation in defect concentration is greatest for transient current measurements near the junction and lowest for capacitance measurements.

We must conclude that the defect responsible for the E_2 level has migrated during irradiation or after toward the junction as a result of the electric field of the junction. Since this field is negative in the n-type depletion side of the junction, the pile-up of E_2 at the junction suggests a net positive charge state for this defect.

Figure 16 shows the DLTS spectra for isochronally annealed p-type material. This sample was irradiated to a thermal fluence of 1.8×10^{16} n/cm^2, a factor of ten higher than the fluence of the n-type sample. An additional peak, E_{15}, not shown, appears at higher reference frequencies. Table 6 lists the energy levels and capture cross sections for p-type hole traps.

The isochronal annealing of the p-type sample is shown in Figs. 17 and 18. The data has been separated for clarity. Of the levels shown, the levels H_2 and H_3 or H_4 have been reported by Barnes[32] as well as H_6 and perhaps H_7.

The annealing of this sample is truly complicated and deserves further study. The behavior of H_3 is remarkable. The concentration of this defect drops an order of magnitude between 300 and 350°C and then rises again at 450°C, then disappears at 500°C. This defect could well be associated with the "600°C defect" seen with resistivity measurements.[15] It may be that we have the same energy level and similar capture cross sections for two different defects.

Identification of the defect structures associated with these energy levels is difficult. Guldberg[30] has recently summarized the known annealing stages in irradiated Si as has Corbett.[35] Kimerling[36] has reviewed DLTS data for electron irradiation. Some similarities occur between these data and our own. From all of these sources we can make some educated guesses as to the nature of the DLTS defects involved.

Our E_1 level appears to be well established as the vacancy-oxygen A-center.[32,36] It should be noted, however, that this level does not anneal as soon as reported in these references. There appears to be some disagreement about the identification of the E_3 level in these references. We believe this is due to the apparent

Table 6. Energy levels (E_T - E_V) and carrier capture cross sections for hole traps in p-type material.

Level	E_T - E_V (eV)	σ (cm^2)
H_1	0.16	6.4×10^{-15}
H_2	0.18	1.6×10^{-16}
H_3	0.35	2.8×10^{-15}
H_4	0.39	9.7×10^{-16}
H_5	0.61	4.1×10^{-13}
H_6	0.28	6.2×10^{-15}
H_7	0.22	1.2×10^{-15}
H_8	0.62	1.2×10^{-10}
H_9	0.19	8.8×10^{-14}
H_{10}	0.25	1.0×10^{-15}
H_{11}	0.20	3.0×10^{-15}
H_{12}	0.37	6.4×10^{-13}
H_{13}	0.45	2.4×10^{-17}
H_{14}	0.51	7.9×10^{-12}
H_{15}	0.19	5.1×10^{-11}

multilevel nature of this defect. The annealing in several steps in Fig. 18 of the E_3 level as well as the shifts in peak position suggest that several different defects have energy levels near the E_3 position. The first annealing stage of E_3 at 100°C is probably related to the dissociation of the vacancy-phosphorus E-center. From 250-400°C the level E_3(III) appears to be a divacancy level as well as the E_2 and H_2 levels.[32,36] All these levels anneal in about the same temperature range.

The growth of the E_4 level can perhaps be related to the EPR P-1 five-vacancy center which grows as the divacancy breaks up. The rise in H_3 should then be related to the growth of the EPR P-4 center which is known to contain oxygen and is thought to be a tri-vacancy oxygen center.[30,35] The rise in E_5 concentration in the same range but to a lower maximum concentration is most likely to be

Table 7. Tentative identification of DLTS levels with known defect structures in Si.

DLTS Level	Energy (eV)	Defect	Structure
E_7	0.52	Surface	?
E_1	0.16	A-Center	vacancy-oxygen
E_3 (< 150°C)	0.42	E-Center	vacancy-phosphorus
E_3(III)	0.45		
E_2	0.24	V-V	divacancy
H_2	0.18		
E_4	0.34	P-1	five-vacancy
H_3 (300-500°C)	0.35	P-4	trivacancy-oxygen
E_5	0.27	P-2	divacancy-dioxygen
H_3 (R.T. - 300°C)	0.35	P-3	four-vacancy
H_4	0.39	P-6	di-interstitial

associated with the P-2 center which is thought to be a divacancy-dioxygen center.[30,35] The other DLTS lines at high annealing temperature probably involve oxygen, however, little more can be said about these centers. The low anneal temperature H_3 line is similar to the annealing of the P-3 four vacancy center while the behavior of H_4 appears to be related to the P-6 di-interstitial.[30,35]

We have summarized our present best guesses as to the nature of the DLTS levels in Table 7. These guesses have been based mainly on annealing temperatures, relative concentrations of known defects in neutron irradiated Si from previous EPR data, and in the case of the A and V-V centers, on other DLTS papers.

SUMMARY AND CONCLUSIONS

A simple Kinchin-Pease model has been presented to illustrate the effects of different reactor spectra on the overall displacement rate. It is argued that most reactors' spectra in which Si is irradiated can be approximated by a K/E_n differential flux and a cut off upper energy to the spectrum. This method works particularly well for Si since the low ionization threshold of Si makes this material relatively insensitive to the exact nature of the spectrum above 1.0 MeV.

A number of vacancy type defects have been observed by optical absorption and EPR. The dominant point defects to be observed are the four-vacancy (P-3), the divacancy (V-V) and the di-interstitial (P-6). The A-center (vacancy-oxygen complex) is a very important sink for vacancies, however, the presence of this defect does not alter the concentrations of P-3 and P-6. It is believed that these two defects are remnants of the original displacement cascade. Consideration of the number of cascades vs. the number of P-3 and P-6 centers for different reactor spectra shows that there are only a few of these defects per damage cascade. Consideration of the energy required to form these defects suggests that they are a result of a displacement spike near the end of the primary knock-on track. Irradiation in a variety of reactor spectra show that these defects are caused by the fast neutron and not by gamma recoils since their concentration correlates to the number of recoils but is independent of the thermal flux.

Raman scattering and EPR both suggest that highly disordered regions, which appear in many ways to be amorphous, also exist. If it is assumed that these regions are amorphous, then most of the defects produced by the cascade must contribute to this type of damage.

A vast array of defect energy levels have been observed by DLTS. One room temperature stable defect appears to migrate toward the junction in the presence of the junction electric field. Tentative defect assignments have been made to the most prominent concentrations of defect traps, however, considerably more experimental evidence will be required to make these assignments definite.

Although the amount of information that has been gathered over the last twenty years about individual defects in Si is indeed vast, there is still considerable new ground to plow concerning the exact nature of the displacement cascade in neutron irradiated Si, the relative concentrations of known defects under different irradiation conditions, cascade overlap, defects above 500°C and the overall effect of these factors on the final minority carrier lifetime in NTD-Si.

ACKNOWLEDGMENTS

We wish to express our thanks to Tom Blewitt of Argonne National Laboratory for providing the CP-5 irradiations and for checking the damage rate calculations in Si at IPNS.

REFERENCES

† University of Missouri Research Reactor, Columbia, Missouri.
* Department of Physics, University of Missouri-Columbia, Columbia, Missouri.
! Department of Electrical Engineering, University of Missouri-Columbia, Columbia, Missouri.
Department of Nuclear Engineering, University of Missouri-Rolla, Rolla, Missouri.

1) G. H. Kinchin and R. S. Pease, Rep. Prog. Phys. 18, 1 (1955).
2) M. W. Thompson, Defects and Radiation Damage in Metals, (Cambridge University Press, Cambridge, 1969).
3) D. R. Olander, Fundamental Aspects of Nuclear Reactor Fuel Elements, TID-26711-P1, (Technical Information Center, ERDA, Washington, D.C. 1976), Chapter 17.
4) J. M. Meese, Neutron Transmutation Doping in Semiconductors, ed. by J. M. Meese, (Plenum Press, New York, 1979), p. 1.
5) H. J. Stein, Ibid., p. 229.
6) O. Malmros, Ibid., p. 249.
7) J. W. Cleland, P. H. Fleming, R. D. Westbrook, R. F. Wood and R. T. Young, Ibid., p. 261.
8) M. A. Kirk and L. R. Greenwood, Ibid., p. 143.
9) J. M. Meese and S. L. Gunn, this conference.
10) J. Lindhard, V. Nielson, M. Scarff and P. V. Thomsen, Mat.-Fys. Med., Dan. Vidensk. Selsk. 33, 1 (1963).
11) Frank A. Valente, A Manual of Experiments in Reactor Physics, (Macmillan Co., N.Y., 1963), p. 69.
12) Handbook on Nuclear Activation Cross-Sections, Tech. Reports Series 156 (I.A.E.A., Vienna, 1974), p. 17.
13) T. Blewitt, Argonne National Laboratory, private communication.
14) R. R. Coltman, Jr., C. E. Klabunde, D. L. McDonald, J. K. Redman, J. Appl. Phys. 33, 3509 (1962); R. R. Coltman, C. E. Klabunde, J. K. Redman, Phys. Rev. 156, 715 (1967).
15) J. M. Meese and P. J. Glairon, Neutron Transmutation Doping in Semiconductors, Ibid., p. 109.
16) G. D. Watkins and J. W. Corbett, Phys. Rev. 134, A1359 (1964).
17) Young-Hoon Lee and J. W. Corbett, Phys. Rev. B 8, 2810 (1973).
18) L. J. Cheng, J. C. Corelli, J. W. Corbett, G. D. Watkins, Phys. Rev. 152, 761 (1966).
19) W. Jung and G. S. Newell, Phys. Rev. 132, 648 (1963).
20) J. W. Corbett, G. D. Watkins, R. M. Chrenko, R. S. McDonald, Phys. Rev. 121, 1015 (1961).

21) W. Kaiser and P. H. Keck, J. Appl. Phys. 28, 882 (1957).
22) H. Y. Lee, N. N. Gerasimenko and J. W. Corbett, Phys. Rev. B 14, 4506 (1976).
23) J. A. Brinkman, J. Appl. Phys. 25, 961 (1954).
24) J. R. Dennis and E. B. Hale, J. Appl. Phys. 49, 1119 (1978).
25) M. H. Brodsky, Light Scattering in Solids, ed. by M. Cardona (Springer-Verlag, 1975), p. 208 and references therein.
26) J. E. Smith, M. H. Brodsky, B. L. Crowder, M. I. Nathan and A. Pinczuk, Phys. Rev. Lett. 26, 642 (1971).
27) R. Shuker and R. W. Gamon, Phys. Rev. Lett. 25, 222 (1970).
28) R. Alben, D. Weaire, J. E. Smith, M. H. Brodsky, Phys. Rev. B 11, 2271 (1975).
29) D. V. Lang, J. Appl. Phys. 45, 3023 (1974).
30) J. Guldberg, J. Phys. D 11, 2043 (1978).
31) B. J. Baliga and A. O. Evwaraye, Neutron Transmutation Doping in Semiconductors, Ibid., p. 317.
32) C. E. Barnes, I.E.E.E. Summary of Papers (abstracts only), Vol. 26A, I.E.E.E. Conf. on Nuclear and Space Radiation Effects, U. of California at Santa Cruz, July 17-20, 1979, p. 57.
33) B. W. Wessels, J. Appl. Phys. 47, 1131 (1976).
34) D. V. Lang, Topics in Applied Physics, Vol. 37: Thermally Stimulated Relaxation in Solids, ed. by P. Braünlich, (Springer-Verlag, New York, 1979), p. 93.
35) J. W. Corbett, J. C. Bourgoin, L. C. Chang, J. C. Corelli, Y. H. Lee, P. M. Mooney and C. Weigel, Radiation Effects in Semiconductors, 1976, ed. by N. B. Urli and J. W. Corbett (Institute of Physics, Bristol, 1977), p. 1.
36) L. C. Kimerling, Ibid., p. 221.

WAFER STABILITY. A COMPARISON OF

NTD-SILICON WITH CONVENTIONAL FZ

O. Malmros, J. Guldberg and K. Lund

Topsil A/S

3600 Frederikssund, Denmark

ABSTRACT

During the irradiation of silicon a large number of crystal defects are created. These defects will give rise to internal stresses in the silicon, which might in turn result in reduced mechanical strength of the wafers and a larger tendency for breakage.

In this paper a comparison is made between the strength of different types of silicon and it is shown that there is no difference between the selected materials.

1. INTRODUCTION.

The impetus for the present investigation were persistent rumours that neutron transmutation doped silicon was weaker than conventionally doped float-zone silicon. It was also reported (1) that CZ wafers possessed better properties strengthwise than FZ wafers. Moreover, a number of investigators had examined the hardening effects of oxygen in the silicon lattice (2). Accordingly, handling of NTD-silicon wafers could possibly require some special precautions.

Cluster formation during neutron irradiation is well documented (3), and after sufficiently high irradiation doses even changes in the lattice constant has been measured (4). It is therefore of interest to investigate to which extent wafer strength is influenced by irradiation doses encountered during ordinary applications and by the subsequent anneal treatments.

The present work had two specific goals:

- development of a suitable experimental technique for comparison of wafer strength, and
- comparison of different types of material.

With respect to the experimental technique we soon realized that any mechanical damage left over on the wafer had a strong detrimental effect on wafer stability during bending. By the same token the amount of material removed from the surface by chemical etching had a pronounced influence on the force which could be applied to a wafer before breakage occurred. Some preliminary work was thus required to define an adequate procedure for wafer preparation.

The measurement technique included recording the forces needed to break a batch of wafers under highly reproduceable conditions at room temperature. This method does not necessarily provide a quantitative assessment of the wafers lost from breakage or slip formation due to handling and high temperature processing - but does give some indication of wafer integrity.

Two series of experiments were incorporated in this analysis.

The first line of experiments served to compare

- unirradiated silicon
- irradiated, but non-annealed silicon
- irradiated and annealed silicon.

A second line consisted in a comparison between silicon produced by floating-zone and crucible growing techniques.

Using two individual crystals to represent each of the above six combinations, a total of 12 ingots were included in the experiments - all of them dislocation free.

2. EXPERIMENTAL PROCEDURES.

2.1. Minimum etching depth

Since microcracks occurring anywhere on a wafer will significantly reduce its ability to withstand externally applied forces, considerable care must be excersized in order to keep the surface and the wafer edge free of mechanical damage.

A 2 inch FZ-ingot with resistivity 5000-8000 Ohmcm, P-type (boron) was ground to a diameter of 50.0 mm in 2 steps. During the first step the rod was ground to 50.3 mm diameter with a diamond cup grinder. In the second operation a beltgrinder equipped with a 400 mesh silicon carbide abrasive was used to bring the diameter down to 50.0 mm. Following this the ingot was etched to 49.5 mm diameter in HF:HNO_3, and finally the ingot was cut into 340 μm thick wafers on a I.D. saw cooled with water.

During all of the above mechanical operations careful precautions were taken not to damage the cylinder surface. From earlier experiments it was known that any residual damage even at the wafer edge would lower the strength considerably of an otherwise perfect wafer.

All wafers were sorted in groups with minimum thickness variation and placed in wafer holders. Using an HF:HNO_3:HAc mixture they were etched to various thicknesses. The etching depth and the resulting final thickness were determined by weighing all wafers in a group together. Again, earlier experience had proven the necessity of using procedures without any surface contacting. Even handling performed during thickness measurements lowered the strength appreciably, also when utilizing a contact less probe.

In what follows etch depths are stated as the sum of material removed from both sides of a wafer i.e. the total thickness change. When calculating the thicknesses the weight per unit length of the silicon ingots was assumed to be

$$0.00183 \times D^2 \text{ grams/mm}$$

where D is the diameter expressed in mm.

It was further assumed that the wafer diameter would be reduced with 1.3 times Δt if the wafer thickness was reduced by Δt.

The strength of a wafer was determined by placing it horizontally on a pedestal consisting of 3 ruby balls (Fig. 1). The rubies had a diameter of 4 mm and formed an equilateral triangle with a side length of 39.04 mm. Thus the pedestal supported the wafer at 3 points, 120 degrees apart, on a concentric circle with diameter 45.08 mm.

During measurements the pedestal was placed on a balance, capable of weighing from 0 to 15 kg with a resolution of 5 grams, and a steel ball with diameter 8 mm was pressed vertically down towards the center of the wafer. Pressure on the ball was increased gradually and the balance reading was noted when the wafer broke.

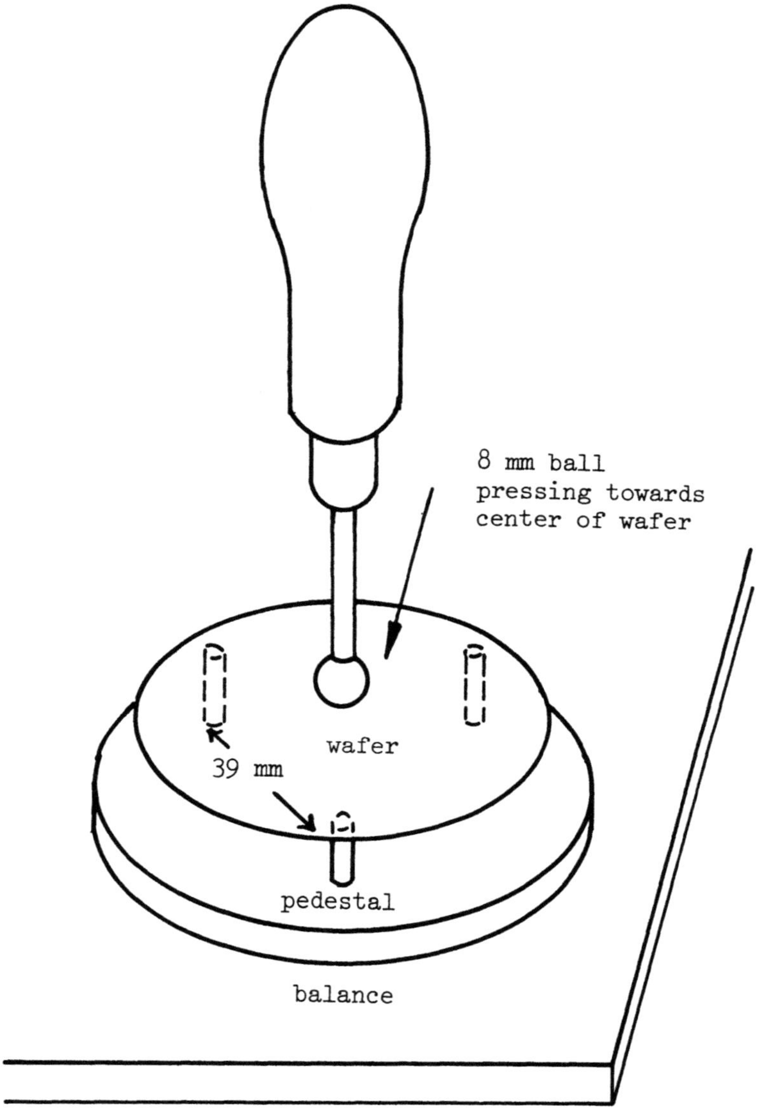

Fig. 1. Sample holder for wafer breakage.

2.2. Crystal damage experiment

Irradiation for this part of the investigation were performed in the Danish reactor, RISOE, with a cadmium-ratio better than 500:1. The resistivity before irradiation was above 600 Ohmcm for all materials, and the irradiations produced a resistivity of 25 Ohmcm, corresponding to a dose of approximately 10^{18} thermal neutrons/cm^2. During the irradiation the crystal temperature was kept below $180^{\circ}C$.

Sample preparation and breakage was performed in the same manner as the above with 2 modifications. Ingot diameter before grinding was $2\frac{1}{2}$", and sawing thickness was 400μm. After etching the thickness ranged from 245 to 270μm, thus more than 125μm was removed in all cases.

2.3. Calculations

The load which can be sustained by a thin infinite plate with a certain thickness supported by 3 points is inversely proportional to the thickness squared. To a first approximation the same is valid for a circular wafer. Hence, we can characterize the strength through the parameter

$$S = \frac{W}{T^2} \quad (\text{gram}/\mu m^2) \qquad (1)$$

where

W (gram) is the balance reading at the moment of breakage
T (μm) is the wafer thickness.

Equation (1) does not hold exactly when very thin wafers are employed. Because of their elasticity they will bow considerably before breaking, therefore the surface will slide on the ruby supports and the surface will become damaged. Consequently a lower strength will be observed on wafers with a thickness less than 200μm.

Strength reduction is primarily caused by mechanical damage on the lower side of the wafer. While under flexure the top of the wafer is under compression and any damage appearing on this side will be inconsequential. The opposite side, however, will experience a tensile stress, and any scratch on this will be torn open due to stress amplification at the exposed micro cracks.

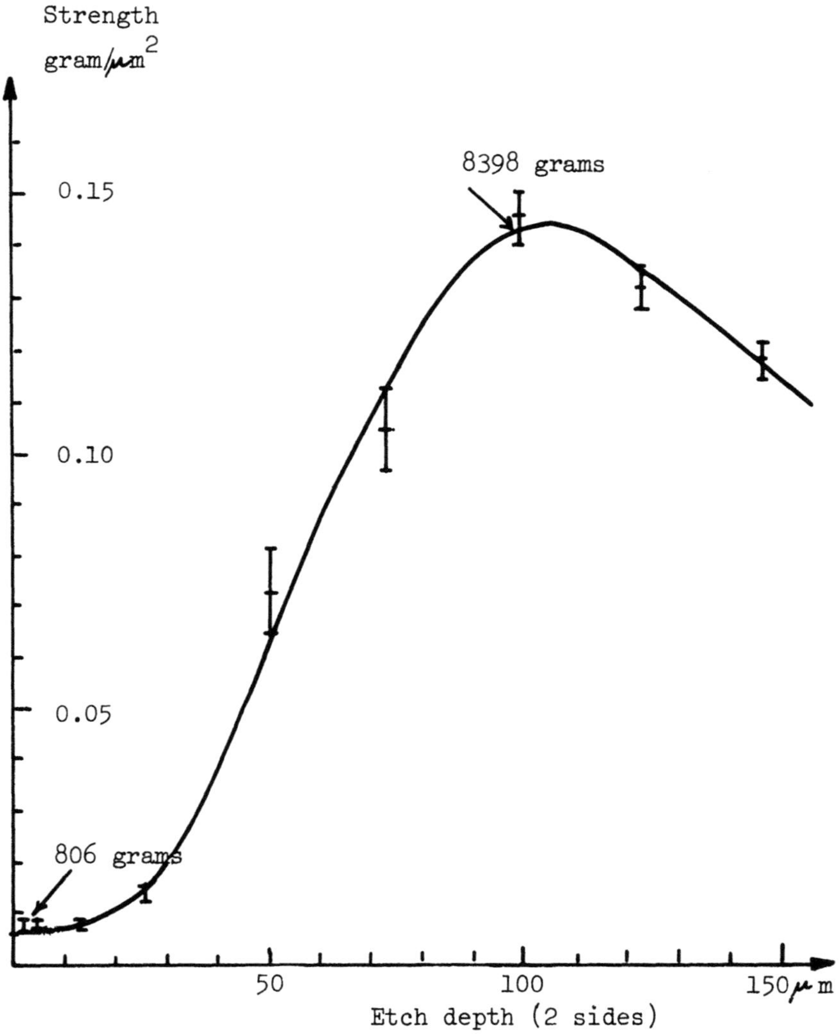

Fig. 2. Strength versus etch depth on wafers originally cut to 340μm thickness.

3. RESULTS AND DISCUSSION.

3.1. Etching experiments

Figure 2 illustrates the measured wafer strength corrected for thickness variations according to equation (1) in relation to the etching depth (total thickness change). From this figure the following observations can be made:

a) The most pronounced feature is the large increase in strength after etching. When all damage is removed wafer strength increased by more than an order of magnitude.

Starting wafers with an average thickness of 344μm required an average of 806 gram to break. After etching to 240μm these wafers required 10 times greater force to break - in average 8398 grams.

b) After 100μm has been etched off the strength ceases to increase. Accordingly the 50μm removed from each side must be related to the depth of damage from the sawing operation. For comparison, a determination of the damage depth by bevel polishing and Sirtl etch on 3 wafers showed the damaged region to extend 10, 12 and 13μm below the surface. Therefore, it is beyond doubt that a Sirtl etch does not reveal the deeper, less severely distorted parts of the crystal. However, this damage can still be discerned in the present type of strength experiment.

c) From the depicted standard deviations in Fig. 2 it can be seen that a very large scatter in the obtained date coincides with the transition from weak to strong wafers. This point is shown more clearly in Fig. 3.

At 50μm the observed scatter exhibits a maximum which further illustrates the statistical fact that some wafers in a batch will reach the defect free condition sooner than others.

d) In the well etched part of the graph (Fig. 2) a slight decrease of strength with thickness can be seen. This may be caused by the damage from the supporting ruby balls as discussed above.

3.2. Crystal damage experiment

Table I compares the strength determined for different types of material, and after various treatments. The values listed in the table are strengths at the point where surface damage is removed

Table 1. Average strength (g/µm^2) for various materials. Standard deviation for each value is .004 gram/µm^2.

Treatment	Crystal			
	CZ 1	CZ 2	FZ 1	FZ 2
As grown	0.146	0.147	0.143	0.146
Irradiated	0.144	0.155	0.157	0.146
Irradiated and annealed	0.143	0.153	0.149	0.150

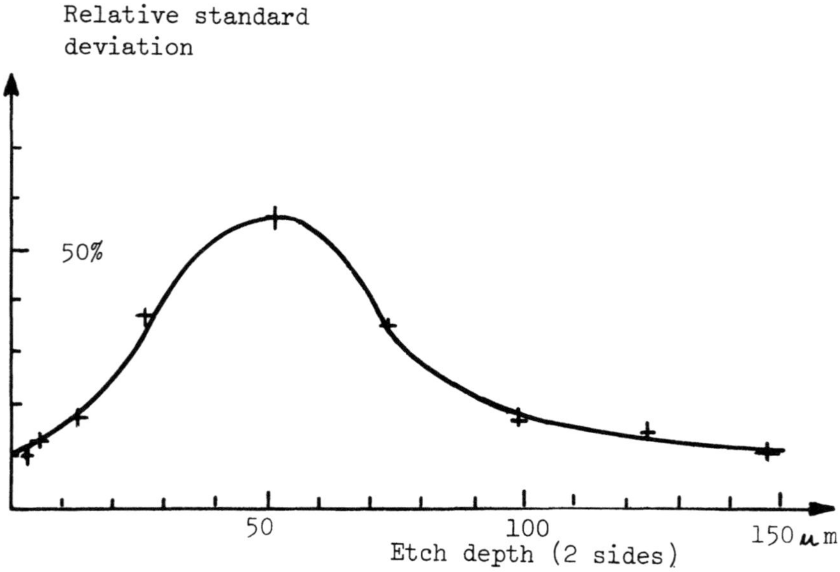

Fig. 3. Relative standard deviation when measuring strength of etched wafers originally cut to 340µm thickness.

completely, but before the thin wafer condition starts to have any influence.

From this table it may be deduced that there is indeed no significant difference in the performance of the various ingots during this test procedure. Moreover, it does not matter whether the silicon is Float-zone or Czochralski grown, nor whether it is "as grown", irradiated or irradiated plus annealed.

Quite evidently there must be a large concentration of cluster type damage sites in the irradiated material which have not yet been annealed, because the temperature during irradiation has not been high enough to cause this type of defect to disappear (5). However, in terms of stresses which the material can endure, there seems to be no difference as compared with unirradiated material.

In this context it may also be mentioned that dislocations, or slip lines, on earlier occasions were found to reduce the wafer strength considerably.

Furthermore the succeeding heat treatment which restores the electrical characteristics of the material appears to have no detrimental effect on the mechanical properties of the wafer, either. In the litterature it has been reported (6) that heat treatments may alter the geometry of unstrained CZ-wafers, however with the temperatures and times adopted for the present purposes no such differences could be observed with the described test procedure.

4. CONCLUSION.

The experiments reported here have lead to a test procedure suitable for testing the mechanical properties of silicon wafers. In particular a measure is established for wafer strength which excludes surface damage.

Through the use of this technique it has been demonstrated that

- Float-zone and Czochralski wafers have the same strength properties, in spite of differences in oxygen content.
- NTD-Silicon - annealed as well an non-annealed - has the same strength as conventional Float-zone Silicon.

However, differences in wafer breakage undoubtedly have been observed in the past. It is quite likely though, that minor differences in ingot/and or wafer finish constituted the real problem to be addressed in this connection.

REFERENCES

(1) D.C. Guidici, Semiconductor International Vol. 1, Aug. 1978, p. 18-19

(2) S.M. Hu, J. Vac. Science, Vol. 14, No. 1, Jan/Feb 1977, p.p. 17-31

(3) H. Stein, in "Neutron Transmutation Doping in Semiconductors" edited by J.M. Meese, Plenum Press, New York 1979, p.p. 229-248

(4) A.R. Chelyadinskii, Sov. Phys. Solid State Vol. 18, No. 3, March 1976, p. 506-507

(5) P.L.F. Hemment and E.M. Gunnersen, Journ. App. Phys., Vol. 37, p. 966, p., 2912. J.M. Pankrantz, J.A. Sprague and M.L. Rudee, Journ. App. Phys., Vol. 39, No. 1, 1968, p.p. 101-106

(6) B. Leroy and C. Plougonven, Journ. Electrochem. Soc., Vol. 127, No. 4, April 1980, p.p. 961-970.

ELECTRON SPIN RESONANCE (ESR) STUDY ON THE THERMAL ANNEALING OF DEFECTS INDUCED IN NEUTRON TRANSMUTATION DOPED SILICON

V. Q. Ho and T. Sugano

Department of Electronic Engineering
The University of Tokyo
7-3-1, Hongo, Bunkyo-ku, Tokyo 113, Japan

ABSTRACT

The effect of thermal annealing on defects induced by transmutation doping process in Si has been studied by electron spin resonance. Two centers have been found for float zone Si irradiated with thermal neutron at the flux of $2x3^{13}n/cm^2sec$ and Cd ratio to fast neutron of 15:1. One center was annealed out after annealing at 420°C but the other center, which was attributed to Si-P_1 center still remained after annealing at 700°C. It was found that vacancy cluster model could be the microscopic structure of defect clusters and the origin of some of those energy levels measured by DLTS.

INTRODUCTION

It has been known that transmutation doping process induces several energy levels in Si band gap, which were measured by Deep Level Transient Spectroscopy (DLTS)[1~3] and defect clusters which were observed by Transmission Electron Microscopy (TEM)[4]. However, the physical model of these levels and microscopic structure of these defect clusters have not been well understood.

Here, we will present our Electron Spin Resonance (ESR) results on the thermal annealing of defects in neutron transmutation doped silicon. It will be shown that the defect center is not simply annealed out at low temperature as reported previously.[5~6] Correlation with results obtained by TEM and DLTS and possible microscopic structure of defect clusters and origin of various energy levels measured by DLTS will be discussed.

EXPERIMENTS

Sample Preparation

Boron doped (111) oriented FZ (purified by the floating zone method) Si wafers with the resistivity of 1.5 KΩcm were choosen as starting material. They were packaged in an aluminum can during the irradiation in JRR-3 reactor of Japan Atomic Energy Research Institute, at the flux of $2x10^{13}$ n/cm^2 sec. The Cd ratio of thermal neutron to fast neutron was about 15. After cooling in several days, irradiated silicon wafers were cut into small chips of 3x10 mm^2 in size for ESR measurements. These chips were etched in a mixture of $HF:HNO_3$ at volume ratio of 1:10 to remove mechanical damages at the surfaces of the chips which give a surface resonant signal appeared at g=2.005 as reported by Haneman.

ESR Measurements

ESR signals at x band were measured at liquid nitrogen and room temperature, with a spectrometer whose sensitivity was $5x10^{10}$ spins /G. Chips were charged into the TE_{011} cylindrical cavity of the spectrometer. In order to minimize noize, the differentiation to the magnetic field of the susceptibility $\frac{d\chi''}{dH}$ were measured. g-values of resonant signals were measured using the relation

$$h\nu = g\beta H$$

where the magnetic field intensity was calibrated using the known g value of a standard DPPH sample, which was assumed to be 2.0036[7] and the microwave frequency was measured using a cavity resonator. The microwave power was kept at 2 mW and the modulation magnetic field intensity at 1 G.

To pursue the variation of the resonance signals with annealing temperature, a marker which was made by thermal diffusion of M_n^{2+} into MgO was inserted in the cavity. M_n^{2+} (I=5/2) has five hyperfine lines in which g value of the fourth line counted from the left is 1.981 and the interval between the 3rd line and the 4th line is about 87 G. The intensity of the 4th line was calibrated using the standard sample of DPPH and used to evaluate the spin density of defects.

Annealing Procedures

Annealing was performed with a resistance heated furnace in Argon atmosphere. Annealing time was choosen to be 20 min for each step since the signal was much the same after prolongating the annealing time to one hour. The chips which were packaged in a quartz holder were slowly pulled out after each annealing step and cooled at the entrance of the reactor for 10 min.

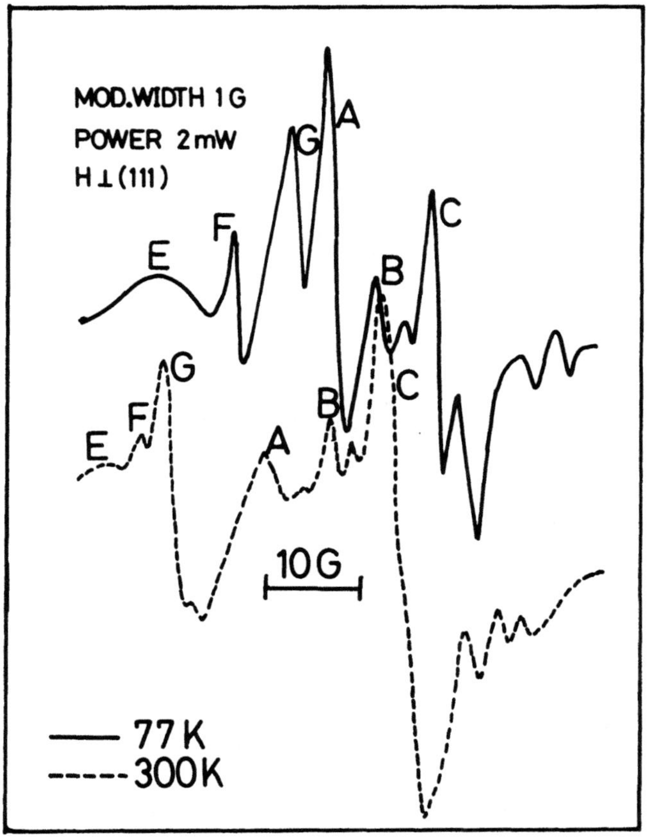

Fig. 1. Resonant signals due to defects in neutron transmutation doped silicon are shown for measurements at 77K and 300K. A, B and C peaks are confirmed to be due to $Si-P_1$ (or Si-N) center. The thermal neutron dose was about $4x10^{17}n/cm^2$.

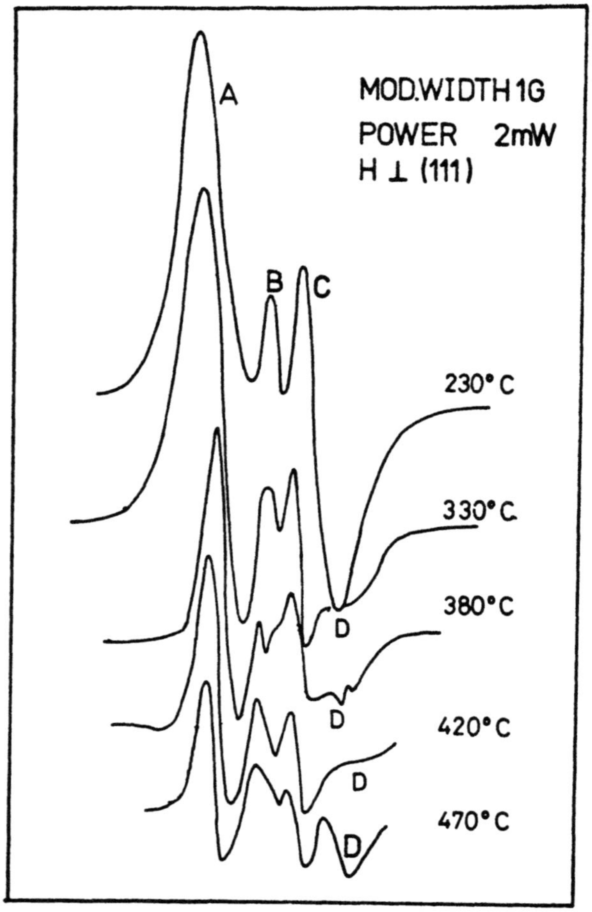

Fig. 2. Variation of resonant signals with annealing temperature is shown for the measurement at room temperature. E, F, G peaks disappeared after annealing at 230°C but a new peak, D is observed after annealing at 470°C.

RESULTS

Figure 1 shows the line shape of the resonant signal measured at liquid nitrogen and room temperature for the case that the magnetic field is in [111] direction.

It is observed that the signal has complicated structure which is composed of 6 dominant peaks and several hyperfine structures. Since g values of A, B, C peaks as in Fig. 1 for Ho // [111] are 2.011, 2.008 and 2.005 respectively, they are attributed to Si-P_1 or Si-N center which was previously reported by Nisenoff and Fan.[7]

The spectra which were measured at room temperature after annealing the chips at various temperatures are shown in Fig. 2. E, F, G peaks in Fig. 1, which may be due to Si-A_2 center disappeared after annealing at 230°C. However, when measuring at liquid nitrogen temperature, it was found that this center still remained after annealing at 330°C and disappeared after annealing at 420°C. Besides, it was found that a new peak, named D, obscured in the broad peak of C, appeared after annealing at 470°C. The decrease in intensity and half width of peak A with increasing temperature shows that some transformation in microscopic structure of Si-P_1 center has happened.

In Fig. 3, the variation of peak to peak height of each resonance peak with annealing temperature is shown. It is seen that the peak A, disappears after annealing at 620°C. This is also confirmed by measuring at liquid nitrogen temperature the decrease of signal intensity with increasing annealing temperature as in Fig. 4 which shows that A peak is annealed at 600°C but other ones still remain after annealing at this temperature. B and C peaks disappeared after annealing at 740°C and D peak at higher temperature.

DISCUSSION

In our experiment, by applying the magnetic field in parallel to [111] direction, the strongest peak of the Si-P_1 center has been separated from other ones so that it was possible to observe the variation of each peak with annealing temperature. The fact that the A peak disappears after annealing at 600°C while other ones still remain after annealing at higher temperature means that some rearrangement of atoms in the vacancy clusters has happened in such a way to remove dangling bonds. Qualitatively, this is in agreement with the finding[4] that the size of defect clusters observed by transmission electron microscopy increases, but their number density decreases with increasing annealing temperature. Moreover, by considering that the size and number density of vacancy cluster which was proposed by Lee and Corbett[8] are in the same order with those

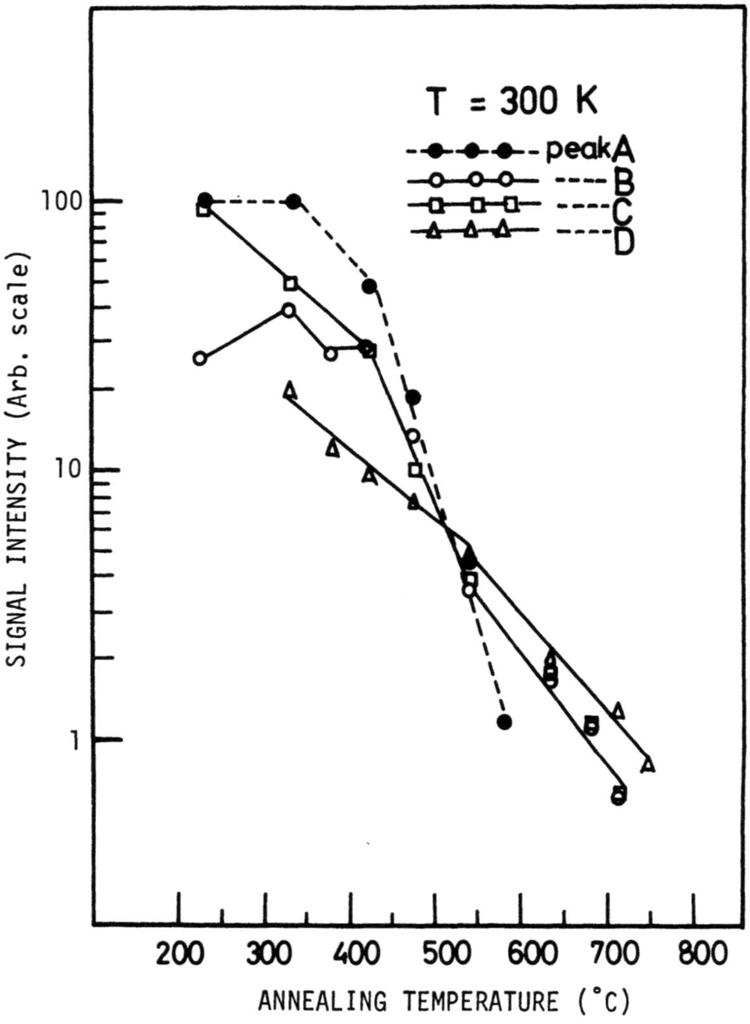

Fig. 3. Variation of signal peak-to-peak intensity with annealing temperature is shown for the measurement at room temperature. The A peak disappeared after annealing at 620°C, while B and C peaks at 740°C but D peak still remained at this temperature.

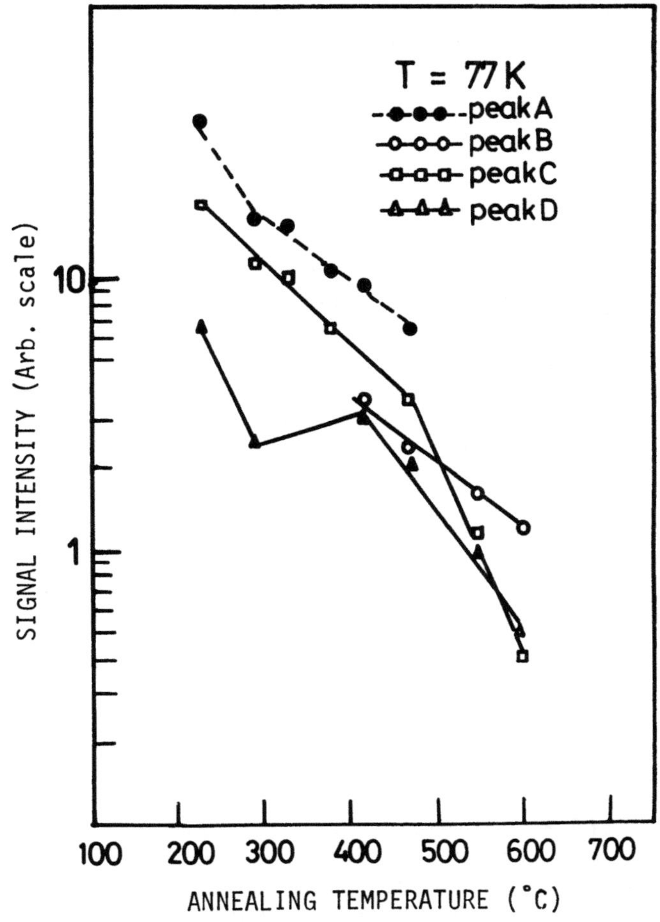

Fig. 4. The variation of signal peak-to-peak intensity with annealing temperature is shown for the measurement at 77k. The A peak of the resonant signal was not observed after annealing at 600°C while other peaks still remained at this temperature.

of defect cluster, it is suggested that vacancy cluster could be the atomic structure of those defect clusters in neutron transmutation doped silicon. Besides, as the anneal behavior of Si-P_1 center is complicated, it is therefore reasonable to assume that such center could result in various energy levels rather than a single one as suggested by Gulberg[9]. Though it is not possible to make clear which levels are belonging to Si-P_1 center at present time since defects which have no dangling bonds would result in no resonance signal detected by ESR, it is thought that the Si-P_1 center is responsible for those levels which still remain after annealing at high temperature.

CONCLUSION

ESR was used to study the effect of thermal annealing on Si-P_1 center in neutron transmutation doped silicon. It has been found that the strongest peak of Si-P_1 center disappears after annealing at 600°C while other ones of smaller intensities still remain at higher annealing temperature. It has been found that this center could result in various energy levels and could be the atomic structure of defect clusters observed by transmission electron microscopy. Annealing at temperature higher than 740°C is therefore necessary to remove these defects completely.

ACKNOWLEDGMENTS

We thank Mr. H. Matsumoto for providing samples for our experiments and to Mr. T. Sakurai for the measurement of microwave frequency.

REFERENCES

1. J. Guldberg, Electron trap annealing in neutron transmutation doped silicon, Appl. Phys. Lett, 31:578 (1977).
2. T. Sugano and H. Matsumoto, Impurity doping of silicon by thermal neutron irradiation, National Convention Japan Society of Applied Physics, 30a p9, March 1979.
3. B. J. Baliga and A. O. Evwaraye, Defect levels controlling the behavior of NTD silicon during annealing, Neutron Transmutation Doping in Semiconductors (1978), ed. by J. M. Messe, Plenum Press, New-York & London, pp. 317-328.
4. B. C. Larson, R. T. Young, and J. Narayan, Defect annealing studied in neutron transmutation doped silicon, ibid. pp.281-290.
5. L. Katz, and E. B. Hale, Electron spin resonance in NTD silicon, ibid. pp. 307-316.
6. W. Jung, and G. S. Newell, Spin-1 centers in neutron irradiated silicon, Phys. Rev., 132:648 (1963).

7. M. Nisenoff, and H. Y. Fan, Electron spin resonance in neutron-irradiated silicon, Phys. Rev., 128:1605 (1962).
8. Y. H. Lee, and J. W. Corbett, EPR studies in neutron irradiated silicon : A negative charge state of a nonplanar five vacancy cluster (V_5^-), Phys. Rev., B8:2810 (1973).
9. J. Guldberg, Electron traps in silicon doped by neutron transmutation, J. Phys., D11:2043 (1978).

OPTICAL STUDIES OF LATTICE DAMAGE IN NEUTRON-TRANSMUTATION-DOPED SILICON*

Noboru Fukuoka**
Department of Physics, College of General Education,
Osaka University, Toyonaka, Osaka 560, Japan

John W. Cleland
Solid State Division, Oak Ridge National Laboratory,
Oak Ridge, Tennessee 37830, U.S.A.

Infrared spectroscopy (IR) techniques in the 1-50 μm wavelength region have been used to investigate the optical properties of neutron-induced lattice defects in high-purity float-zone (FZ)-refined Si containing $\leq 10^{16}$ interstitial oxygen (O_i) cm^{-3} and in high purity Czochralski (Cz)-grown Si containing 2.4×10^{18} O_i cm^{-3}. The measurements were made following $\leq$ 100°C fast neutron ($\geq$ 1 MeV) irradiations and after 20 min isochronal annealing at 50°C intervals from 150 to 900°C.[1] All samples were irradiated in a reactor in-core position with a thermal-to-fast ($\geq$ 1 MeV) neutron ratio of ∿ 5, so that the lattice damage introduced by thermal neutron absorption and (n,γ) recoils was negligible compared to that introduced by fast neutron collisions with Si atoms.[2]

An absorption band at 1.8 μm due to the introduction of divacancies was observed in all samples after irradiation with a low (1.5×10^{19} n_f cm^{-2}) fast neutron fluence, but was not observed in more heavily irradiated ($\geq 10^{20}$ n_f cm^{-2}) samples, due to the introduction of very strong near-band edge absorption above about 1.0 μm. At least 90% of the near-edge absorption was removed in each sample by annealing up to 450°C, irrespective of neutron fluence, and complete removal was obtained by annealing the more heavily irradiated samples up to 750°C.

No irradiation-induced IR-active centers were observed in either the FZ or Cz Si samples in the 5-10 μm wavelength region; however, single-phonon absorption bands were observed in every sample of FZ or Cz Si after irradiation in the 15-30 μm range and the absorption coefficient of these bands increased with neutron fluence. Higher-order bands were also observed in all FZ or Cz Si samples from about 8 to 15 μm after irradiation and annealing at 300 to 500°C, and the absorption coefficient of these bands decreased with neutron fluence.

Infrared measurements of the 8.87 μm absorption band can be used to measure the concentration of electrically inactive O_i in Si and heat treatment of Cz Si at temperatures above 300°C has been shown to form electrically active oxygen related donors, that can be removed by annealing at higher (≥ 500°C) temperatures. In the present experiments, the absorption coefficient of the 8.87 μm band of an unirradiated Cz Si sample decreased on annealing from 150 to 500°C as a consequence of oxygen donor formation and then increased on annealing up to 850°C as a consequence of oxygen donor dissociation. The absorption coefficient of the 8.87 μm band of a Cz Si sample, irradiated with 1.5 x 10^{19} n_f cm^{-2}, decreased on annealing to 300°C, increased on annealing to 550°C, and then decreased again on annealing to 700°C. The relatively large increase in the apparent O_i concentration in the irradiated sample above 300°C may have been due to the release of oxygen from radiation-induced lattice defects, and the apparent decrease in O_i concentration above 550°C may have been due to the formation of new oxygen-defect complexes, since this effect was not observed in the unirradiated sample.

Additional samples of FZ or Cz Si were irradiated with several different neutron fluences and the absorption coefficient was measured after annealing for 20 min at 350°C. The results were such as to indicate the introduction of six distinct absorption bands at 12.1, 12.0, 11.9, 11.2, 10.8, and 9.87 μm in the Cz Si samples only, and two additional bands were also observed in the Cz Si samles at 9.62 and 9.49 μm after higher temperature (> 500°C) annealing. The fact that similar bands were not observed in the FZ Si samples under identical irradiation and annealing conditions can be taken as an indication that these distinct absorption bands in Cz Si are due to some type of interaction between O_i and fast-neutron induced lattice defects.

Previous investigators have identified an absorption band at 12.0 μm in Cz Si as due to an oxygen-vacancy pair (the A center) and another band at 19.5 μm as due to the presence of oxygen. In the present experiments the absorption coefficient of the A center was larger in neutron irradiated Cz Si than that of the six other distinct absorption bands below an annealing temperature of 350°C. All of the A centers were removed by annealing at 500°C, but the 19.5 μm band was still present after annealing at 900°C.

Infrared absorption measurements of conventionally doped FZ or Cz Si have indicated a band at 31.8 μm as introduced by neutral, n-type, P dopant. In the present experiments, an absorption band was observed at that wavelength in FZ or Cz Si samples after irradiation with a moderate (4.5×10^{19} n_f cm^{-2}) fluence after annealing at 650°C and a similar band was observed at that wavelength in FZ Si after irradiation with a large ($1.3\text{-}7.6 \times 10^{-20}$ n_f cm^{-2}) fluence after annealing at 700 to 750°C. Electrical property measurements on companion samples of FZ or Fz Si indicated a large increase in donor concentration after annealing at these temperatures, which would suggest that the P was dissociated from other defect complexes.

The primary results of these studies are that divacancy, near-edge absorption and single-phonon bands were observed after fast neutron irradiation of FZ or Cz Si, irrespective of O_i concentration, and the absorption coefficient of these IR active defect centers increased with fluence. Higher-order bands were formed in FZ or Cz Si after irradiation and annealing to 300°C and the intensity of these bands decreased with increasing fluence. The complete removal of all of the absorption introduced by these bands following annealing for 30 min at 750°C, even in very heavily irradiated samples, and the observation of the anticipated carrier concentration and mobility due to P, is taken as evidence that the absorption was not the result of a very complex defect or defect-impurity structure. The formation of at least six distinct absorption bands in the 9.1-12.1 μm wavelength region in Cz Si, after irradiation and partial annealing, may have been associated with the formation of different oxygen-defect related centers, but their complete removal on annealing to 750°C is presented as evidence that there was no special oxygen-lattice defect complex that remained in Cz Si above 750°C under these irradiation and annealing conditions.

ACKNOWLEDGMENTS

The authors wish to thank P. H. Fleming for technical assistance, R. T. Young for many valuable discussions, and D. H. Lowndes, G. E. Jellison, Jr., and F. W. Young, Jr. for a careful review of the manuscript.

REFERENCES

* Research sponsored by the Division of Materials Sciences, U.S. Department of Energy under contract W-7405-eng-26 with the Union Carbide Corporation.

** Experiments performed at the Oak Ridge National Laboratory as a guest scientist.

1. N. Fukuoka and J. W. Cleland, Radiat. Eff. <u>51</u>, 215 (1980).
2. J. W. Cleland and N. Fukuoka, Proceedings of this Conference.

A FACILITY AND PROGRAM AT IPNS TO STUDY DEFECTS PRODUCED BY FAST NEUTRONS IN SEMICONDUCTORS*

M.A. Kirk, R.C. Birtcher, T.L. Scott and T.H. Blewitt

Materials Science Division
Argonne National Laboratory
Argonne, Illinois 60439

ABSTRACT

The Radiation Effects Facility at the Intense Pulsed Neutron Source (IPNS), under construction at Argonne National Laboratory, will offer a genuinely unique opportunity to study the effects of fast-neutron bombardment of semiconductors. This user-oriented Facility will be described and the results of recent measurements[1] of spallation neutron flux, flux gradients, and energy spectrum will be presented. A close similarity with the fast component of the neutron spectrum in a typical fission reactor will be demonstrated. Two unique features of a pulsed spallation neutron source will be exploited in experiments to be performed on semiconductors: the low γ flux and the pulsed nature. The unusually low production of γ-rays in a spallation neutron source will make it possible to perform experiments on semiconductors which in the past have been performed only on metals. For example, low-temperature defect production and annealing will be measured in situ by electrical resistivity techniques. The pulsed nature of the neutron source will be used to study the kinetics of defect annihilation and agglomeration. The structure and character of defects in Si will be investigated by techniques including resistivity, Hall coefficient, DLTS and TEM. Early experiments are being planned in cooperation with co-workers at the University of Missouri, Columbia. Other experimenters with an interest in this Facility are encouraged to contact the authors.

---

[1]To be published in J. of Nucl. Materials.

*Work supported by the U. S. Department of Energy.

DEFECTS IN NEUTRON-IRRADIATED EXTRINSIC P-TYPE SILICON

M. H. Young* and O. M. Stafsudd

Department of Electrical Sciences and Engineering
University of California at Los Angeles
Los Angeles, CA 90024

K. L. Brower

Sandia Laboratories
Albuquerque, NM 87185

O. J. Marsh and R. Baron

Hughes Research Laboratories
3011 Malibu Canyon Road
Malibu, CA 90265

INTRODUCTION

At the 1978 International Conference on Transmutation Doping in Semiconductors, we reported the initial results of a study of radiation defects in neutron-irradiated, low-oxygen extrinsic p-type silicon.[1,2] In that initial study we observed two new shallow acceptor levels denoted A_1 and A_2 located near the valence band edge in the forbidden gap of Si. The activation energies for A_1 and A_2 were determined from measurements of Hall effect vs. temperature to be $\sim$0.03 eV and $\sim$0.04 eV, respectively, in irradiated Si:Ga. These defect centers were observable after annealing at temperatures between $\sim$575°C (below which temperature the centers were overcompensated by excess damage-produced donors) and $\sim$625°C. Damage-produced donors became observable at an anneal temperature of $\sim$500°C with concentrations well in excess of the total concentration of initial donors

* Present address, Hughes Research Laboratories, 3011 Malibu Canyon Road, Malibu, CA 99265.

(measured) plus transmutation-produced phosphorus. These donors annealed away with increasing anneal temperature, and the concentration became too small to observe at an anneal temperature of ∿700°C. Annealing at 700°C to 850°C removed all observable radiation effects — both shallow acceptors and radiation-produced compensating donors. We also observed the shallow acceptors in low-temperature photoconductive spectral response measurements in our initial work, but no extensive study of the optical properties of these centers was carried out. These early results provided the motivation for a new electron paramagnetic resonance (EPR) study of defects in the same Si material, as well as a more thorough investigation of the electrical and optical properties and annealing characteristics of the A_1 defect level in Si:Ga. A_1 is the more readily observable acceptor defect and the more interesting, since not only is it much shallower than other reported radiation defects in silicon (of which most are associated with deep energy levels, typically near the center of the band gap), but it is also shallower than any of the normal Column III acceptors in silicon.

The silicon material studied was especially grown for infrared (IR) photoconductive detector applications. For p-type detectors, the intentional dopant chosen is most often gallium or indium, but there is inevitably some residual boron content, typically in the range from 10^{12} to 10^{13} B/cm^3, even in the best detector material. For proper detector operation, it is necessary that the boron be overcompensated by an n-type dopant such as phosphorus. Neutron transmutation is an ideal doping technique for this purpose, since the phosphorus concentration can be set in detector material to precisely the desired value with good spatial uniformity over the large areas needed for detector array applications. We have therefore irradiated especially grown and carefully characterized p-type IR detector silicon with thermal neutrons under conditions appropriate for phosphorus counterdoping the residual boron and then studied the defects associated with this doping process.

STUDIES OF THE A_1 LEVEL

Experimental Procedures

The irradiation facility and conditions were the same as reported in our previous work.[1,2,3] We used a neutron flux of 3×10^{17} neutrons/cm^2 which produced 5×10^{13} P/cm^3 in the silicon samples. From an analysis of damage mechanisms appropriate to our irradiation conditions, we estimate[4] the displacement density to be approximately 1.2×10^{18} displacements/cm^3. The primary displacement mechanism is hard sphere knock-on reactions with the small fraction of energetic neutrons present in the flux. A small contribution is also present from recoiling atoms undergoing transmutation.

Analysis of Hall effect measurements as a function of temperature from ∿18K to 325K allowed us to determine the concentration of intentional Column III impurities, A_1 defects, and compensating donors, as well as the electronic energy level positions of the acceptors. The Hall effect measurement procedure and analysis method used was reported in our earlier work.[1,2]

Photoconductive spectral response measurements were made using a Beckman IR 12 spectrometer which is adapted to provide a focused monochromatic beam to illuminate the sample. We measured photoconductivity as a function of the wavelength of incident IR radiation. Reliable spectral measurements could be made out to ∿40 μm, which is somewhat short of the expected cutoff wavelength ($\lambda_{cutoff} = hc/E$) for the A_1 defect level (∿ 50 μm). The measurement temperature was 5K. The samples were irradiated Si:Ga doped with ∿$7x10^{16}$ Ga/cm^3 and measured to contain ∿$2x10^{14}$ A_1 centers/cm^3 after a 600°C, one-hour anneal.

RESULTS AND DISCUSSION

Our previously reported[1,2] electrical and optical investigations have dealt primarily with irradiated Si doped with $3x10^{16}$ to $8x10^{16}$ Ga/cm^3. In the present work we have expanded our study of Si:Ga to cover a doping range from $2x10^{15}$ to $5x10^{17}$ Ga/cm^3 in order to determine with confidence whether the production of shallow defects depends on Column III atom doping, as was tentatively indicated in the initial study. As shown in Fig. 1, a clear dependence of N_1, the concentration of A_1 centers, on Column III atom doping is indeed observed. This figure compares samples with various Ga doping levels and identical irradiations for two anneal states. Results from measurements on a Si:In and a Si:Aℓ sample irradiated under the same conditions as Si:Ga samples are also plotted in Fig. 1 and show that the N_1 concentration produced in Si doped with other Column III impurities fits well on an empirical N_1 vs. $N_{Col\ III}$ curve determined from the Si:Ga data. Straight lines of unity slope are drawn for a reference to indicate that a linear dependence of N_1 on N_{Ga} fits the data reasonably well.

A lightly doped Si:Ga sample containing $2.5x10^{15}$ Ga/cm^3 was also included in the study. We observed no detectable shallow defect concentration in this sample. This result was expected, since from the results of Fig. 1 only $1.2x10^{13}$ A_1/cm^3 would be present after a 600°C anneal, which is less than the lower limit on the donor concentration of $5x10^{13}$ cm^{-3} set by the concentration of transmutation-produced phosphorus. To observe A_1 centers by temperature-dependent Hall effect measurements, A_1 must be undercompensated, so that N_1 must exceed the concentration of donors.

The dependence of N_1 on N_{Ga} provides an important clue to the nature of the A_1 center, since it indicates that the defect responsible for the center contains a Column III atom and is not an

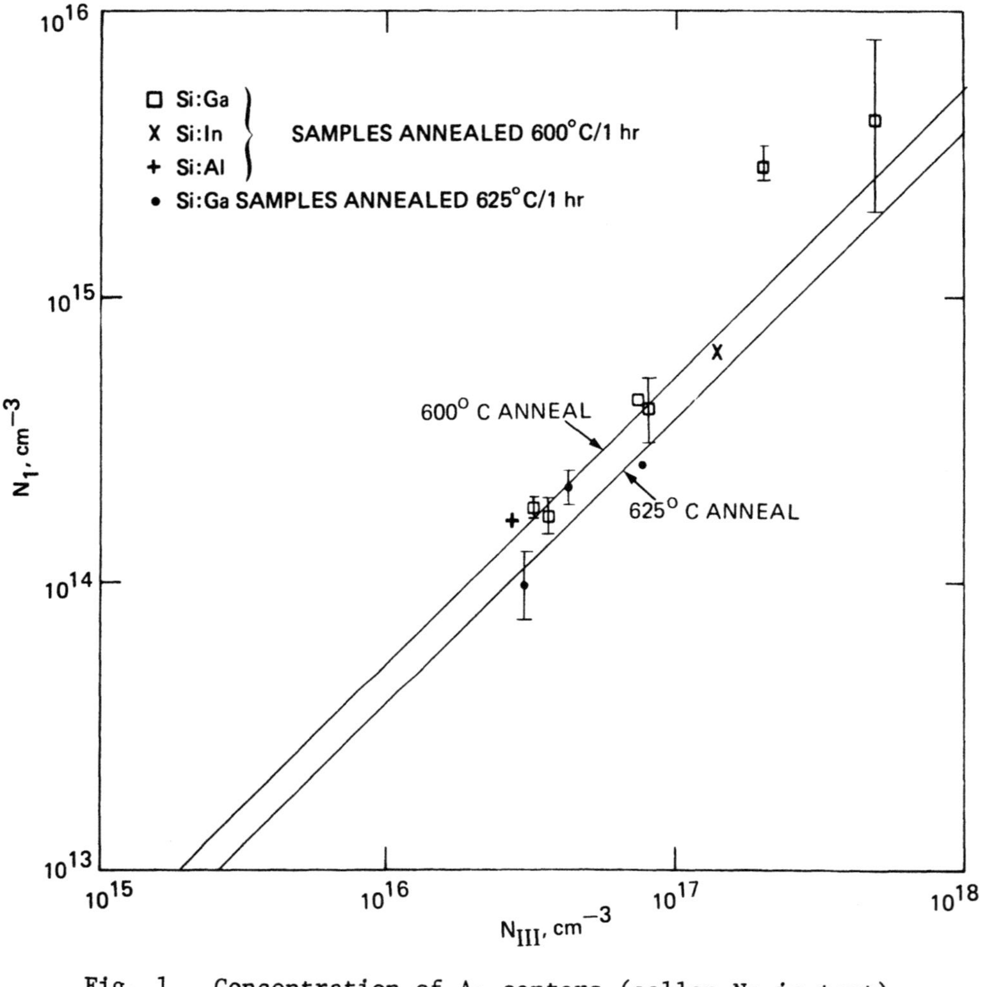

Fig. 1. Concentration of A_1 centers (callen N_1 in text) generated vs. sample Column III doping concentration at two different anneal temperatures.

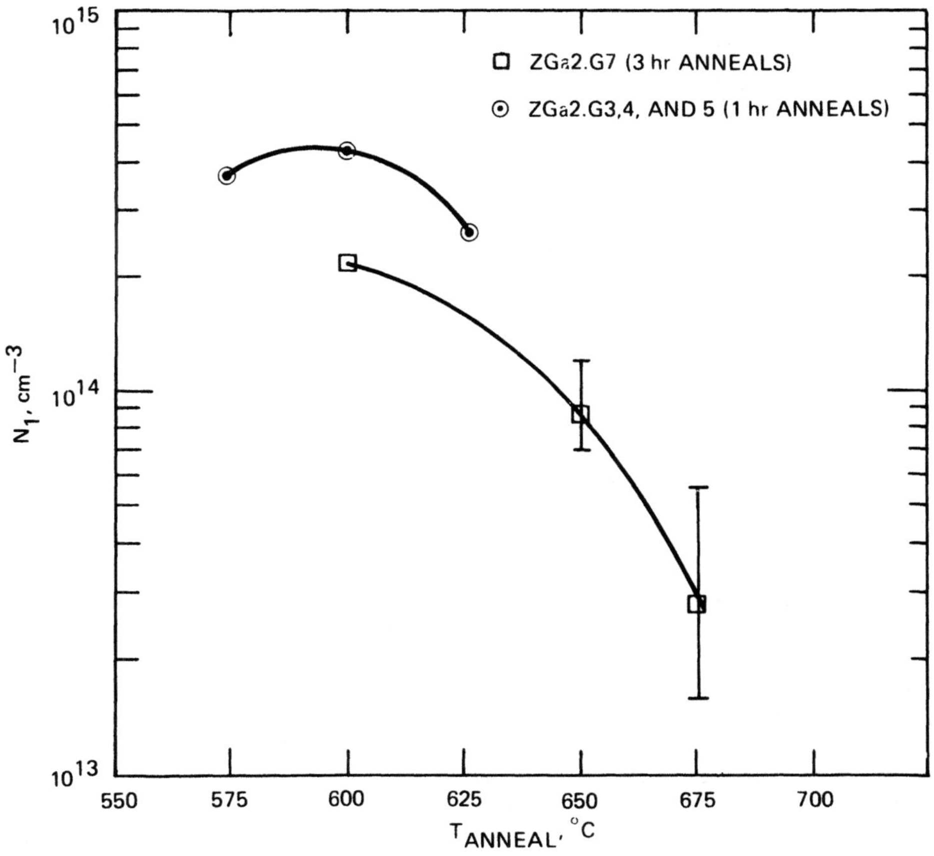

Fig. 2. Concentration of A_1 defects as a function of anneal temperature with constant anneal times.

intrinsic radiation defect. It should be noted that the N_1 concentration in irradiated Si:Ga is less than 1% of N_{Ga}, so that the concentration of Ga atoms tied up in these defect structures is negligible compared to the Ga doping. Such a small decrease in substitutional Ga could not be detected by the experimental techniques used here.

From the results of measurements of the activation energy of A_1 in sixteen Si:Ga Hall samples, we determined the energy level to be located at $E_v + 0.025 \pm 0.003$ eV. From a more limited study of other irradiated, float-zone grown extrinsic p-type Si, we measured the very shallow A_1 level in Si:Aℓ to be located at about $E_v + 0.03$ eV, and at a similar, but measurably deeper, position in Si:In of about $E_v + 0.031$ to 0.032 eV. Thus it appears that a family of A_1 defect complexes involving Column III impurity atoms is formed in low-oxygen, extrinsic, p-type silicon.

Some key results of a study of the annealing behavior of A_1 are summarized in Fig. 2. The defect is readily observable after anneals of about 600°C. At anneal temperatures much lower than 600°C the shallow A_1 acceptors are overcompensated by excess radiation-produced donors and are therefore not observable in the electrical measurements. At higher anneal temperatures (∿675°C) the defect complex begins to break up and disappear.

An example of a photoconductive spectral response scan of the long wavelength region appropriate to the study of the optical properties of the Si:Ga-A_1 defects is shown in Fig. 3. The Ga impurity level excitation spectrum is superimposed on the shallow level response and is readily observable. The positions of the deepest absorption peaks in the Ga excitation spectrum, as given by Onton, et al.,[5] are marked on the figure. A dip at 519 cm^{-1} (19.3 μm) corresponding to the energy of an optical phonon in silicon is seen in all spectra and is marked in Fig. 3 by an arrow. This structure is due to resonant interaction between a hole in the ground state of an impurity-defect located at less than 0.064 eV from the valence band and the 519 cm^{-1} optical phonon.[6,7] The only conventional substitutional Column III impurity that could interact in this way with the optical phonon is boron with energy level at $E_v + 0.0457$ eV.[8] Since the boron concentration in these irradiated float-zone grown Si is very low ($\lesssim 10^{13}$ B/cm^3), we conclude that the spectral structure at 519 cm^{-1} in these samples results from interactions of the shallow defect level ground states with the optical phonon.

From measurements of the peak photoconductive response in numerous spectra at 5K due to the A_1 defect at ∿35 μm, which is beyond the cutoff wavelengths of Ga (16.8 μm) and the A_2 level (∿32 μm), we determined the peak optical absorption cross section, σ_1, for the A_1 shallow defect to be 1.6×10^{-14} cm^2. We obtained this result by

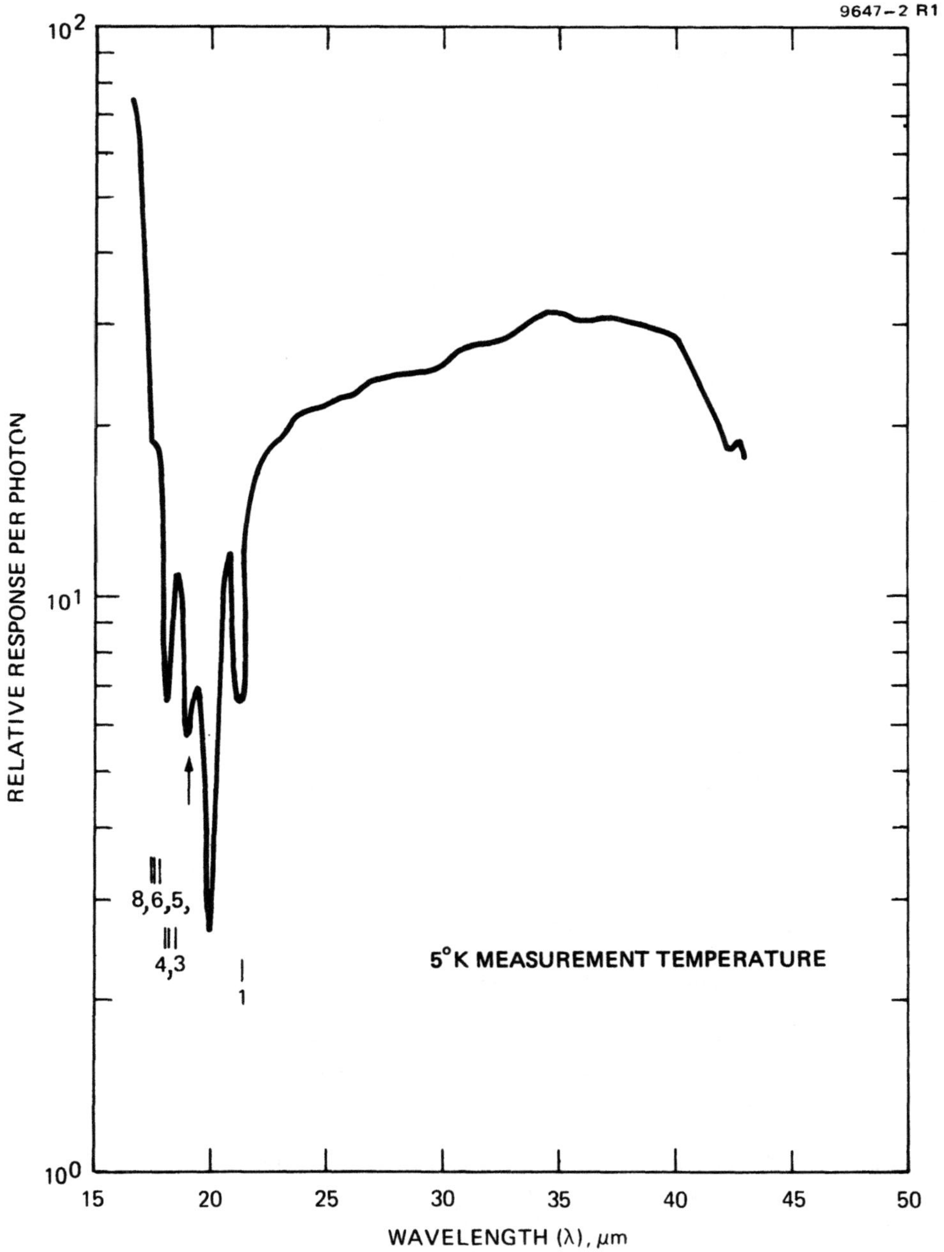

Fig. 3. Spectrum of relative photoconductive response per photon for irradiated Si:Ga sample.

determining the quantum efficiency for the A_1 center, η_1, from the relation

$$\eta_1 = \frac{R(35\ \mu m)}{R(16\ \mu m)} \frac{c_1}{c_{Ga}} \eta_{Ga}$$

where each R is the measured response per photon at the respective peak wavelength, c_1 and c_{Ga} are the recombination coefficients for A_1 and Ga, and η_{Ga} is the calculated quantum efficiency for Ga at its peak wavelength of 16 μm. We assumed $c_1 = c_{Ga}$ in the calculation. The value of η_{Ga} was calculated from the usual relationship

$$\eta_{Ga} = \frac{(1-r)\left(1-e^{-N_{Ga}\sigma_{Ga}\ell}\right)}{1-re^{-2N_{Ga}\sigma_{Ga}\ell}} ,$$

and σ_1 was then determined from

$$\eta_1 = \frac{(1-r)\left(1-e^{-(N_1-N_D)\,\sigma_1\ell}\right)}{1-re^{-2(N_1-N_D)\sigma_1\ell}} ,$$

where r is the single surface reflectivity of Si. The concentrations were determined from Hall measurements. The measured values from various investigations of peak optical absorption cross section for impurities or defects in Si are plotted in Fig. 4 as a function of impurity activation energy. Our measurement of the peak absorption cross section for A_1 is also plotted in Fig. 4, and clearly fits well on this empirical curve.

All of our investigations of the properties of the A_1 defect have been in low-oxygen silicon. In an initial study of high-oxygen (Czochralski-grown) Si:In, we could not detect the presence of A_1 centers. If they are present in the Czochralski material they are overcompensated by a rather large concentration of radiation-produced, oxygen-related donors commonly found in irradiated Czochralski-grown silicon.[13] There is an indication, however, that the production of A_1 centers is actually inhibited by the presence of oxygen. Even in a silicon sample containing 3.4×10^{17} In/cm^3, and thus expected to produce a large Si:In-A_1 concentration, no shallow centers were seen.

EPR STUDIES

A survey of the paramagnetic defects in this same thermal-neutron-irradiated, extrinsic p-type silicon was made. In particu-

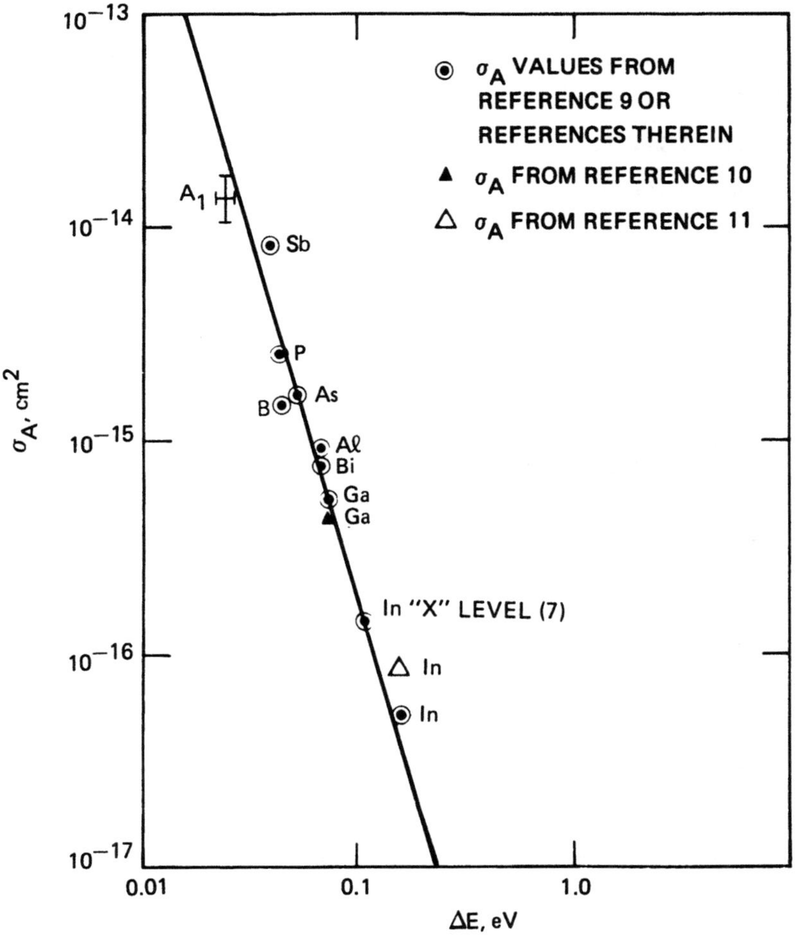

Fig. 4. Peak optical absorption cross section as a function of impurity activation energy for impurities or defects in silicon. ΔE values for Column III impurities from Reference 8. ΔE values for Column V impurities from Reference 12. In "x"-level ΔE value from Reference 7.

Table 1. Doping Characteristics of Float-Zone Silicon Samples Irradiated with Thermal Neutrons to a dose of ∿ $3x10^{17}$ neutron/cm^2.

Sample Number	Doping
(Si:Ga ZGa3.L)	3 - $4x10^{16}$ Ga/cm^3
(Si:Ga ZGa2.G)	8 - $10x10^{16}$ Ga/cm^3
(Si:Ga ZGa4.V)	$2x10^{17}$ Ga/cm^3
(Si:Aℓ ZAℓ1.D)	$3x10^{16}$ Aℓ/cm^3
	$2x10^{13}$ B/cm^3

lar, four samples which had been doped as indicated in Table 1 and irradiated with $3x10^{17}$ neutrons/cm^2 were examined with electron paramagnetic resonance (EPR) after 20 min. isochronal anneals in ∿50°C steps from room-temperature to 600°C. The dominant EPR spectra which we observe, shown in Fig. 5, correspond to the neutral (110) planar 4-vacancy (P3),[14,15] the positive divacancy (G6),[16] the negative T-shaped 5-vacancy (P1),[17] and the <001> split interstitial (B3).[18] The P3 and P1 centers are commonly observed with EPR in neutron-irradiated and ion-implanted silicon.[19,20] It is interesting to note that we now observe the B3 center in Ga and Aℓ-doped silicon in concentrations which exceed the background boron concentration by a factor of 10-100. Previous studies[18] had observed the B3 spectrum only in boron-doped silicon, and there was some concern that boron possibly might be a part of the B3 structure. The present results indicate that the B3 center is impurity-independent. There are some weak lines near 7300 G which we have not identified, whose intensity increases with increasing Ga concentration after the 400 to 500°C anneals.

In Fig. 6 we have plotted the concentrations of specific defects versus annealing temperatures. Since the concentrations for each kind of defect were observed to be nearly the same among the four samples we studied, we have plotted their mean concentrations. The specific concentrations for each defect and sample usually[21] fall within the error bars indicated in Fig. 6. Although the relative concentrations indicated in Fig. 6 are accurate to ± 15%, the absolute

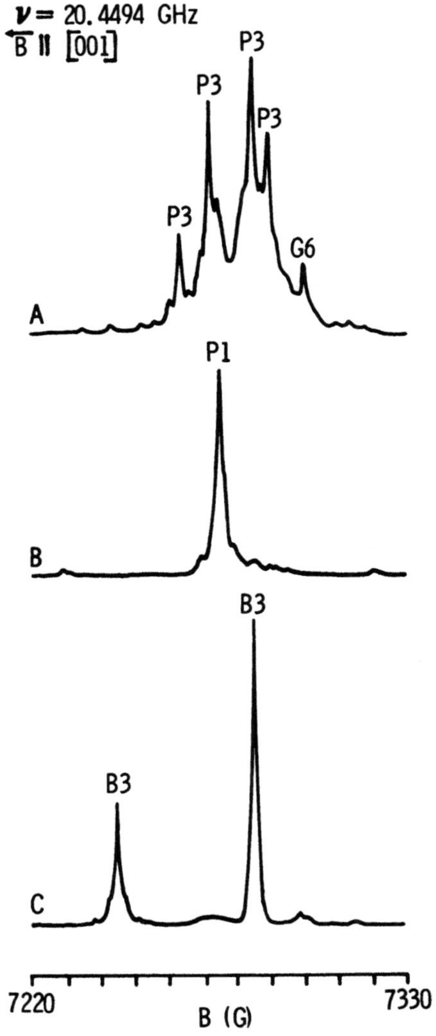

Fig. 5. EPR spectra of thermally annealed, thermal-neutron-irradiated, p-type silicon. A) $2x10^{17}$ Ga/cm^2 with 60°C, 20 min. anneal; B) $3x10^{16}$ $A\ell/cm^3$ and $2x10^{13}$ B/cm^2 with 300°C, 20 min. anneal; and C) $8\text{-}10x10^{16}$ Ga/cm^3 and 400°C, 20 min. anneal.

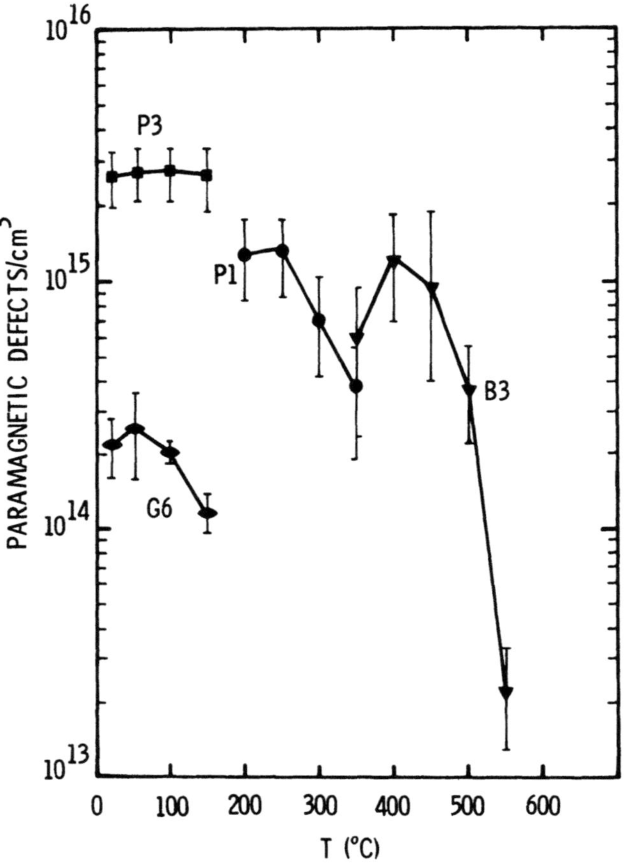

Fig. 6. Concentration of the dominant paramagnetic defects in p-type silicon thermal-neutron-irradiated with $\sim 3\text{x}10^{17}$ neutrons/cm^2 versus annealing temperature.

concentrations are estimated to be accurate to within a factor of 5. This uncertainty in the absolute concentration arises because the spectra were observed in fast passage.

We had hoped to correlate EPR spectra with the A_1 and A_2 centers observed with Hall measurements; however, for reasons not understood, the A_1 and A_2 centers do not give rise to observable EPR spectra.

DISCUSSION

No measurable EPR spectra were evident in the extrinsic p-type samples after anneals at a temperature of about 600°C — the temperature at which the Si:Ga-A_1 was most readily observable in electrical and optical measurements. The precise nature of the very shallow A_1 therefore cannot be determined conclusively from this work. However, from the marked dependence of the observed A_1 concentration on Column III atom doping, it appears that the defect contains a Column III atom (Ga, In, or Aℓ). It is likely that the Column III atom is complexed with an intrinsic defect such as a vacancy or interstitial complex. The indication that large concentrations of oxygen in Czochralski-grown samples may inhibit the formation of A_1 defects suggests that the A_1 defect very well may contain vacancies, since oxygen has been identified as a particularly effective trap for vacancies.[22,23,24] Vacancy-oxygen centers form so readily that other impurity-vacancy defects such as the E-center (Column V atom-vacancy pair)[25] have been observed only in low-oxygen silicon.

Multi-vacancy complexes are observed to form readily in neutron-irradiated silicon, as demonstrated in the EPR study results presented here. Impurity-vacancy defects have been observed to form at rather elevated anneal temperatures in neutron-irradiated silicon as these vacancy clusters break up, releasing vacancies and divacancies which can then complex with impurity atoms present.[26] We observe the A_1 defect to anneal at a relatively high temperature in this work. It may very well be that this defect is formed following the breakup of the vacancy clusters (P3, G6, P1) observed in our EPR measurements.

A Column III atom-vacancy pair has been observed by EPR techniques,[27] but the energy level associated with this defect has been determined by deep level transient spectroscopy (DLTS) techniques[28] to be a deep level located at E_v + 0.48 eV. Therefore, the very shallow A_1 defect cannot be associated with a simple vacancy-impurity atom pair. We speculate that the A_1 defect complex may contain more than one single vacancy in its configuration.

REFERENCES

1. M. H. Young, O. J. Marsh, and R. Baron, "Neutron Transmutation Doping in Semiconductors," ed. by Jon M. Meese (Plenum

Press, N.Y., 1979) p. 335.
2. M. H. Young, O. J. Marsh, and R. Baron, J. Appl. Phys. 50, 3755 (1979).
3. All irradiations were performed at the Texas A&M University Nuclear Science Center with supervision of Prof. R. R. Hart.
4. We acknowledge Prof. Hart's contributions to the calculation of these displacements.
5. A. Onton, P. Fisher and A. K. Ramdas, Phys. Rev. 163, 686 (1967).
6. G. D. Watkins and W. Beall Fowler, Phys. Rev. B16, 4524 (1977).
7. R. Baron, M. H. Young, J. P. Baukus, O. J. Marsh, and T. C. McGill, Proceedings of the IRIS Specialty Group on Infrared Detectors (Infrared Information and Analysis Center, Ann Arbor, Mich., 1978), p. 281.
8. A. Baldereschi and N. O. Lipari, Proc. of the 13th Int. Conf. on Semiconductors, ed. by L. G. Fumi (Tipographic Marves, Rome, 1977) p. 595.
9. R. Baron, M. H. Young, J. K. Neeland, and O. J. Marsh, Appl. Phys. Lett. 30, 594 (1977).
10. M. H. Young, R. Baron, and O. J. Marsh, unpublished measurement of σ_A of Si:Ga.
11. J. P. Baukus, R. Baron, and S. D. Allen, unpublished measurement of σ_A of Si:In.
12. P. Fisher and A. K. Ramdas, Physics of the Solid State, ed. by S. Balakrishna, M. Krishnamurthi, and B. Ramachandra Rao (Academic, N.Y. 1969), p. 149.
13. H. M. Liaw and C. J. Varker, Proc. of the Third International Symposium on Silicon Materials Science and Technology, ed. by H. R. Huff and E. Sirtl (The Electrochemical Society, Inc., Princeton, N.J., 1977), p. 116.
14. K. L. Brower, Radiation Effects 8, 213 (1971).
15. Y. H. Lee and J. W. Corbett, Phys. Rev. B9, 4351 (1974).
16. G. D. Watkins and J. W. Corbett, Phys. Rev. 138, A543 (1965).
17. Y. H. Lee and J. W. Corbett, Phys. Rev. B8, 2810 (1973).
18. K. L. Brower, Phys. Rev. B14, 872 (1976).
19. W. Jung and G. S. Newell, Phys. Rev. 132, 648 (1963).
20. K. L. Brower, F. L. Vook, and J. A. Borders, Appl. Phys. Lett. 15, 208 (1969).
21. Although a defect may exist over a particular temperature range, it may be observable only with EPR over a smaller temperature interval which is contingent on the position of the Fermi level. For example, the B3 center forms between 50° and 175°C but does not become paramagnetic in these samples until they are annealed near 350° to 400°C.
22. G. D. Watkins and J. W. Corbett, Phys. Rev. 121, 1001 (1961).
23. J. W. Corbett, G. D. Watkins, R. M. Chrenko, and R. S. McDonald, Phys. Rev. 121, 1015 (1961).
24. G. D. Watkins, Lattice Defects in Semiconductors, 1974 International Conf. Series Number 23 (The Institute of Physics,

London, 1975), p. 1.
25. G. D. Watkins and J. W. Corbett, Phys. Rev. 134, A1359 (1964).
26. R. E. Whan, J. Appl. Phys. 37, 3378 (1966).
27. G. D. Watkins, Phys. Rev. 155, 802 (1967).
28. L. C. Kimerling, Proc. of the Int. Conf. on Radiation Effects in Semiconductors, 1976 Conference Series No. 31 (Institute of Physics and Physical Society, London, 1977), p. 221.

THE HEALTH AND SAFETY ASPECTS OF NEUTRON DOPED SILICON

T.G.G. Smith

Atomic Energy Research Establishment,
Harwell,
England.

INTRODUCTION

At the present time, the neutron irradiation of mono-crystal silicon ingots for the purpose of "doping" with phosphorous is an established production process. Each year tens of tonnes of silicon ingots are being transported from producers to nuclear reactor centres, are then irradiated; transported back to the producers, machined, sliced and polished and the slices eventually incorporated in electronic devices with which the general public have contact and free access.

In common with all materials, silicon, when irradiated with neutrons in a reactor, becomes radio-active; however, it is a fortunate fact of nature that the half lives involved are very short and the decay of radio-activity down to negligible levels, which are completely safe, is very rapid. To those who are knowledgeable in the subject that is a convincing statement and one which can be proved by a simple mathematical calculation. To those less knowledgeable, the fact that the material has been irradiated in a reactor can cause concern, particularly if they are required to work with the material, or to be in contact with devices containing the material. The object of this paper is to demonstrate how safe the material is and to do this in simple terms which can be understood by those having no special knowledge of physics or of nuclear processes.

In the overall safety of the process there is another aspect which requires consideration and this is the safety of the irradiation process itself. There is no advantage in producing a better electronic material if in doing so the reactor operators receive large radiation doses or, large quantities of radio-active waste are produced.

THE IRRADIATION PROCESS

The neutron irradiation is carried out in research reactors and these can be divided broadly into two types:

(i) The 'Pool' Type in which the reactor core is situated in an open pool of water which allows easy access to the neutron flux regions and also permits post irradiation storage in the pool whilst the initial high residual radio-activity of the silicon and its container decay to safe levels.

(ii) 'Closed Tank' Type in which access to the neutron flux is more difficult and the silicon must be removed from the reactor immediately the desired amount of irradiation has been achieved and the silicon and the container must then be placed in suitable shielded storage adjacent to the reactor until it is safe to handle.

At Harwell, the reactors used for silicon irradiation are of the tank type and therefore the silicon container must be removed to storage at the conclusion of irradiation. These same reactors have been used for many years for the production of large quantities of radio-isotopes, and on-power load and unload of materials has long been a daily routine operation by the reactor operators; techniques and equipment have merely been adapted for silicon irradiations. Nevertheless, even very low levels of radiation encountered during such operations can lead to measurable doses when some hundreds of silicon irradiations are undertaken each month. This is an aspect which is continually monitored and as quantities continue to increase, we are about to bring new equipment into operation which will have a two-fold purpose: firstly, it will be virtually radiation dose free and, secondly, increase the production capacity.

RADIO-ACTIVE WASTE ARISINGS

In reactors with very large experimental and production programmes, it is important to limit the quantity of radio-active waste material to a minimum. This is important because handling and storage of waste is expensive and reflects in the price of the irradiations. In the case of the silicon work, we achieve this by a maximum re-usage of the equipment and components and by choosing metals of low neutron capture cross section and short half-life. The assemblies on which the silicon, in its container, is introduced to the reactor do, with time, become radio-active, but these are continuously re-used and remain in operation for several years.

The containers are made from aluminium, which in itself has good nuclear properties, but lacks mechanical strength. Therefore, normally when aluminium is considered in an engineering sense, it is an alloy rather than a pure metal. The alloying elements are responsible for much of the induced radio-activity and it is therefore necessary to compromise between impurities which give rise to radio-activity and the mechanical strength of the container. In practice we have found that aluminium of 99.95% purity is best suited to this application and containers made from this material survive 300 - 400 irradiations. A further source of induced radiation which can be avoided,is that arising from human contact when contaminants such as sodium are transferred from operators' hands to can surfaces and then become active when irradiated. Cleanliness and the wearing of gloves, etc. are of prime importance when loading containers with silicon.

LEVELS OF SILICON RADIO-ACTIVITY

It is often stated in the literature that a few days after irradiation in a reactor, silicon is safe to handle. Whilst this is true as a general statement, it is necessary to establish how safe is safe.

Fig. 1 shows the typical build up of radio-activity in silicon with time; this shows that for average resistivity material (about 35 Ω cms) 20 hours of irradiation are required in a flux of 10^{13}n cm^{-2} sec^{-1} and the material has reached its maximum activity of about 20 curies/Kg. It should be noted that it is only possible to reduce this value if the irradiation time is very short < 5 hours, but this will result in material of high resistivity; hundreds of Ω cms.

In contrast to Fig. 1, Fig. 2 shows the rate at which the radio-activity decays and this clearly shows the two constituents which contribute; ^{31}Si with its short half-life of 2.6 hours, and ^{32}P with its longer half-life of 14.5 days. These activities are total values, but as the radiation is almost exclusively β, then only a small proportion escapes to the surface of the ingot, the rest being absorbed within the ingot itself.

TRANSPORT

Up until the time that the silicon leaves the reactor station and starts on its return journey to the producer, it has been in an environment in which there are specialists accustomed to handling radio-active materials and expensive and complicated equipment for measuring and analysing radio-activity.

Once the silicon leaves the reactor station, it is handled by non-experts, particularly during its transport to the producer when

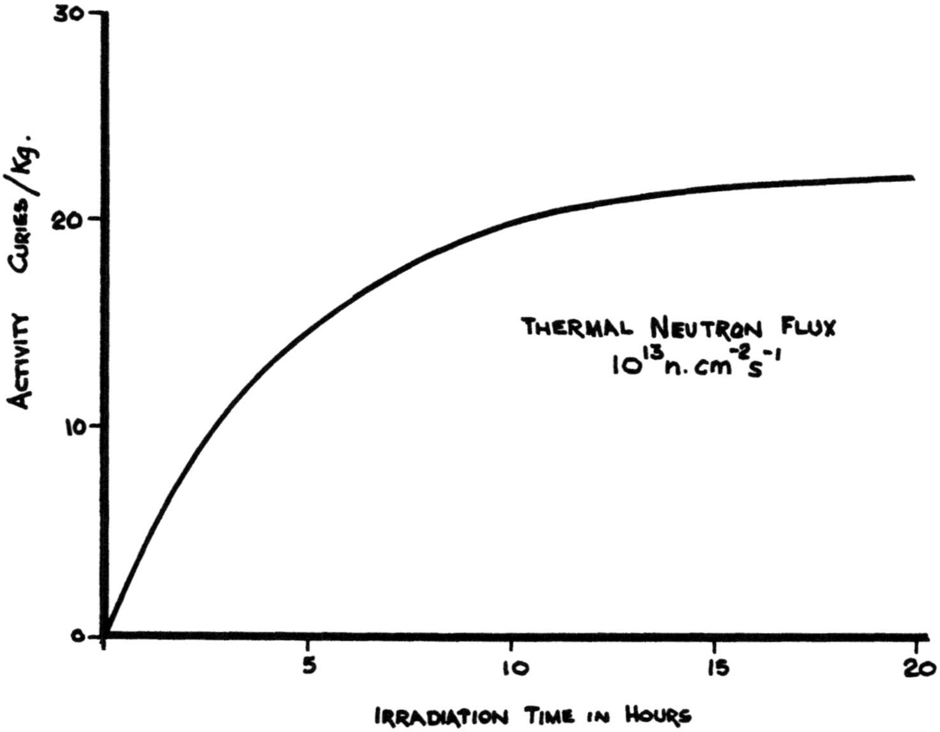

Fig. 1. Build-up of Radioactivity in Silicon during Irradiation.

road, rail or air transport may be used. The regulations for the international transportation of such materials are clearly defined by the I.A.E.A. in 'Regulations for the Safe Transport of Radioactive Materials (1977)' and these specify the radiation levels and non-fixed contamination levels which are not to be exceeded at the surface of a package if it is to qualify as 'Exempt' or safe material. These state that the radiation level at any point on the external surface shall not exceed 0.5 mRem/hr and that the non-fixed contamination shall not exceed 10^{-4} μCi cm^{-2}. These are limits which must be complied with, but it should be noted that they apply to the outside of a package and not to the silicon ingots which the package contains. At Harwell we naturally check and comply with these regulations, but of greater concern to us is the radiological state of each individual ingot or crystal.

INDIVIDUAL INGOT CHECKS

In setting the levels of radioactivity with which each ingot must comply, we have borne in mind that when returned to the producer, the material will be handled by non-radiation workers, that it will

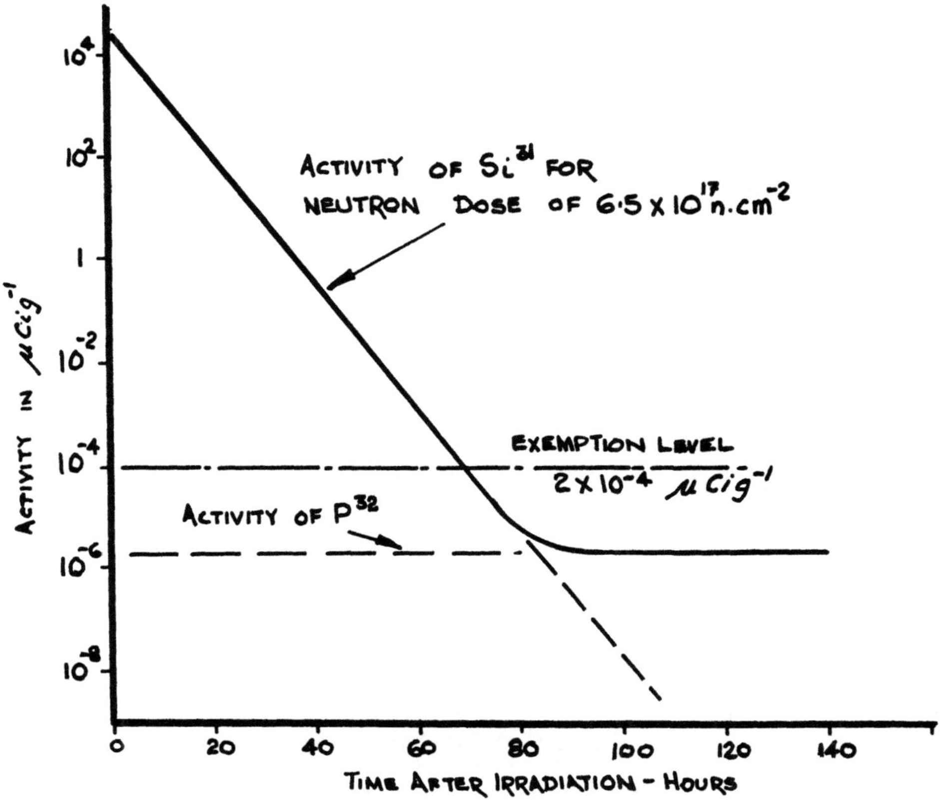

Fig. 2. Decay of Silicon Isotopes after Irradiation.

be subjected to grinding and polishing processes and that during these processes, particles of silicon will be liberated from the centre of the ingots. Additionally, these processes will be carried out in workshops and laboratories which will not be equipped with instruments for the detection of air-borne radio-active dust. In our view, there is an unquestionable responsibility on the reactor operator to ensure that the material will be safe and free from radiological hazards for the remainder of its life. In this respect, the half-life of the material acts in the direction of safety since with time the radio-activity, however small, will continue to decay through all time.

Within the United Kingdom there is legislation which defines "exemption" limits for particular nuclides and for total radio-aotivity of materials. These are:

The Radio-active Substances Act 1960 and the Ionising Radiations (Unsealed Radio-active Substances) Regulations 1968.

In addition, as the silicon is to be transported and distributed on an international basis, it is necessary that it complies with the regulations laid down by the I.A.E.A. under; 'Regulations for the Safe Transport of Radio-active Materials (1973)'

From these regulations, the lowest level of radio-activity which will allow the material to be classified as "EXEMPT" has been determined and this is $2 \times 10^{-4}\mu Ci g^{-1}$. This is a level at which national and international advisory and regulatory bodies consider a material is completely safe and cannot cause harm to those coming in contact with the material.

To measure a value of $2 \times 10^{-4}\mu Ci g^{-1}$ directly on a silicon ingot is extremely difficult and a more convenient measurement is that of the counts per second at the surface of the ingot. Making due corrections for the type of detector which is used and for its geometry relative to the ingot, gives a count rate equivalent for $2 \times 10^{-4}\mu Ci g^{-1}$ of 23 counts per second. As a further measure of

Certificate № 3699 HARWELL

Certificate of Radiological Monitoring of Silicon Ingots

It is certified that in the case of Silicon Ingot No. ....

1. The ^{32}P specific β-activity is less than $2 \times 10^{-4} \mu C.g^{-1}$.
2. The total specific activity from all γ-emitting nuclides is less than $10^{-5} \mu C.g^{-1}$.
3. The removable surface contamination is less than $10^{-4} \mu C\, cm^{-2}$.

It is therefore exempt from the provisions of the Radioactive Substances Act 1960 (United Kingdom) and the Ionising Radiations (Unsealed Radioactive Substances) Regulations 1968 (United Kingdom) and is an exempt radioactive material as defined in the IAEA Regulations for the Safe Transport of Radioactive Materials (1973).

For dispatch to: ....

Signed .... Date ....
Health Physics Operations Group.
Building No. ....
Environmental and Medical Sciences Division,
AERE Harwell, Didcot, Oxon OX11 0RA. England

Fig. 3. Harwell's Certificate of Radiological Monitoring of Silicon Ingots.

safety and to ensure beyond doubt that the exemption limit is complied with, the acceptable count rate is reduced to not more than 10 c.p.s. for Harwell clearance tests.

Having set the standard, it is necessary to ensure that it is rigorously applied and to this end we subject each ingot to two separate checks. Each ingot is placed in a machine and is rotated whilst a counter is traversed along its length. The counter is set to alarm at a level of 10 c.p.s. and to stop the machine, thus indicating any spot which exceeds the set limit. Such spots are usually quite small and are invariably due to contaminants present on the ingot prior to irradiation. Swabbing with a suitable solvent is sufficient to remove the contamination.

In the many thousands of ingots which have been tested in this fashion, there have been a small number which indicate how searching the test is. In one, a minute area contaminated with gold was found and it was suspected that this had been transferred from a ring worn by someone who had handled the ingot. In another, classical fission products were detected and we suspect that these arose from thorium in the silicon which, being a fertile material, was transmuted in the reactor to uranium 233 which, being fissile, gave rise to the fission products. These are isolated incidents and in each case only a few atoms of the 'foreign' material were present, but they do serve to demonstrate the extreme sensitivity of the equipment and the searching nature of the test.

The first of the two tests is carried out by the operators responsible for the handling and the irradiation of the silicon. The second test is carried out by members of a group who are responsible for radiological safety at Harwell and are entirely independent from the reactor operating staff and the silicon team.

Fig. 3 shows the certificate which is issued with each ingot and this is completed and signed by the independent inspectors at the conclusion of the second test. Every ingot irradiated at Harwell is dispatched to the customer with a certificate stating that it complies with U.K. and I.A.E.A. regulations.

HOW SAFE IS SAFE?

Every human being spends the whole of his or her life in a radioactive environment, subjected to bombardment by radiation. The radiation is of cosmic origin, from the everyday materials around us, and from the food we eat. Fig. 4 shows the average total dose of radiation which every individual receives and the sources from which it arises. The total average dose is about 90 mRem per year. These values are average for the United Kingdom and will be increased if we live at higher altitudes where the atmosphere provides less

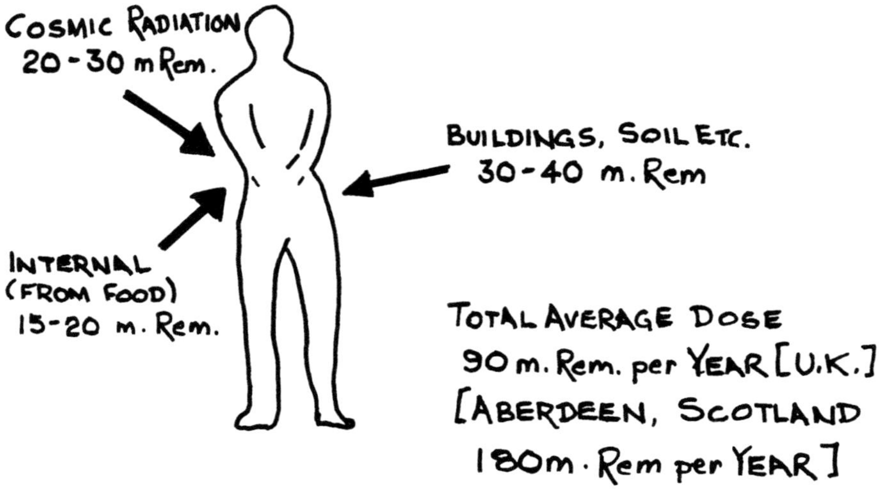

Fig. 4. Natural Sources of Radiation.

protection from cosmic irradiation, or if we live in an area of higher than normal background.

Aberdeen in Scotland is an interesting example, where the buildings are constructed from local granite which has a relatively high level of radio-activity and an inhabitant of that city will receive an annual dose twice that of the average, i.e. 180 mRem.

There are other sources of radiation to which we are subjected and some of these are shown in Fig. 5. Normal aircraft flights give rise to a dose of 1 mRem per hour of flight, so a holiday flight from northern Europe to the Mediterranean would contribute about 5 mRem, and a return atlantic flight to Europe would add about 10 - 12 mRem.

Another common source of radiation dose is medical treatment and few people have not had one or more X-ray examinations during their life. A normal X-ray picture of the chest contributes 30 - 40 mRem to the individual. So we see that most individuals receive an average of about 100 mRem per year, but a large number receive considerably more without ill-effects.

We can now consider the effect of working with neutron doped silicon. If we assume, pessimistically, that on its return to the producer's works, a large silicon ingot is placed on a table at which an operator is sitting, that the operator remains this close to the ingot for 8 hours of every working day and that the ingot is replaced

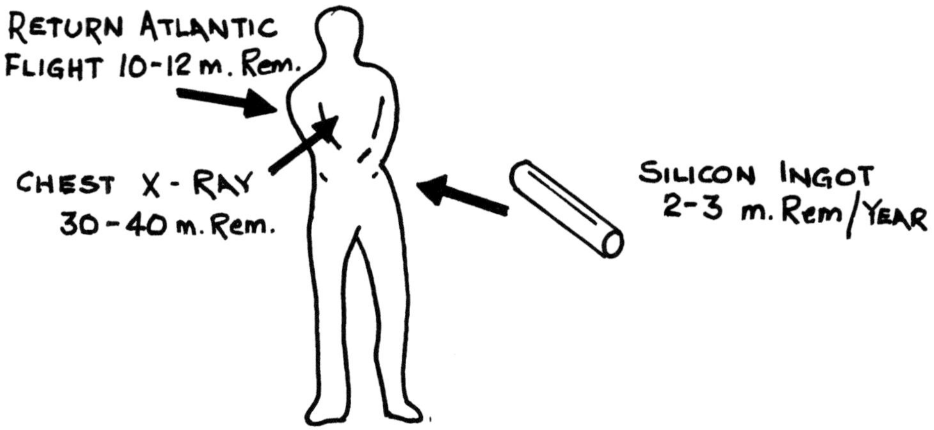

Fig. 5. Other Sources of Radiation.

by a new freshly delivered ingot each day, then in one year the operator would receive a radiation dose of at most 2 - 3 mRem.

If we consider the possible effects from slices or devices and take account of the fact that the radioactivity of the silicon is reducing by one half in every 14 - 15 days, and that the amount of material is some hundreds of times less than in an ingot, then the resultant amount of radiation becomes so infinitesimally small as to be completely meaningless.

CONCLUSIONS

From the foregoing, it can be seen that in the production of neutron doped silicon, there are areas in which radiological hazards can arise, but these are confined exclusively to the reactor stations where the work is carried out under strict radiological control by radiation workers.

For those handling the material outside the reactor stations, i.e. the transporters, the silicon producers and the general public, there is no danger and the material is much safer than many materials in general use by the public.

PRECISION AND ACCURACY OF NTD SILICON PRODUCTION

BASED ON CALORIMETRIC NEUTRON DOSE CONTROL

K. Heydorn and Kirsten Andresen

Isotope Division
Risø National Laboratory
DK 4000, Roskilde

INTRODUCTION

Industrial neutron irradiation of silicon at the Risø National Laboratory began in 1974 in co-operation with the Danish company Topsil A/S in Frederikssund[1]. A modest production of T∿silicon* was based on irradiation in the thermal column of the Danish reactor DR 2, a tank-type light-water reactor operating at 5 MW.

In 1975 a unique facility was designed specifically for the irradiation of silicon crystals and installed in the Danish reactor DR 3, which is a heavy-water reactor operating at 10 MW. Commercial production gained momentum during 1976, and in 1977 additional irradiation facilities were constructed and installed in the DR 3's graphite reflector.

A considerable effort was made in the design of these facilities to ascertain an accurate control of neutron dose, so that NTD silicon semiconductors could be produced to close tolerances. Results of almost 3 years of continual operation indicate that deviations from the nominal value on a routine basis have a standard deviation of less than 5%.

SILICON IRRADIATION FACILITY

The Danish reactor DR 3 is a PLUTO-type materials testing reactor, cooled and moderated with heavy water and fueled with 93% enriched uranium. A vertical section of the reactor is shown in

* Registered trade mark.

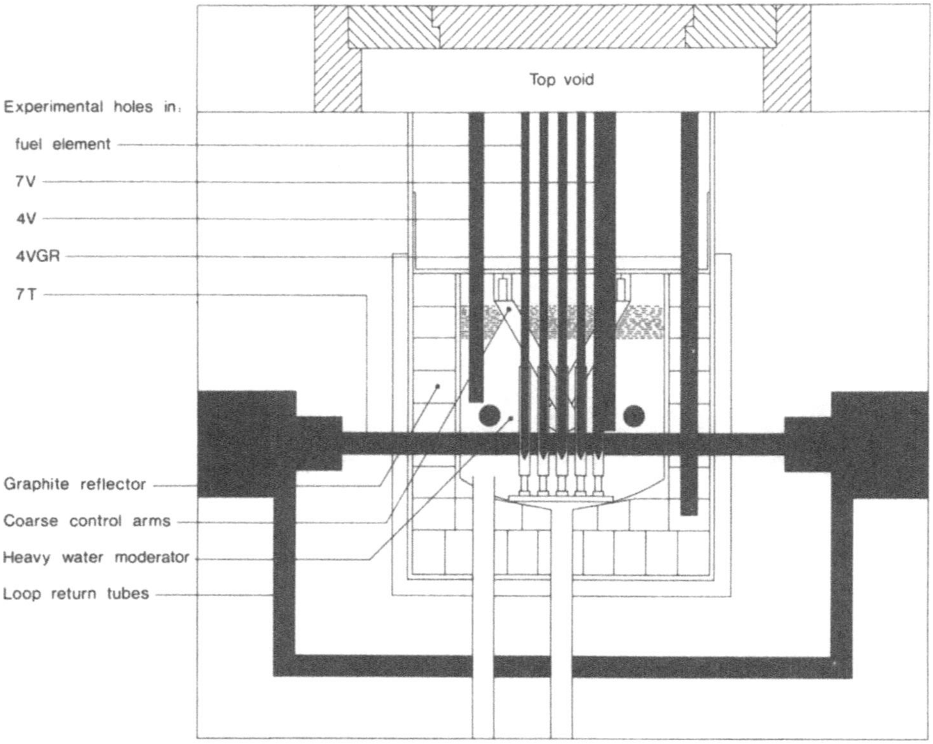

Fig. 1. Vertical section of Danish reactor DR 3. The reactor core height is 600 mm, and the thickness of the graphite reflector is 300 mm. Altogether 6 irradiation tubes in the reflector lie on a 2.5 m diameter circle.

Fig. 1 illustrating the position of the vertical irradiation tubes in the graphite reflector relative to the reactor core. All 4VGR tubes have a diameter of 100 mm, and the maximum thermal neutron flux density is approx. $4 \times 10^{12} n/(cm^2\ s)$ or 60 $nmol/(m^2\ s)$ at a power level of 10 MW.

The temperature of the graphite reflector is about 185°C, and the maximum heat generation from γ-radiation is approx. 20 W/kg of aluminium. The normal reactor operating schedule is based on a 4-week cycle with 23 days of continuous operation followed by 5 days shut down.

A schematic drawing of the irradiation rig designed[2] for the neutron transmutation doping of silicon crystals with a diameter up to 78 mm or 3 inches is shown in Fig. 2. Crystals with a total length up to 590 mm can be irradiated in aluminium cans with a wall thickness of 0.5 mm.

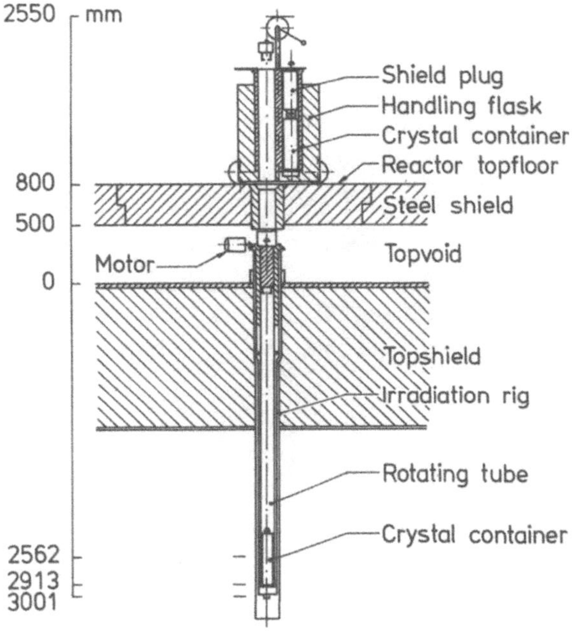

Fig. 2. Rig for the irradiation of 3" silicon crystals in a 100-mm ⌀ vertical irradiation tube in the DR 3 graphite reflector.

The spatial variation in neutron flux density within the irradiation volume of almost 3 liters exceeds 50% and is assymetrical. The effect of the radial flux gradient is, however, practically eliminated by rotating the irradiation container at approx. 2 revolutions per minute during the entire irradiation period.

The vertical flux variation along the axis of rotation is depicted in Fig. 3, and its effect is reduced by the installation of graded absorbers of stainless steel around the position of maximum flux. By reducing the peak neutron flux to approx. 50 $nmol/(m^2\ s)$ or $3 \times 10^{12} n/(cm^2\ s)$, the variation is lowered by a factor of almost 2 over the 500-mm length of the absorber, as can be seen in the figure. The vertical flux profile, however, changes during the course of the reactor period within the shaded area between the first and last flux distribution.

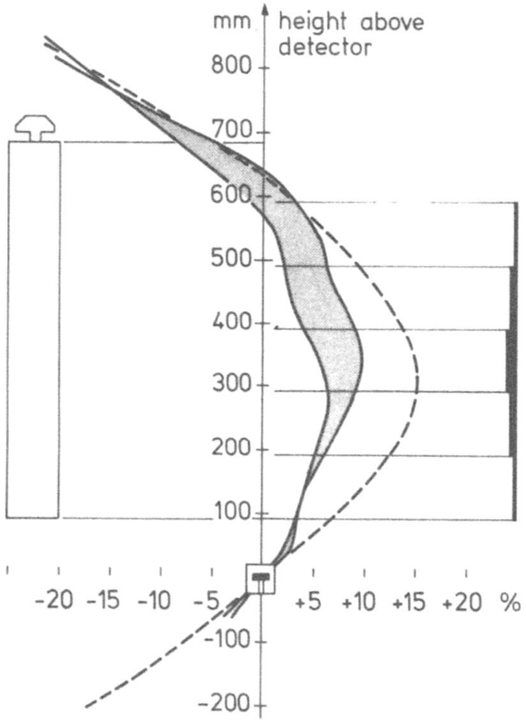

Fig. 3. Axial variation of neutron flux density before and after installation of stainless steel absorbers. During an operating period the flux distribution varies within the shaded area.

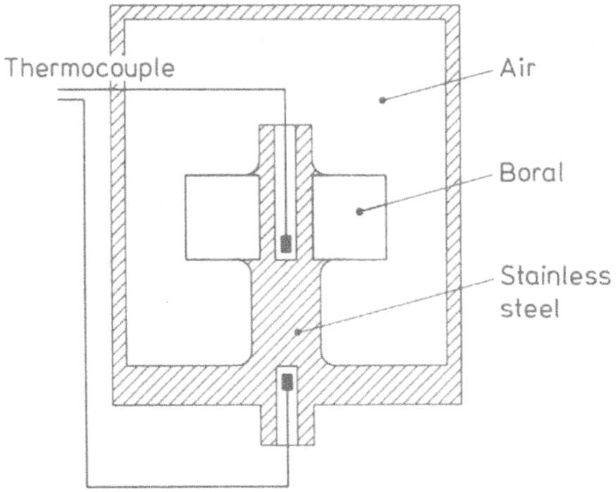

Fig. 4. Calorimetric thermal neutron flux detector. A Boral disc 15 mm ∅ is mounted in a 26-mm ∅ stainless steel capsule of total length of 30 mm.

NEUTRON DOSIMETRY

Reactor power level is usually kept constant to within 1% during the entire 23 days of operation. However, this does not mean that the neutron flux density in a particular irradiation position remains constant. Changes in fuel element configuration, burn-up, control rod position, as well as the presence of other experiments in the reactor, results not only in changes from one operating period to the next, but also during the course of a single period.

This temporal variation necessitates continuous flux monitoring in order to control the neutron transmutation doping with sufficient accuracy. This is achieved with a detector[3] based on a measurement of the heat produced by the reaction $^{10}B(n,\alpha)^{7}Li$, and shown in Fig. 4.

A small Boral disc is mounted in a stainless steel capsule in a way such that the heat generated by neutron absorption is conducted to the surroundings via a 5-mm ∅ steel rod. The temperature difference between the ends of the rod is measured by chromel/alumel thermocouples and is proportional to the neutron flux density. The thermocouple signal is converted to a frequency of pulses, and the number of pulses recorded is a measure of the thermal neutron fluence. The detector is located in the bottom of the rotating tube close to the irradiation can, as indicated in Fig. 3.

The calorimetric dosimeter has to be calibrated in a known neutron flux density, and its calibration should be checked under different, prevailing reactor conditions. This calibration is performed by means of neutron fluence monitors of 1.19% cobalt in aluminium as 1-mm ∅ wire.

A length of 20 mm - approximately 40 mg - is placed in the bottom of the irradiation can during actual neutron transmutation doping of 3" silicon crystals. Used in this way they provide quality control for the continuous measurement of neutron dose, as well as a lasting record of actual exposure of individual crystals.

The irradiated monitors contain 5-15 µCi or 0.2-0.5 MBq of ^{60}Co, and they are measured at approx. 25 cm from a 3" x 3" NaI(Tl) scintillation detector relative to a ^{60}Co reference standard source[4] of similar strength.

The measured activity is corrected to the end of irradiation, using a half-life of 5.26 years, and the calculation of the neutron flux is based on a thermal neutron cross section of 37.5 barns.

Measurements are carried out with a precision of approx. 0.5%, and systematic errors in the flux determination are estimated to be less than 2%.

Quality Control Data

Consistent quality in the production of NTD silicon is based on the precision and stability of the calibration of the neutron dosimetry system, which is a prerequisite for the precision and accuracy of the resistivity of the final product.

The primary quality control system therefore compares the reading of the calorimetric dosimeter with the induced activity in a cobalt monitor placed at the bottom of the irradiation can in a fraction of the samples irradiated. Other data recorded with the sample include the irradiation time, time after reactor start-up, date and time of completing the irradiation, magnitude of the thermocouple signal, and the position of the reactor control arms.

A total of 456 sets of data from the 6 different installations belonging to 25 different reactor operating periods over the years 1977-1980 are available for a detailed study. An investigation of the stability of calibration can be carried out by using the logarithm of the ratio of ^{60}Co activity to dosimeter reading as the dependent variable in a multiple regression on time, control rod position and irradiation position.

First of all, the variability within the 23-day reactor operating period was studied separately for each facility. Here a small, but significant influence of the reactor control arm position was found in all cases. This is illustrated for the rig installed in position 6 in Fig. 5, which shows the deviations of 88 observations from a reference value of 20°CCA position as a function of control arm position. The shape of the regression line was determined in all cases, and 4 facilities were indistinguishable in this respect and could be pooled, while the remaining had to be treated separately throughout the calculations. Values given in Table 1 refer to the pooled datasets, and the residual standard deviations reflect the precision of the measurements.

The variation of the actual neutron flux densities in the course of the reactor period was determined for the various installations in exactly the same way by using the ratio of ^{60}Co

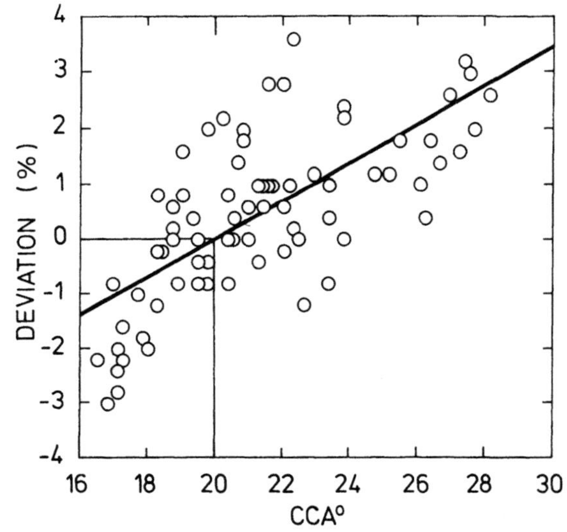

Fig. 5. Influence of control arm position in °CCA on the calibration of the calorimetric dosimeter in the 4VGR6 irradiation position. Deviations in % refer to the value at 20°CCA.

activity to irradiation time as the dependent variable. Even after correction for the control arm position the residual variability had a standard deviation of about 6%.

Based on the reference values at 20°CCA the ratios between the calibrations of the various installations may be calculated from the data. These results are presented in Table 1 relative to the calibration of installation no. 6 in 4VGR6, and the precision of these experimentally determined calibration factors is based on the results of the multiple linear regressions discussed above.

The long-term stability was investigated by calculating a weighted, average calibration value for Installation no. 6 based on all measurements in the period by means of the data from Table 1. This mean value and its associated standard deviation for each period is now compared with another value determined 13 or 26 periods later, corresponding to 1 or 2 years ahead. In the present data set 9 such comparisons can be made, and the stability of the equipment can now be tested from these duplicate results by the Analysis of Precision in the same way as analytical quality control is carried out[5]. Results are shown in Table 2, including the test statistic T which is approximately a chi-squared distribution with 9 degrees of freedom, and shows that no additional sources of variation can be detected; the long-term stability is therefore very satisfactory and does not contribute to the overall uncertainty.

Finally the seasonal variation of calibration can be determined from 13 consecutive period values, including the weighted mean values of the results given in Table 2.

Table 1. Experimentally determined calibration factors for silicon irradiation facilities in DR 3

Installation No	Ratio to VGR6	Precision SEM	Regression %/°CCA	Residual SD
1	0.965	1.1%		
2	1.135	1.1%	0.17%	1.4%
3	1.011	1.1%		
4	1.008	1.0%		
5	0.858	1.2%	0.34%	1.2%
6	1.000	-		

Table 2. Long-term stability of calibration values for calorimetric neutron dosimetry

Initial calibration	Second calibration	Interval
mmol neutrons/m^2 per 10^6 counts ± SEM		years
39.61 ± 0.34	40.01 ± 0.18	+ 2
40.24 ± 0.24	40.23 ± 0.19	+ 2
39.90 ± 0.24	39.82 ± 0.20	+ 2
39.54 ± 0.18	39.18 ± 0.17	+ 2
39.28 ± 0.18	39.27 ± 0.15	+ 2
40.01 ± 0.19	39.01 ± 0.40	+ 1
39.29 ± 0.21	39.06 ± 0.27	+ 1
39.50 ± 0.20	39.58 ± 0.59	+ 1
39.66 ± 0.19	39.32 ± 0.30	+ 1

T = 9.768 9 d.f.
$P(\chi^2 \geq T) = 0.37$

From these data the mean calibration value for installation no. 6 over a period of one year is found to be 40.2 mmol of neutrons/m^2 per 10^6 counts, corresponding to 41313 counts per 10^{17} neutrons/cm^2. The standard deviation of the 13 individual results is 0.8% which must be included in the overall uncertainty of calibration values.

In comparison it can be mentioned that the standard deviation of the average neutron flux densities for the sequence of reactor opening periods is 5 times higher or approx. 4%.

RESULTS AND DISCUSSION

The quantitative evaluation of individual components of the overall uncertainty in the determination of neutron fluence, based on the quality control data, permits the estimation of actual calibration values with a known precision.

The calibration value for a particular installation is affected by a seasonal variation with a standard deviation of 0.8%, and after correction for CCA position other sources of variation are responsible for the precision of the calibration ratios given in Table 1. The standard deviation of an estimated calibration value is thus 1.1-1.2%.

The actual measurement of integrated dose is carried out with a precision of 1.2-1.4% according to Table 1, and this means that the overall standard deviation of neutron dose imparted to a silicon crystal near the bottom of the irradiation container is approx. 1.6%.

The dose to other parts of the crystal is, however, influenced by the vertical neutron flux density variation shown in Fig. 3, and as an example we calculated the resulting dose distribution for crystals with a length of 400 mm.

During the course of a reactor operating period the vertical variation changes, as shown in Fig. 3, so that the maximum flux difference becomes 7½%. With a target value approximately 3% higher than the base value, deviations of up to ±1.25% yield an axial homogeneity of better than ±5% in all cases.

Greater deviations give axial homogeneities better than ±5% in fewer cases, and with a standard deviation of 1.6% it was calculated that ±5% material was achieved in 85% of the cases; this is equivalent to a standard deviation of 3.5% for the axial variation of NTD silicon crystals of up to 400 mm length.

Based on irradiation time the overall standard deviation of estimated dose is not less than 7%, and this excludes the use of time as the controlling factor for neutron transmutation doping in DR 3.

Resistivity

After the decay of ^{31}Si and suitable annealing the final resistivity ρ of the n-type material is measured at the centre of the ingot end face close to the cobalt monitor. These measurements are carried out by Topsil A/S in connection with their comprehensive inspection of incoming NTD material[6], and serve to ascertain the accuracy and reliability of the system as a whole. The uncertainty of the resistivity measurements used in this context has been estimated at approximately 1.5%, which means that the contribution from resistivity measurements before irradiation ρ_o can be neglected at the normal levels of doping, i.e. doping factors of 30-40.

The quality of the final product is, however, not determined by the quality of the neutron transmutation doping alone, but also by the initial homogeneity of the starting material[7].

From the relation*

$$\frac{1}{\rho} = \frac{1}{\rho_o} + \frac{1}{\rho_{NTD}} \quad (1)$$

* Only n-type material is used.

ρ_{NTD} can be calculated from the experimentally determined values of resistivity, ρ and ρ_o; but ρ_{NDT} can also be determined from the expression

$$\frac{1}{\rho_{NDT}} = C_p \mu e \quad (2)$$

where C_p is the number of phosphor atoms per cm^3 of silicon produced by neutron capture,

$$C_p = N_o \gamma \sigma \int \phi dt \quad (3)$$

The same relation applies to the activity of the cobalt monitor

$$A = k \times \int_o^t \phi dt \quad (4)$$

and neutron fluence is therefore determined by the specific activity of ^{60}Co which amounts to 96.3 TBq/kg for a fluence of 1 mol thermal neutrons/m^2; this corresponds to an activity of 2.13 μCi for a monitor containing 0.492 mg of cobalt exposed to $10^{17} n/cm^2$.

With $N_o = 4.997 \times 10^{22}$ atoms/cm^3 and $\gamma = 3.09\%$ Eqs. (1) - (4) may be rearranged to express the value of $\sigma\mu$ as a function of experimentally determined results

$$\frac{1}{\rho} - \frac{1}{\rho_o} = \sigma \mu e \frac{\gamma}{k} N_o A$$

and with $e = 1.602 \times 10^{-19}$C the mean values of

$$\overline{\sigma\mu} \cdot \overline{A\rho_{NTD}} = 8.60 \times 10^{-20} \quad (5)$$

with A in μCi and ρ in Ω cm.

The products of the corresponding values of A and ρ are used to determine the practical uncertainty of resistivities in the range of 18 to 180 Ω cm. The distribution of 238 values yields a standard deviation of 3% for the resistivity measurements, which reflects not only the reproducibility of measurement, but also the effects of inhomogeneities in the starting material and the possible long-term variation in instrumental calibration, etc.

Inserting the mean value in Eq. (5) gives a mean value of

$$\overline{\sigma\mu} = 1.681 \times 10^{-22}$$

With a mobility of 1396 $cm^2/(V\ s)$[7], the thermal neutron cross section of the $^{30}Si(n,\gamma)^{31}Si$ reaction becomes

$\sigma = 0.120$ b

which is in good agreement with other recently reported results[8].

CONCLUSION

An investigation of possible sources of variability in the production of NTD silicon in the Danish reactor DR 3 has been made on the basis of quality control data for the last 3 years of operation.

Calibration values for the calorimetric neutron dosimeter were in statistical control when compared with values taken 1 or 2 years later. This demonstrates the absence of significant, unknown random sources of error in the determination of neutron fluence.

Calculation of the distribution of axial variations in neutron dose based on the vertical flux density distribution in the irradiation facilities gave an approximately normal distribution with a standard deviation of 3½%.

Corresponding values of resistivity and neutron dose were used to calculate the cross section for the NTD reaction, and satisfactory agreement with other recent data indicates the absence of significant systematic errors in the production control.

REFERENCES

1. H.M. Janus and O. Malmros, Application of Thermal Neutron Irradiation for Large Scale Production of Homogeneous Phosphorus Doping of Floatzone Silicon, I.E.E.E. Trans. Electr. Dev. ED-23: 797 (1976)
2. Kai Hansen, Facility for Irradiation of Large Silicon Crystals in the Danish DR 3 Reactor, IR no. 494, Risø Engineering Dept. (1978)
3. K. Haack, A Calorimetric Thermal Neutron Dosimeter, Risø-M-2247 (1980)
4. Source no. 1274, The Radiochemical Centre, Amersham
5. O. Malmros, Precision Resistivity Measurements on NTD-Silicon, in: "3rd International Conference on Neutron Transmutation Doped Silicon", J. Guldberg, ed., Plenum Press, New York (1981)
6. K. Heydorn and K. Nørgaard, Analysis of Precision of Activation Analysis Methods, Talanta 20: 835 (1973)
7. J.M. Meese, D.L. Cowan, and M. Chandrasekhar, A Review of Transmutation Doping in Silicon, I.E.E.E. Trans. Nuc. Sci. NS-26: 4858 (1979)

8. N.W. Crick, The Optimisation of Nuclear Parameters Used for Silicon Irradiations in the Harwell Research Reactors, in: "3rd International Conference on Neutron Transmutation Doped Silicon", J. Guldberg, ed., Plenum Press, New York (1981).

THE SELECTION OF STARTING MATERIAL

FOR NEUTRON-TRANSMUTATION DOPED SILICON

Horst G. Kramer

Electronics Division
Monsanto Industrial Chemicals Company
St. Peters, Missouri 63376

The selection of starting material for neutron-transmutation doped (NTD) silicon has been described by Janus and Malmros[1] for the case where the starting material is entirely of one or the other conductivity type. After defining homogeneity factors for the starting and for the NTD material, the authors derive an expression for the doping factor, which essentially sets a fixed limit on the maximum dopant concentration of the starting material for NTD material of a required homogeneity.

The portion of material neglected by this analysis, namely that containing p-n junctions, is obviously an excellent starting material for NTD, since junctions in undoped crystals generally occur whenever the base-level impurities of the polycrystalline silicon, both p and n type, are present in only very small quantities. A general analysis, encompassing starting material with p-n junctions, seems hence desirable. As will be shown in the next paragraph, this general analysis leads to the establishment of two selection criteria for the starting material - a maximum dopant difference, which depends only on the desired homogeneity of the NTD material; and a maximum average dopant concentration, which is a function of the homogeneity of both the starting and the NTD material.

We define the dopant differences in starting and NTD materials respectively as

$$\Delta C_D = C_D \max - C_D \min \qquad (1)$$

$$\Delta C_s = C_s \max - C_s \min \qquad (2)$$

where the C's refer to dopant concentrations, the subscript D's to the NTD material, and the subscript s's to the starting material. For convenience, we define donor dopant concentrations as positive, and acceptor dopant concentrations as negative numbers.

Following Janus and Malmros, we assume that neutron-transmutation doping itself is completely homogeneous, and that the homogeneity of the doped silicon is determined by the distribution of impurities in the starting material. We hence have

$$\Delta C_s \leq \Delta C_D \tag{3}$$

We define average dopant concentrations as

$$\overline{C}_D = 1/2\ (C_D \max + C_D \min) \tag{4}$$

$$\overline{C}_s = 1/2\ (C_s \max + C_s \min) \tag{5}$$

Equation (2) sets no limits on the maximum or minimum dopant concentration in the starting material, only on the difference between them. Given the limitations of the NTD process, we must set

$$\overline{C}_s \leq \overline{C}_D \tag{6}$$

We define normalized homogeneity factors

$$a_D = \frac{\Delta C_D}{\overline{C}_D} \tag{7}$$

$$a_s = \frac{\Delta C_s}{\overline{C}_s} \tag{8}$$

Then the allowable dopant difference in the starting material is given by

$$\Delta C_s \leq a_D \overline{C}_D \tag{9}$$

If we define the doping factor F as

$$F = \frac{\overline{C}_D}{\overline{C}_s} \tag{10}$$

then, for $\overline{C}_s \geq 0$, that is, for material that is essentially n-type or intrinsic,

$$F \geq \frac{a_s}{a_D} \tag{11}$$

and,

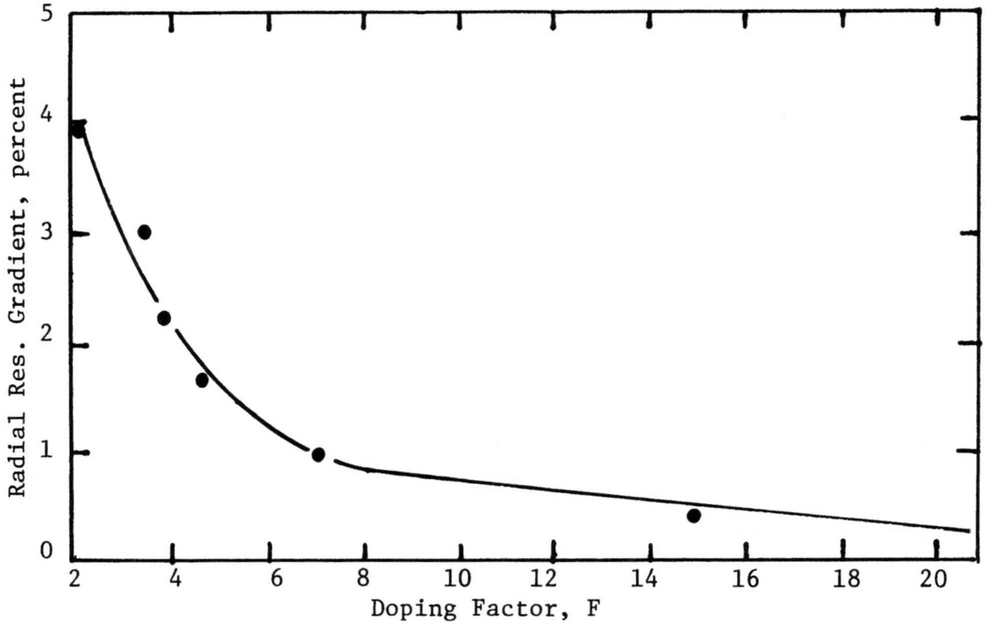

Fig. 1: NTD Material Homogeneity vs. Doping Factor

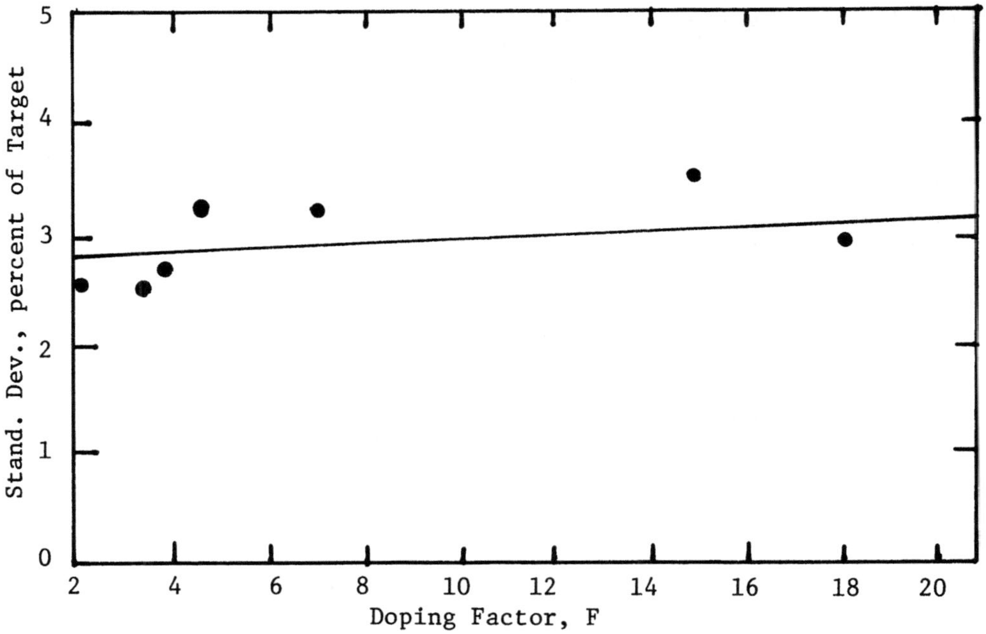

Fig. 2: Doping Precision vs. Doping Factor

$$\overline{C}_s \leq \frac{a_D}{a_s} \overline{C}_D \tag{12}$$

For C_s <0, that is, for p-type material,

$$F \leq \frac{a_s}{a_D} \tag{13}$$

and,

$$\overline{C}_s \geq \frac{a_D}{a_s} \overline{C}_D \tag{14}$$

In the case of p-type material, this selection of the minimum average dopant concentration in the starting material may not be realistic, and may be altered by such considerations as the amount of allowable dopant compensation in the NTD material, or the additional irradiation costs of such compensation.

The results obtainable with these two selection rules are shown in Figures 1 and 2. Figure 1 shows the homogeneity of the NTD material (measured in this case as the radial resistivity gradient) as a function of the doping factor F. The solid curve shows calculated values; the dots represent experimentally obtained values. The agreement between predicted and actual values is very close. Figure 2 shows the doping precision obtainable when batches of starting material with various average doping factors are irradiated. The doping precision is expressed by the standard deviation of individual ingot doping concentrations about the batch mean doping concentration, divided by the mean doping concentration. The least-squares line fitted to the data seems to indicate an improved precision as the doping factor becomes smaller; however, this trend is statistically not significant.

[1] H. M. Janus and O. Malmros, Application of Thermal Neutron Irradiation for Large Scale Production of Homogeneous Phosphorus Doping of Float Zone Silicon, IEEE Transactions on Electron Devices, Vol. ED-23, No. 8, August]976, pp. 797-802.

THE OPTIMISATION OF NUCLEAR PARAMETERS USED FOR SILICON IRRADIATION IN THE HARWELL RESEARCH REACTORS

N.W. Crick

Materials Physics Division
AERE, Harwell,
Oxfordshire, England

INTRODUCTION

Irradiations of semi-conductor silicon for the purposes of neutron doping have been carried out in DIDO and PLUTO, the materials-testing reactprs at Harwell, since 1975. The reactors are fuelled with highly enriched uranium in a core which is cooled and moderated with heavy water. At the present time irradiations are carried out in the outer graphite reflector where the neutron spectrum is well thermalised, resulting in relatively low damage rates from fast neutrons. (FIG. 1) Reactivity of the core is partly controlled by a combination of burnable poisons so that flux changes caused by the movement of control absorbers are kept to a minimum.

The purpose of this paper is to show how a resistivity/neutron dose relationship has been optimised by using a "best fit" neutron cross section for Si^{30}. Other nuclear parameters are taken into account so that accurate allowances can be made for flux depressions of crystals of different size. A detailed analysis of a small series of irradiations is included so that the main sources of error can be identified.

RESISTIVITY/NEUTRON DOSE RELATIONSHIP

Assuming that both ending and starting material is n-type silicon, the relationship between the dopant concentration of P^{31} after irradiation C_E and the resistivity ρ_E can be written as:-

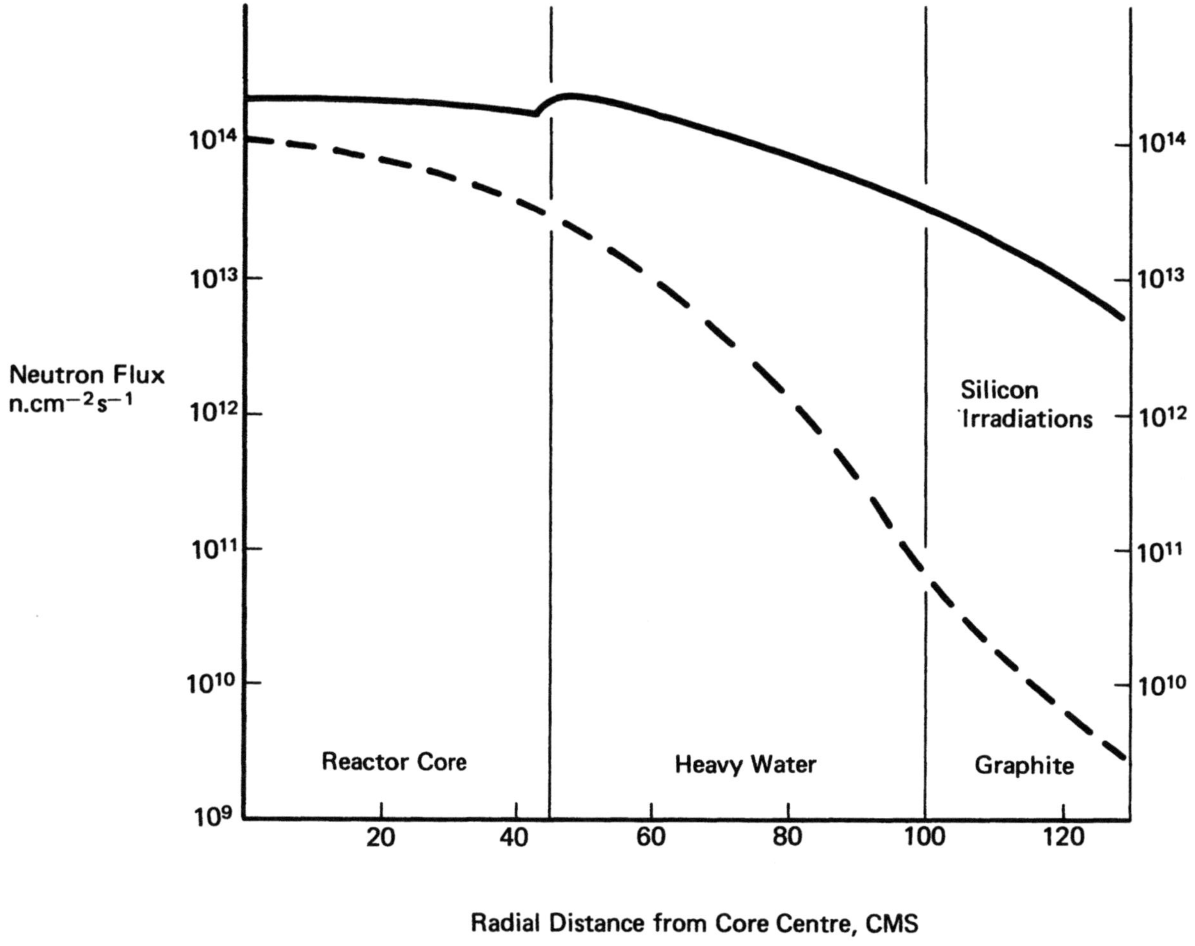

Fig. 1. Comparison of thermal and fast fluxes in Pluto Reactor.

$$\rho_E = (C_E \mu\ \varepsilon^{-1}),\ \Omega\ \text{cm} \tag{1}$$

where μ is the lattice drift mobility of electrons which for this work has been taken as 1350 $cm^2\ Vsec^{-1}$ and ε is the electronic charge, 1.6 x 10^{-19} coulombs so that $C_E = \rho_E^{-1}\ 4.624 \times 10^{15}$ atoms cm^{-3}.

To allow for the effect of dopant concentration present in the starting material, it is assumed that:-

$$C_E = C_D + C_S \tag{2}$$

Where C_S is the dopant concentration in the starting material

$$\text{i.e. } C_S = \rho_S^{-1}\ 4.624 \times 10^{15} \text{ atoms cm}^{-3} \tag{3}$$

and C_D is the dopant concentration of P^{31} added during the irradiation

$$\text{i.e. } C_D = N_o\ \sigma\ \phi t \text{ atom cm}^{-3}. \tag{4}$$

Where N_o is the atoms Si^{30}, cm^{-3} in the starting material. For silicon with a specific gravity of 2.33 $g.cm^{-3}$, atomic weight of 28.086 and a Si^{30} isotopic abundance of 3.09%.

N_o = 1.544 x 10^{21} atoms cm^{-3}

σ is thermal neutron absorption cross section for Si^{30}, cm^2

and ϕt is the neutron dose, $n.cm^{-2}$.

As the resistivity measurements from irradiated silicon became available, it was possible to derive a "best fit" value for the Si^{30} absorption cross-section. The resistivity measurements were made with a conventional four-point probe at several points across both ends of each crystal. The mean measured resistivity, ρ_m, for each irradiation could then be related to the mean neutron dose, (ϕt), over the full length of each irradiation. It was then possible to derive a value for the Si^{30} absorption cross section for which the average value of ρ_m/ρ_E was unity. The "best fit" value from the results for over 1000 irradiations was 0.118 x 10^{-24} cm^2.

Substituting the values for No and σ in equation 4:-

$$C_D = 1.822 \times 10^{-4}\ \phi t \text{ atoms cm}^{-3}$$

By substituting values of C_E, C_D, and C_S into equation (2), it can be shown that the relationship between neutron dose and resistivity is:-

$$\phi t = 2.54 \times 10^{19}\ (\rho_E^{-1} - \rho_S^{-1})$$

THERMAL NEUTRON ABSORPTION CROSS SECTION OF Si^{30} BY NEUTRON ACTIVATION

In an effort to confirm the value derived from resistivity measurements the thermal neutron absorption cross section of Si^{30} has been measured using a neutron activation technique. Small samples of silicon were irradiated in one of the reactor positions normally used for silicon irradiations so that after chemical preparation the absolute activity as the Si^{31} decayed could be measured using a 4π β counter. Counting normally took place within ten hours of irradiation and continued until background level had been reached. (FIG. 2).

It was assumed that the activity of the Si^{31}

$$A = N_o \sigma \phi (1 - e^{-\lambda t}) e^{-\lambda T}$$

where N_o is the atoms of Si^{30} cm^{-3}, using the data from equation 4.

ϕ is the thermal flux, measured during the irradiation
λ is the decay constant for Si^{31}
t is the irradiation time normally about 1 hour,
and T is the time after the irradiation had ended.

The data from these measurements which included ten separate irradiations gave the following mean values:-

$\lambda(Si^{31}) = 7.237\ 10^{-5}\ s^{-1}$ (equivalent half-life 2.66 hrs)
$\sigma(Si^{30}) = 0.118 \times 10^{-24}\ cm^2$

This cross section compares very closely with the value derived from resistivity measurements, but it should be noted that the parameter common to both values is the measurement of neutron flux.

It is interesting to note that the absorption cross-section of 0.118 barns for Si^{30} is close to the mean of values used by other reactors operators,

i.e. 0.11 barns used by Meese et al (1) and reported by Hughes and Schwartz in 1958 (2),
0.12 barns used by Cleland (3)
and 0.13 barns used by Janus and Malmros (4)

NEUTRON FLUXES DURING IRRADIATIONS

The axial thermal flux for a typical silicon irradiation is shown in FIG. 3. A typical mean value is 8×10^{12} $n.cm^{-2}\ s^{-1}$ and the axial shapes have been improved by the use of neutron shields

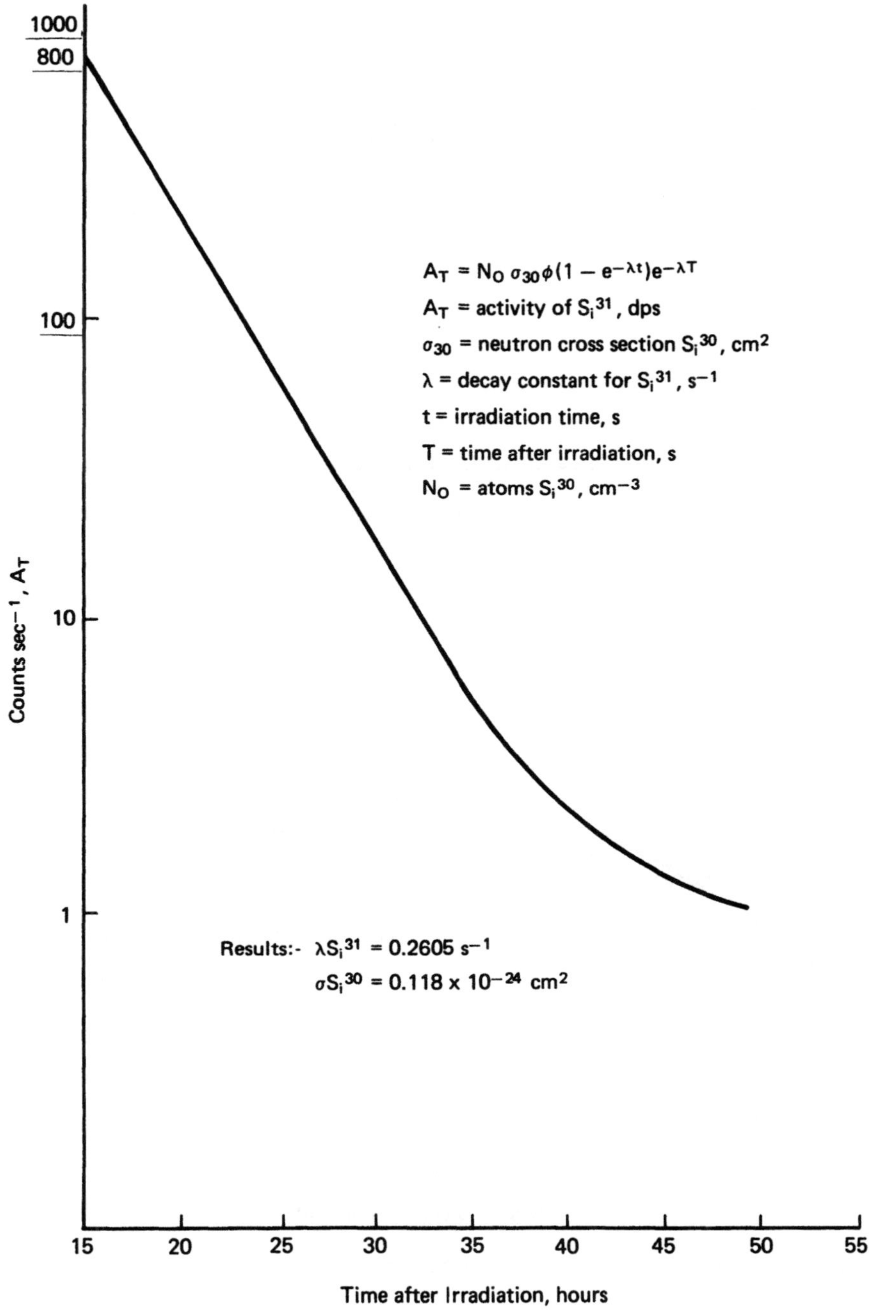

Fig. 2. Radioactive decay of S_i^{31}

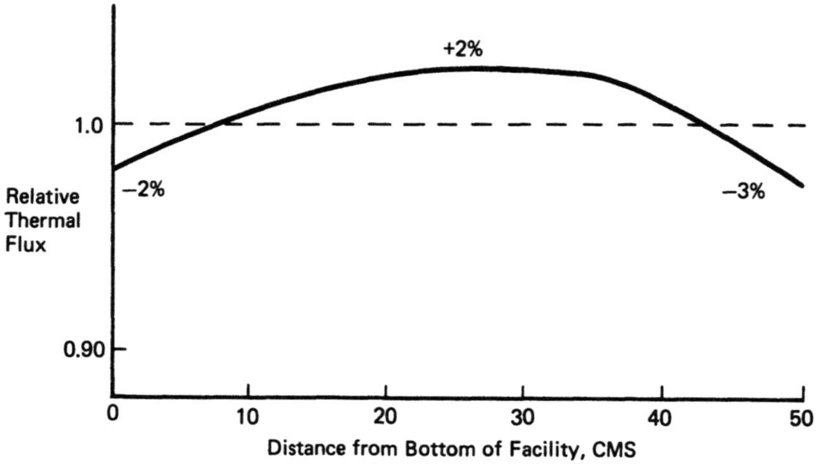

Fig. 3. Typical thermal flux shape in the Harwell reactors

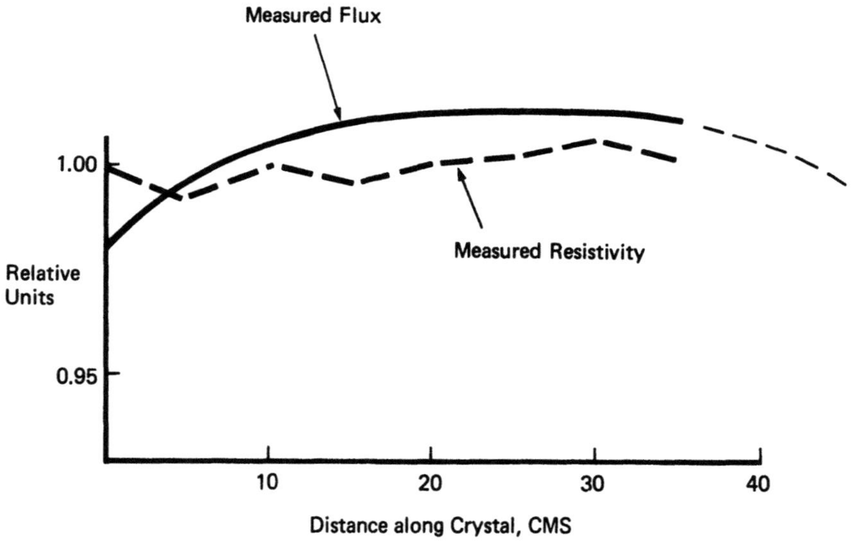

Fig. 4. Comparison of measured flux with resistivity

so that variations from the mean value over a length of 500 mm are + 2% at the centre and -2 to -3% at either end.

The variations due to shape of the axial flux can be minimised by positioning the silicon crystals so that the highest starting resistivity is located in the region of highest flux. A result using this procedure is shown in FIG. 4 which shows that the measured variation of resistivity is less than the measured variation in the axial flux.

The neutron flux is normally measured inside an empty irradiation facility so that different flux depressions can be calculated for crystals of different sizes or for combinations in which several crystals are irradiated side by side. The calculations used to develop a relationship between crystal size and flux depression (FIG. 5) assumed the generally accepted neutron absorption cross section for natural silicon of 0.17 barns. Recent

Data	Thermal Cross Section x 10^{-24} cm^2	Isotopic Abundance V_o
S_i^{28}	0.170	92.23
S_i^{29}	0.270	4.68
S_i^{30}	0.118	3.09

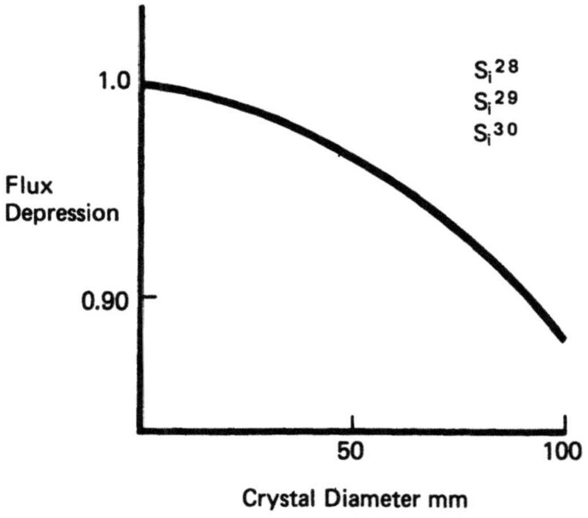

Fig. 5. Flux depression/crystal size relationship

measurements of neutron fluxes on the surface of silicon crystals of various sizes agree with the theoretical predictions of flux depression. Also the cross section of natural silicon has been confirmed at Harwell in the GLEEP oscillator.

RESULTS

A distribution of results from about 1000 irradiations is shown in FIG. 6 as a frequency distribution of the ratio:-

$$\frac{\text{mean measured resistivity for each irradiation}}{\text{calculated resistivity}}$$

The mean measured resistivities have been obtained as described in section 5 and the calculated resistivities computed using equations 1 to 4. The histogram shows that the ratios are normally distributed with a standard deviation of $\pm$ 2.9% and that the spread of results is very similar to that reported by other reactor operators.[1] It is considered that the contributions to the total error are approximately, 2% due to flux variations, 2% due to inhomogeneity of the starting material and 1% from resistivity measurements.

A more detailed analysis of a small series of irradiations can be used to illustrate the magnitude of errors which can occur. Such a series, as shown in TABLE 1, was carried out in the same reactor facility over a period of about eight days. The series included irradiations in which crystals were positioned side by side and covered a wide range of neutron doses. In this case the neutron flux was calibrated three times during the irradiations and remained constant to within $\pm$ 1% compared with the normal flux variation of $\pm$ 2%. A close examination of the results shows that even though the variations in the neutron dose are kept to a minimum, errors of 5-6% are not uncommon. This strongly suggests that major contributions to the total error are inhomogeneity of the starting material[3] and the accuracy with which resistivity measurements can be made.[1] Because of these effects any further reduction in the observed standard deviation will be small.

SUMMARY

A resistivity/neutron dose relationship has been developed using a "best fit" neutron absorption cross section for Si^{30} derived from resistivity measurements. The value of 0.118×10^{-24} barns is within the range of values used by other reactor operators and has been confirmed independently by neutron activation techniques. A relationship between thermal flux depression and crystal size is used to correct the unperturbed flux and an analysis of results shows that the error distribution has a standard

TABLE 1

Results of Silicon Irradiations made in the same reactor facility

Irradiation No	Time days	Dose $n.cm^{-2}$ $\times 10^{17}$	Target Ω cm.	Diameter mm	Average Measured ρ / Calculated ρ	Comments
1	0.94	5.77	43	4 x 35	1.000	Flux calibrated
2	1.74	4.93	50	4 x 35	0.999	
3	2.50	4.60	50.5	2 x 40	1.011	
4	3.34	4.50	54.5	80	1.007	
5	4.00	4.14	54.5	80	0.964	
6	4.40	2.47	100	3 x 40	0.997	Flux recalibrated
7	4.96	3.48	65	80	0.952	
8	5.60	3.92	63	80	1.006	
9	6.24	3.90	63	80	1.023	
10	6.80	3.39	65	80	0.961	
11	7.36	3.36	75	4 x 35	0.980	Flux recalibrated
12	7.54	1.04	220	80	0.940	
13	8.07	3.30	70	80	1.002	

Average 0.998
$\sigma =$ 0.026

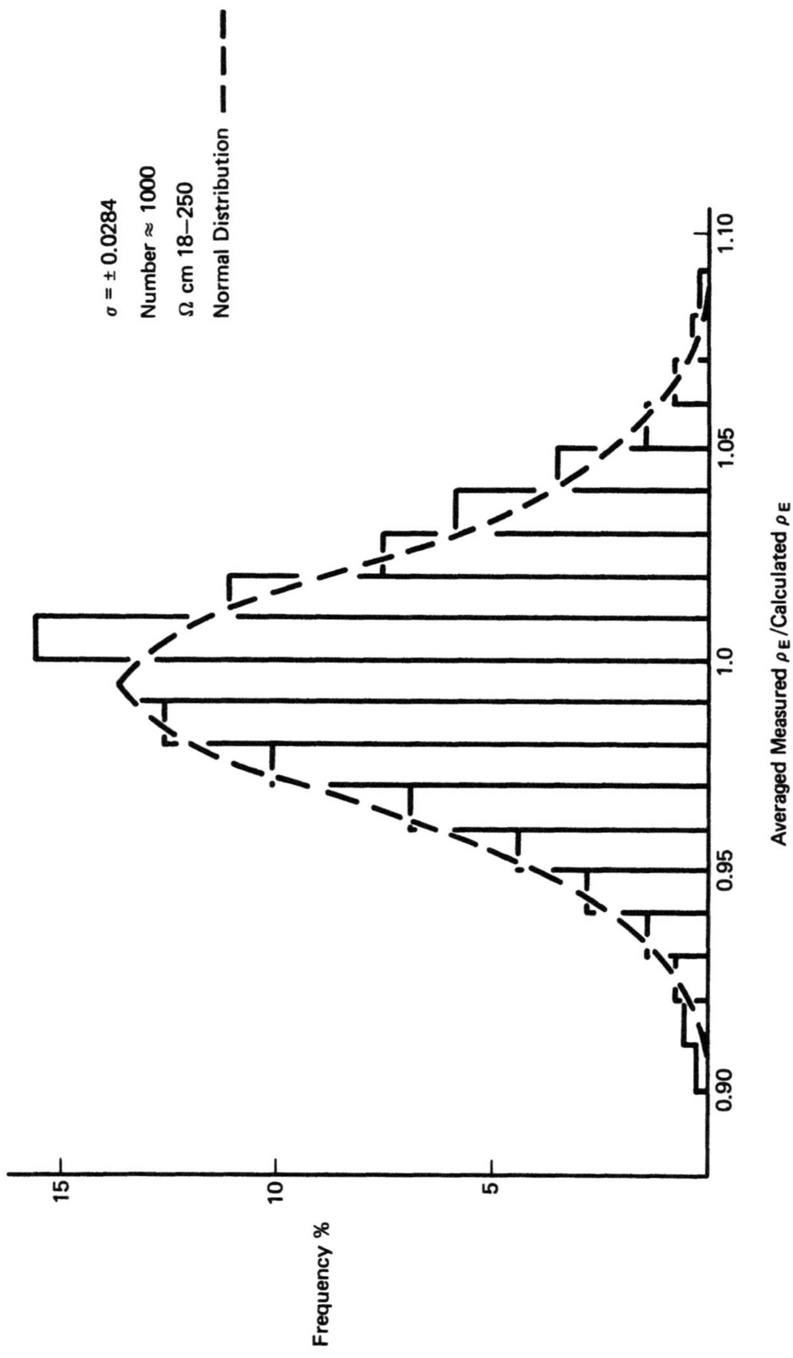

Fig. 6. Results of Silicon irradiations in Dido and Pluto

deviation of approximately ± 3%. A detailed examination of the results from a series of irradiations carried out in the same reactor position suggests that the nuclear component of the total error is small compared to those caused by material effects.

References

1. J.M. Meese and P.J. Glairon. Resistivity Fluctuations in Highly Compensated NTD Silicon. 2nd Int. Conf. of Neutron Transmutation Doping in Semiconductors. Un. of Missouri (1979).

2. D.S. Hughes and R.B. Schwartz. Neutron Cross Sections. U.S. Gov't. Printing Office (1958).

3. J.W. Cleland. Proc. Int. School of Physics - Course XVIII - Radiation Damage in Solids (Academic Press 1962).

4. H.M. Janus and O. Malmros. Application of Thermal Neutron Irradiation for Large Scale Production of Homogeneous Phosphorus Doping of Floatzone Silicon. IEEE Trans. on Electronic Devices, Vol. ED-23 No. 8. Aug. 1978.

FACTORS AFFECTING PHOSPHORUS PRODUCTION RATE IN NTD SILICON

J. M. Meese[+] and S. L. Gunn

University of Missouri Research Reactor

Columbia, MO 65211

ABSTRACT

We will discuss the many factors which determine the overall experimentally observed phosphorus production rate in NTD-Si. This production rate can vary considerably for experiments run in various reactors. The errors associated with the determination of this production rate are the result of our inexact knowledge of the thermal neutron capture cross section, the epi-thermal resonance integral, the thermal and epi-thermal fluences and the phosphorus concentration in both the starting and final NTD material. Other possible errors include mis-typing of starting material, contamination during both irradiation and annealing, and incomplete annealing. It is important for both the silicon manufacturer and the reactor service engineer to understand these factors if the accuracy of doping inherent in the NTD process is to be realized.

INTRODUCTION

Although the transmutation doping process can potentially be used to control phosphorus concentration to an accuracy of better than 1% in small samples; a number of factors related to the actual phosphorus production rate per unit fluence, K, as well as factors related to the accuracy of determining the resultant phosphorus concentration tend to make the whole NTD calibration process somewhat circular. In practice, the irradiation calibration procedure is usually based on the starting and final resistivity and some method of thermal fluence measurement rather than on an actual determination of phosphorus concentration. In effect, fluence measurement errors, cross section errors, epi-thermal neutron doping, and errors associated with the estimation of phosphorus concentration from resistivity values tend to be lumped into a single production rate parameter, K'. It is, therefore, not unusual to obtain different production rate parameters, K', for different reactors, different positions in a single reactor or for different silicon manufacturers irradiating in the same reactor position who use differing conversions between resistivity and phosphorus concentration. We will analyze to some extent the consequences of the assumption usually made that the actual production rate, K, is equal to the observed production rate parameter, K'. We will discuss the accuracy with which the nuclear parameters such as cross sections, fluence and the relative abundances of the isotopes are known in order to estimate the overall accuracy of our knowledge of the actual production factor, K. We will also discuss errors associated with various methods of determining the actual phosphorus concentration such as resistivity and Hall effect measurements, junction capacitance measurements, neutron activation analysis and optical measurements. Finally, we will discuss the errors to be expected from using a lumped parameter production rate, K', as determined by resistivity measurements.

PRODUCTION RATE DISCREPANCIES

The concentration of phosphorus, N_P, produced by the reaction

$$^{30}Si(n,\gamma)^{31}Si \rightarrow {}^{31}P + \beta^- \quad (T_{1/2} = 2.62\ hr) \tag{1}$$

is given by the well known relationship

$$N_P^{(th)} = \sigma_c N_T \psi t = K\Phi \tag{2}$$

where σ_c is the thermal neutron capture cross section for the isotope ^{30}Si, N_T is the concentration of target isotope (^{30}Si) and $\Phi = \psi t$ represents the thermal neutron fluence (flux times irradiation time). The parameter, K, is the phosphorus production rate per unit fluence. Equation (2) tells us that $K = N_P/\Phi$ can be determined experimentally if the phosphorus concentration and fluence can be accurately measured or that it can be calculated from atomic and nuclear constants using the expression

$$K = \sigma_c N_T = \sigma_c (R.A.) L_o\ \rho/A \tag{3}$$

where (R.A.) is the relative abundance of the ^{30}Si isotope in Si, L_o is Avogadro's constant, ρ is the density of Si and A is its atomic weight. Hopefully, these two methods should yield the same value for the phosphorus production rate. That this is not always the case is illustrated in Table 1. This table gives the 2200 m/sec ^{30}Si capture cross section from two reference sources over a period of five years. In comparison, Table 1 also lists apparent capture cross sections determined by experiment in the NTD literature over the last twenty years. It is clear that the NTD literature values are about 7% higher on the average than the accepted reference values![1-12] This is to be compared with the stated accuracy of the 2200 m/sec cross section (± 1.9%).

From Eq. (3), the capture cross section observed in an NTD experiment can be in error only if the concentration of targets of ^{30}Si is in error or if the experimental production rate, K, is in error. Table 2 lists those parameters which are needed to determine the number of target atoms, N_T.

Table 1. Thermal neutron absorption cross sections (2200 m/sec) for the ^{30}Si isotope from standard references and from the NTD literature.

Standard References:

Cross section (barns)	Source + Date	Reference
0.107 ± 0.002	Brookhaven (1973)	(1)
0.108 ± 0.002	Table of Isotopes (1978)	(2)

mean 0.1075 ± 0.0020 (1.9%)

NTD Literature:

Cross section (barns)	Source + Date	Reference
0.12	Schweiner (1959)	(3)
0.11 ± 0.01	Tanenbaum + Mills (1961)	(4)
0.11 ± 0.01	Messier et al. (1964)	(5)
0.11	Kharchenko + Solov'ev (1971)	(6)
0.110	Herman + Herzer (1975)	(7)
0.13	Hass + Schnöller (1976)	(8)
0.13	Janus + Malmros (1976)	(9)
0.110	Herzer (1977)	(10)
0.12	Benière + Agrawal (1978)	(11)
0.102	Agrawal + Benière (1979)	(12)

mean ± avg. deviation 0.115 ± 0.009 (7.8%)

Table 2. Constants used to determine the concentration of ^{30}Si target isotopes per cm^3, N_T.

Parameter	Value	Reference
Atomic wgt. of Si (A)	28.0855 ± 0.0003	(13)
Density of Si (ρ)	2.33 ± 0.01 gm/cm^3	(13)
Avogadro's Constant (L_o)	$(6.022045 \pm 0.000031) \times 10^{23}$ atoms/mole	(13)
^{30}Si Rel. Abundance (R.A.)	$3.10 \pm 1\%$	(2)
^{30}Si Rel. Abundance (R.A.)	3.09	(13)

$N(Si) = L_o\,\rho/A = 4.996 \times 10^{22}\ cm^{-3}$

$N_T = (R.A.)N(Si) = 1.549 \times 10^{22}\ cm^{-3}$

One notices immediately from Table 2 that the largest errors in the determination of N_T are in the relative abundance of ^{30}Si ($\pm$ 1%) and in the Si density ($\pm$ 0.43%). The reason that these numbers have not been measured to better precision is that the relative abundance of the ^{30}Si isotope is not constant in the earth's crust![14] We must hold out the possibility that the process of crystal growth by the float zone process can enhance the relative abundance of ^{30}Si in some way. To our knowledge, this possibility has not been checked experimentally.

It is much more likely, however, that either the accepted 2200 m/sec cross sections are in error, or that systematic errors are to be found in the experimental determination of K, which are related to either fluence measurement errors or to phosphorus concentration measurement errors. It should be mentioned that the first transmutation doping experiments in Ge thirty years ago by Cleland, Lark-Horovitz and Pigg[14] eventually led to a revision of the capture cross sections in this material. In Si, however, there has been a systematic reduction in σ_c (see Ref. 4 for an early value) which, in fact, has widened the gap between accepted 2200 m/sec cross sections and those observed in NTD experiments.

It should be mentioned that Eq. (2) is only approximately true. Because the decay of ^{31}Si to ^{31}P has a 2.62 hr half-life,[2] there is a build up during irradiation of ^{31}Si which has not decayed and which can be further transmuted to ^{32}Si. Since ^{32}Si has a half-life of 280 yr[2], this process represents a potential loss of phosphorus from the process illustrated in Eq. (1). Furthermore, for irradiation times which are several times the 2.62 hr half-life of the decay of ^{31}Si to ^{31}P, the build up of ^{31}P represents an additional target which can be transmuted by beta decay to ^{32}S. This process is the primary source of activity in NTD-Si irradiated to high fluences and represents an additional loss of phosphorus concentration from Eq. (2). It can easily be shown, however, that both these processes as well as the burn up of our original target material (^{30}Si) are much too small to be observed by the usual methods of phosphorus determination in Si.

Another factor which could perhaps account for the enhanced NTD phosphorus production rate is epi-thermal and resonance neutron capture.[8] These effects are dependent on the neutron energy spectrum used to irradiate the silicon. Figure 1 shows the neutron energy spectrum at the beryllium reflector of a typical light water reactor. The dotted curve was obtained by measurement while the solid line represents a calculated spectrum taken from the reactor design data. The neutron distribution up to energies of 0.5 eV are dominated by a Maxwell-Boltzman distribution in which the neutrons are in approximate thermal equilibrium with their environment. The density of neutrons dn with velocity v in the range dv is given by

$$dn = \frac{4n}{v_o^3 \sqrt{\pi}} v^2 \exp(-v^2/v_o^2)\, dv \tag{4}$$

where $v_o = (2kT/m)^{\frac{1}{2}}$ is the most probable velocity of this distribution ($\sim$ 2200 m/sec).[15] Since the flux distribution is $d\psi = d(nv)$, then the flux distribution is given by[15]

$$d\psi = \frac{4n}{v_o^3 \sqrt{\pi}} v^3 \exp(v^2/v_o^2)\, dv \ . \tag{5}$$

This distribution can readily be converted to $d\psi/dE$ using $E = \frac{1}{2}mv^2$.

At high neutron energies (E > 500 keV in Fig. 1), the neutron distribution is dominated by a fission spectrum which has been empirically fit to relationships of the form[15]

$$dn = k_1 e^{-E} \sinh\, (2E)^{\frac{1}{2}}\, dE \ . \tag{6}$$

Here, k_1 is a proportionality constant used to adjust the strength of the source and E is in MeV. Since the 1 MeV capture cross section of ^{30}Si is of the order of 1 mb[16] and the 1 MeV flux is several orders of magnitude lower than the peak thermal flux, fast neutron capture can be ignored.

In the energy range between the thermal and fission spectra all reactors exhibit a down scattered spectra with E^{-1} energy dependence as shown in Fig. 1. Since the probability for elastic iso-

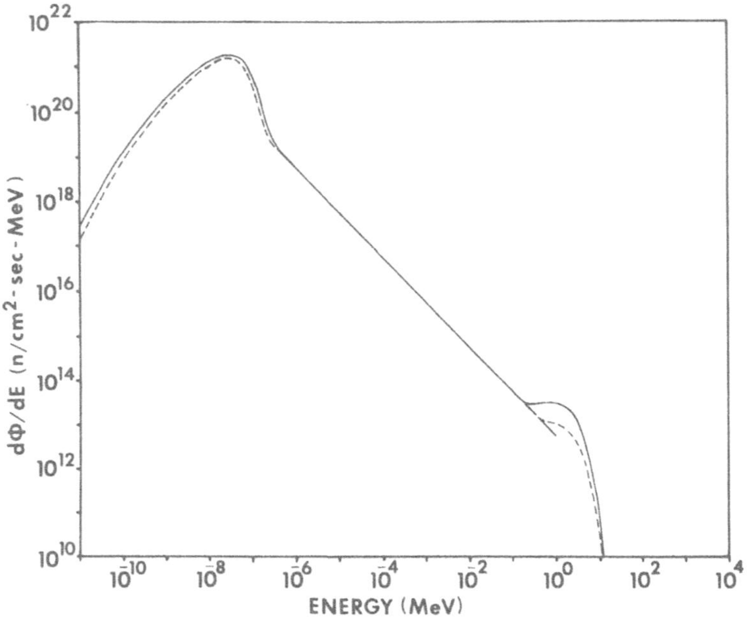

Fig. 1. Neutron differential flux spectrum of MURR at the edge of the beryllium reflector before appreciable thermalization in the graphite reflector.

tropic scattering is proportional to E^{-1}, this region of the spectrum can be represented by

$$d\psi = k_2 \, dE/E \ . \tag{7}$$

Although the spectrum of Fig. 1 is for a light water reactor, all reactor spectra exhibit these same factors to a greater or lesser degree. In a heavy water reactor with a Cd ratio of 1000:1, the thermal spectrum dominates, however, in a light water reactor with a Cd ratio of 10:1, this "resonance" portion of the spectrum must be considered. It is from this E^{-1} epi-thermal range that additional neutron captures occur.

We have tried to estimate the importance of this region of the spectrum for phosphorus production in Si for a typical row-2 irradiation position in our own light water reactor. For this position, the flux distribution is well represented by

$$\frac{d\psi_r}{dE} = \frac{k_2}{E}, \quad (0.5 \text{ eV} \leq E \leq 3.27 \text{ MeV}) \tag{8}$$

where $k_2 = 7.8 \times 10^{11}$ n/cm^2/sec.

The total integrated resonance flux is obtained from

$$\psi_r = \int_{0.5\text{eV}}^{3.27\text{MeV}} \frac{d\psi_r}{dE}\, dE = 1.2 \times 10^{13} \text{ n/cm}^2\text{/sec.}$$

The thermal flux in this same position is $\psi_{th} = 4 \times 10^{13}$ n/cm^2/sec.

The total phosphorus production at these energies is composed of resonance capture (i.e., capture of neutrons in regions of enhanced capture cross section) and by the 1/v capture of epi-thermal neutrons. The phosphorus concentration produced by resonance capture is given by

$$N_P^{(r)} = N_T t \int_{0.5\text{eV}}^{\infty} \frac{d\psi_r}{dE}\, \sigma_r(E)\, dE$$

$$= N_T t\, k_2 \int_{0.5\text{eV}}^{\infty} \frac{\sigma_r(E)}{E}\, dE\ . \tag{9}$$

Notice that Eq. (9) is just a spectral average of Eq. (2). The last integral above is known as the resonance integral, I_r, whose value for ^{30}Si is given as 0.106 ± 0.012b.[1] Also using Eq. (1) for the thermal neutron production, we can find the ratio of resonance to thermal phosphorus production as

$$\frac{N_P^{(r)}}{N_P^{(th)}} = \frac{I_r}{\sigma_c}\, \frac{k_1}{\psi_{th}} = 0.0193$$

for the light water spectrum represented by Eq. (8). We therefore predict that the thermal production rate is enhanced by less than 2%. The epi-thermal 1/v contribution to phosphorus production can be obtained from the 1/v dependence of neutron capture in this energy region. Since $\sigma_c \propto v^{-1} \propto E^{-\frac{1}{2}}$,

$$\sigma_c(E) = \sigma_o (E_o/E)^{\frac{1}{2}}$$

where σ_o is the 2200 m/sec cross section and $E_o = 0.025$ eV. The epi-thermal phosphorus production is then given by

$$N_P^{(epi)} = N_T t \int_{0.5eV}^{\infty} \sigma_c(E) \frac{d\psi^{(r)}}{dE} dE$$

$$= N_T t \, \sigma_c \, k_2 \sqrt{E_o} \int_{0.5eV}^{\infty} E^{-3/2} \, dE \, . \quad (9)$$

Using the spectrum of Eq. (9), we find the epi-to-thermal ratio of phosphorus produced to be

$$N_P^{(epi)}/N_P^{(th)} = 8.77 \times 10^{-3} \, . \quad (10)$$

From the above, the increase in the phosphorus production rate due to resonance and epi-thermal capture will not exceed 2% in a light water reactor. For a heavy water reactor with a high Cd ratio, we would expect these two processes to be completely negligible. Therefore, production rates for transmutation doping should not vary from reactor to reactor by more than 2%.

We must conclude from the above that the discrepancy between calculated and experimental production rates is most likely to be caused by the experimentally determined production rate, $K = N_P/\Phi$. In the next section, we will investigate errors associated with the determination of thermal neutron fluence, Φ. This will be followed by a discussion of the errors which commonly occur in the determination of phosphorus concentration, N_P.

ERRORS IN FLUENCE MEASUREMENT

Many methods have been devised to obtain the relative fluence for an NTD irradiation based on the time integration of a signal from a neutron detector which is proportional to the neutron flux.[18-21] Such integration systems can reproduce the fluence at a particular irradiation position to an accuracy of better than ± 1%[18] but must be calibrated in counts per unit fluence in order to make the measurements absolute. The activity induced in a standard flux wire is used to calibrate the integration system.

Ideally, the flux foil or wire should have a 1/v cross section dependence so that the flux measurement is independent of neutron velocity. To see this, consider an "ideal" 1/v neutron absorber. The capture cross section is then given by

$$\sigma(v) = k/v \, . \quad (11)$$

For a neutron velocity, v_0 = 2200 m/sec, the above equation becomes $\sigma(v_0) = k/v_0$. Solving for k and substituting into Eq. (11) we find that

$$\sigma(v) = v_0\sigma(v_0)/v \ . \tag{12}$$

Using the fact that the flux $\psi = nv$, where n is the neutron density, we see that the foil activity for a 1/v absorber is given by

$$\begin{aligned} A &= N_T n v \sigma(v) \\ &= N_T n v_0 \sigma(v_0) \\ &= N_T n k \ . \end{aligned} \tag{13}$$

We note that the activity actually measures the neutron density, not the flux, and that a 1/v absorber activity is independent of the neutron velocity. The results are modified if the energy dependence of the cross section is not given by Eq. (11).

A number of different materials have been recommended for flux calibration[22] including In, Dy, Cu and Mn, however, the most commonly used are Au and Co. Although Au is often used, it has a strong resonance capture at 4.9 eV which must be accounted for by cadmium difference measurements if the flux is to be measured correctly.[22] Also the high capture cross section of Au can cause flux depression near the foil so that corrections for foil thickness must be made.[22] We have found at MURR that Co doped Al flux-wire (% Co = 0.116%) is nearly ideal for NTD calibration and is readily available from the U.S. National Bureau of Standards as standard reference material 953. The % Co in this wire is certi-

Table 3. Thermal neutron capture cross section (2200 m/sec) and resonance integrals (including 1/v contribution) for Cobalt and Gold from Ref. 23.

isotope	σ_0(2200 m/sec)	R.I. (with 1/v contribtion)
^{59}Co	37.5 barns	75.0 barns
^{197}Au	98.8 barns	1550.0 barns

fied by NBS to an accuracy of ± 1.7%. Care must be taken in using this wire, however, due to the presence of ^{24}Na (from an (n,α) reaction on ^{27}Al) and from the presence of Au contamination at a level of ∿ 5 ppb. The percentage cobalt in these dilute wires is sufficiently low as to avoid flux depression.

Table 3 gives the 2200 m/sec and resonance integrals (including the 1/v contribution) for Co and Au.[23] It is apparent that ignoring the resonance integrals can lead to substantial errors in the flux measurement. We have estimated the ratio of activity induced by the resonance flux to the thermal flux for a light water reactor using the methods of the previous section. Using $k_2 = 7.8 \times 10^{11}$ n/cm^2/sec in Eq. (8) and $\psi_{th} = 4 \times 10^{13}$ n/cm^2/sec, we find that $A^r/A^{th}(^{59}Co) = 0.039$ while $A^r/A^{th}(^{197}Au) = 0.306$. The resonance integral for a Au flux wire and a light water reactor spectrum therefore contributes an additional 31% to the foil activity. The resonance contribution is 3.9% if Co flux wires are used. For very high Cd ratios found in heavy water reactors, these resonance integral effects can sometimes be neglected.

In addition to the above effects, other factors which must be considered when using flux wires for calibration are:[24]

1. Calibration of multichannel analyzers
2. Dead time corrections
3. Counting statistics
4. Detector efficiency
5. Background corrections
6. Measurement of flux wire mass
7. Corrections for irradiation time and decay before and during counting.

Any of these can lead to significant error in the flux calibration if care is not taken to understand these factors and make the appropriate corrections. As an indication of typical laboratory flux wire counting errors which are encountered even when using NBS flux wires, we have listed a typical calibration run of ten separate irradiations to the same fluence as determined by the flux integration system (10,000 integrator counts). The mean deviation in Table 4 is just about as expected for ± 1% counting statistics and ± 2% deviations in % Co in the NBS wire. It should be noted, however, that several irradiations deviate by twice the mean error. For any single irradiation where the flux is determined by flux wires only, experimental errors as large as 7 to 8% can be expected even if the resonance integrals have been taken into account. Flux integration systems, if carefully designed and calibrated, can reduce this scatter substantially.

Table 4. Typical flux wire calibration run using NBS standard reference material 953 Al (0.116% Co) flux wire.

Fluence per Integrator count ($n/cm^2/ct.$)	Deviation from mean (%)
1.446×10^{13}	-2.56
1.456	-1.89
1.491	+0.47
1.501	+1.15
1.431	-3.57
1.433	-3.44
1.529	+7.27
1.584	+6.74
1.487	+0.20
1.417	-4.51
mean = 1.484	$\lvert$mean$\rvert$ = ±3.2%

METHODS OF PHOSPHORUS DETERMINATION

The other factor needed in order to determine the production rate given by Eq. (2) is the phosphorus concentration after transmutation doping and annealing. By far the most popular method is to measure resistivity with four-point probes and then use a conversion factor to obtain phosphorus concentration. This method is independent of the degree of initial compensation in the sample. To show this, consider an initially p-type sample which is transmutation doped to n-type. Then, the initial and final carrier concentrations (assuming that the intrinsic carrier concentration, n_i, can be neglected), are given by

$$p_o = N_A^{(0)} - N_D^{(0)} , \text{ (initial)}$$

and

$$n_f = N_P + N_D^{(0)} - N_A^{(0)} \ , \text{ (final after NTD)}$$

where $N_A^{(0)}$ and $N_D^{(0)}$ are the initial acceptor and donor concentrations and $N_P = K\Phi$ is the concentration of phosphorus added by transmutation doping. The production rate is then given by

$$K = N_P/\Phi = 1/\Phi \ (n_f + p_o)$$

$$= 1/\Phi \ (K_n/\rho_f + K_p/\rho_o) \ , \ (p \rightarrow n) \tag{14}$$

where $K_n = (e\mu_n)^{-1}$ and $K_p = (e\mu_p)^{-1}$. Similar expressions are obtained for initial n-type material or final p-type material and are given as

$$K = K_n/\Phi \ (\rho_f^{-1} - \rho_o^{-1}) \ , \ (n \rightarrow n) \tag{15}$$

and

$$K = K_p/\Phi \ (\rho_o^{-1} - \rho_f^{-1}) \ , \ (p \rightarrow p) \ . \tag{16}$$

It is even possible to use a similar scheme for those cases where the intrinsic carrier concentration cannot be neglected as in the close compensation of detector material.[25]

Several errors can be introduced using the above procedure. It is essential, first of all, that the NTD material be completely annealed so that 100% electrical activity of the phosphorus is achieved. Secondly, annealing can introduce additional donors or acceptors by contamination or electrical activation of interstitial oxygen. These effects can be detected to some extend by annealing unirradiated samples as control specimens along with the NTD material.

It should also be recognized that four-point probe resistivity measurements contain inherent systematic errors which increase with resistivity.[26] At 100 Ω-cm, resistivity errors of ± 3% are usually to be expected while at 1000 Ω-cm ± 13% errors are common.[26] In fact we have obtained an approximate expression for the expected four-point probe error as a function of resistivity as defined by A.S.T.M. standards[26] which is given by

$$\text{Percent error} \approx 0.1 \ \rho^{0.71} .$$

However, we have also found through our own "round-robin" and by comparison with van der Pauw measurements, that this error can be substantially reduced if care is taken to prevent sample heating and eliminate thermal emf's.[27]

Finally, there is considerable difficulty in obtaining accurate values of the conversion factors, $K_n = (e\mu_n)^{-1}$ and $K_p = (e\mu_p)^{-1}$. These factors depend on the drift mobility, not Hall mobility. It has been our experience that different silicon manufacturers use values for these factors which differ by as much as 15%! Relatively few studies of drift and Hall mobility comparisons have been made using the high quality Si available today. From a survey of several review papers on Si, we have estimated that drift mobilities have not been determined to better than ± 5% while the ratio of Hall to drift mobilities is not known to better than approximately ± 10%.[28,29] NTD technology can ultimately contribute to a better understanding of these mobility errors. Using the data in Refs. 28 and 29, we have obtained the following conversion factors:[25]

$$K_n = 89.5 \text{ ppb} - \Omega\text{-cm}$$
$$K_p = 253 \text{ ppb} - \Omega\text{-cm} \qquad (17)$$

These factors can be converted into concentrations using 1 ppb = 5×10^{13} cm^{-3}. The estimated accuracy of these factors is about ± 5%.[25]

Other methods by which the NTD phosphorus concentration can be measured have not thus far yielded phosphorus concentrations with sufficient accuracy to challenge the accepted cross sections. Although the resistivity method above leaves one with the uncomfortable feeling that the carrier mobilities should be measured, Hall effect measurements lack absolute accuracy because of the mobility ratio inaccuracies mentioned previously.

The neutron activation analysis technique which is generally quite powerful for trace substance analysis, can be used to determine phosphorus in one of two ways. One can try to measure phosphorus directly by the reaction $^{31}P(n,\gamma)^{32}P$, however, this method is hampered by interfering effects from the transmutation process.[30] The other method is to measure the 1.266 MeV gamma which is emitted in approximately 0.07% of the decays of ^{31}Si to ^{31}P.[31] This latter method is free of interference effects, however, Hart, et al., have shown that the 0.07% branching ratio is apparently in error.[31] After a careful comparison with Hall effect measurements, they have estimated the error in phosphorus concentration to be ± 10% using this method.

Junction capacitance measurements appear to be an attractive method of determining N_P if it can be demonstrated that the diode behaves as an abrupt diode. The NTD phosphorus concentration is given by

$$N_P = \frac{2(V_{Bi} + V)}{A^2 q\varepsilon} \left\{ [C_J^{(f)}]^2 - [C_J^{(o)}]^2 \right\} \quad (18)$$

where V_{Bi} and V are the built-in and external biases, A the junction area, q the electronic charge and ε the dielectric constant of Si. The junction capacitances before and after doping are given by $C_J^{(o)}$ and $C_J^{(f)}$ respectively. The most inaccurately known quantity in Eq. (18) is the junction area. Conceptually, this could be eliminated by successive transmutation doping of equal fluence steps on the same diode. In practice, most diode structures will not survive the 850°C anneal needed to activate the NTD phosphorus electrically. The technique is also hampered by contamination problems and oxygen activation. Nevertheless, this technique appears to be promising and should be investigated further since it yields a direct measure of the dopant concentration.

Perhaps the most promising techniques are optical measurements.[32-34] At the present time, they have usually been limited in accuracy by a cross calibration with electrical measurements, however, they have the inherent virtue of measuring the phosphorus concentration directly and are insensitive to most contamination problems. Of all the optical techniques, luminescence appears to have the most sensitivity for small samples. Several papers at this conference discuss this method in detail.[32-33] To obtain phosphorus production rates to high accuracy, the optical techniques require a substantially phosphorus free starting material in order to establish absolute sensitivity factors. An estimate of the present accuracy for phosphorus determination using free and bound exciton luminescence intensities is given in Ref. 32 as ± 7%. It is felt that this accuracy can be improved by at least a factor of two with further work.[35]

MOBILITY FACTOR EFFECTS ON APPARENT PRODUCTION RATES

In this section we will discuss the consequences of using incorrect values for the parameters $K_n = (e\mu_n)^{-1}$ and $K_p = (e\mu_p)^{-1}$. We shall show that even though our knowledge of these parameters is very inexact, a simple procedure will insure that target resistivities can still be achieved to accuracies of close to ± 1% assuming we can obtain accurate fluence and resistivity measurements.

We shall assume that our starting material is p-type while after transmutation doping and annealing the material will be n-type. The equations presented will eventually be generalized to n-type starting material. According to Eq. (14), the fluence actually needed to obtain a final target resistivity, ρ_T, is

$$\Phi = \frac{1}{K}\left(\frac{K_n}{\rho_T} + \frac{K_p}{\rho_o}\right) \tag{19}$$

if the correct proportionality constants K_n and K_p and the correct production rate K are used.

When incorrect constants K_n' and K_p' are used to determine target fluence, a situation we have argued is likely, the target fluence used to irradiate the sample is given by

$$\Phi_T = \frac{1}{K}\left(\frac{K_n'}{\rho_T} + \frac{K_p'}{\rho_o}\right) . \tag{20}$$

From Eqs. (19) and (20), using incorrect values K_n' and K_p' has produced a fractional error in the fluence used to reach the target resistivity given by

$$\frac{\Phi_T - \Phi}{\Phi} = \frac{(K_n' - K_n) + (\rho_T/\rho_o)(K_p' - K_p)}{K_n + (\rho_T/\rho_o)\ K_p} \tag{21}$$

where we notice that this error is a function of the resistivity ratio ρ_T/ρ_o.

To obtain the corresponding equations for n-type starting material, we notice from Eqs. (14), (15), (19) and (20) that all we must do is perform the following substitutions:

$$\begin{cases} \text{Replace } K_p \text{ with } -K_n \\ \text{Replace } K_p' \text{ with } -K_n' \end{cases} \tag{22}$$

The fluence error for n-type starting material is then easily shown to be

$$\frac{\Phi_T - \Phi}{\Phi} = \frac{K_n' - K_n}{K_n} . \tag{23}$$

The error in this case is independent of resistivity and depends only on the error in the constant K_n.

To illustrate Eqs. (21) and (23), and all those equations which will follow, we have assumed the values given in Table 5. It should be noted that the values for K_n and K_n' and K_p and K_p' differ by approximately a factor of $3\pi/8$, a factor sometimes used for the Hall ratio if better values are not available. Values for K_n and K_p are calculated using the carrier mobilities from Sze.[36] These values[36] differ appreciably from those given in Eq. (12), again further illustrating our inadequate knowledge of these parameters.

Table 5. Constants used for the calculations in this section.

Assumed Correct Values	Assumed Erroneous Values
K_p = 208.3 ppb Ω-cm	K_p' = 248 ppb Ω-cm
K_n = 83.3 ppb Ω-cm	K_n' = 96 ppb Ω-cm
$K = 3.355 \times 10^{-18}$ ppb/n/cm^2	
1 ppb = 5×10^{13} cm^{-3}	

Plots of the fluence errors given by Eqs. (21) and (23), using the constants given in Table 5, are shown in Fig. 2. The actual fluence errors are found to be between 15 to 16% for the parameters used. These errors are typical of the Hall to drift mobility ratio errors we have assumed.

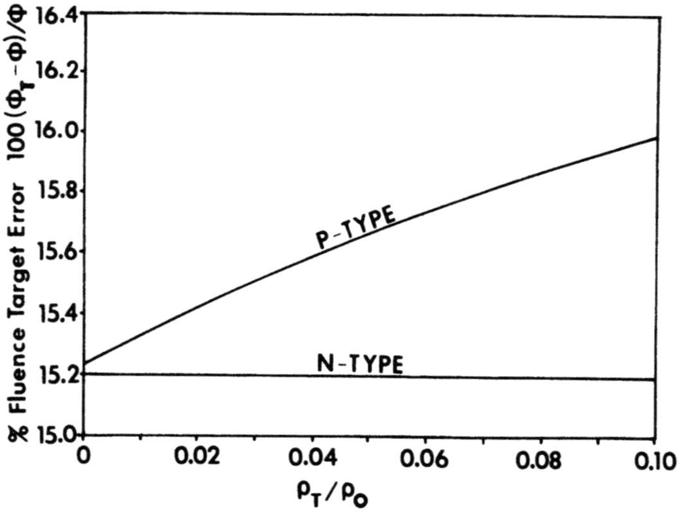

Fig. 2. Percentage error in target fluence induced by incorrect mobility factors K_n' and K_p' of Table 5.

We now wish to calculate the error in target resistivity that the above fluence error will actually produce. Using Eq. (14), we find that the final resistivity, ρ_f, which we actually produce using the erroneous fluence, Φ_T, is

$$\rho_f = \frac{K_n \rho_o}{K\,\Phi_T\,\rho_o - K_p} \tag{24}$$

Substituting Eq. (20) into (24) we find that

$$\rho_f = \frac{K_n}{\dfrac{K_n'}{\rho_T} + \dfrac{K_p' - K_p}{\rho_o}} \tag{25}$$

Therefore, the fractional error in resistivity following our initial irradiation is given by

$$\frac{\rho_f - \rho_T}{\rho_T} = \frac{(K_n - K_n') - \dfrac{\rho_T}{\rho_o}(K_p' - K_p)}{K_n' + \dfrac{\rho_T}{\rho_o}(K_p' - K_p)} \tag{26}$$

Similarly, for n-type starting material

$$\frac{\rho_f - \rho_T}{\rho_T} = \frac{1 - \rho_T/\rho_o}{\rho_T/\rho_o - \left(\dfrac{K_n}{K_n' - K_n}\right)} \; . \tag{27}$$

These two equations are plotted on Fig. 3. The observed resistivity error induced by using incorrect mobility factors, K_n' and K_p', is between 12 to 17% for the case considered for doping ratios up to 10:1.

At this point it is typical for the silicon manufacturer to assume that the ^{30}Si capture cross section is in error. The reactor personnel tend to agree since they are generally unaware of how poorly the values of K_n and K_p are known. They also have reason to believe, based on past experience with other irradiations, that the fluence measurements, cross sections and epi-thermal neutron capture could conspire to produce a 15% error in isotope production. There is also a tendency to adopt a policy in which the customer is always right.

Under these conditions, it is inevitable that the production rate parameter, K, will be adjusted to produce the desired target resistivity rather than adjusting erroneous values of K_n' and K_p'. Since the observed error in target resistivity given by Eqs. (26) or (27) is a function of the doping ratio ρ_T/ρ_o, it is clear that an adjustment of the production rate factor will produce desired results until the doping ratio is changed. We analyze this situation below.

We assume that the initial experiments have misled us into readjusting the production rate parameter from K to K'. It is assumed, then that our first experiment produced a phosphorus concentration given by

$$N_P = K' \Phi_T$$

since Φ_T was the fluence used in the experiment. By Eq. (20), we believe that

$$N_P = K' \Phi_T = \frac{K_n'}{\rho_f} + \frac{K_p'}{\rho_o} . \tag{28}$$

What actually happened is that the phosphorus concentration produced was determined by

$$N_P = K \Phi_T = \frac{K_n}{\rho_f} + \frac{K_p}{\rho_o} . \tag{29}$$

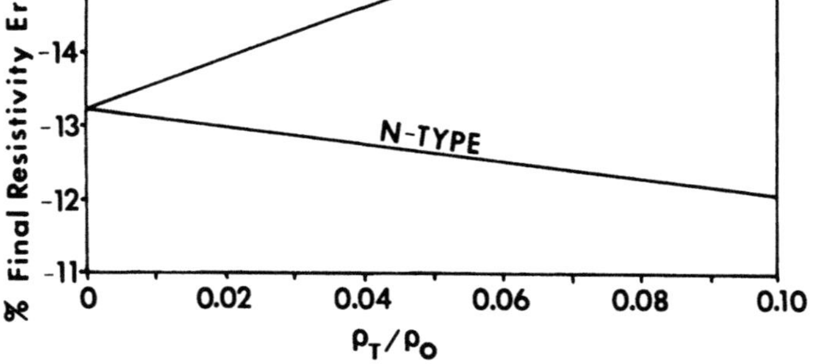

Fig. 3. Percentage error in final resistivity after first irradiation.

Taking the ratio of Eqs. (28) to (29) we can estimate the observed shift in phosphorus production rate to be

$$K' = K\left[\frac{K_n' + K_p'(\rho_f/\rho_o)}{K_n + K_p(\rho_f/\rho_o)}\right] \tag{30}$$

for initially p-type material. For n-type starting material

$$K' = K(K_n'/K_n) \ . \tag{31}$$

These effective production rates, K', are plotted in Fig. 4 assuming that the actual rate is $K = 3.355 \times 10^{-18}$ ppb/n/cm^2. We have supposedly observed an effective production rate which is 15 to 16% higher than the rate determined by calculation using the capture cross section. Smaller errors in K_n' and K_p' would produce values of K' which are correspondingly closer to those expected. It is clear that such effects are more than adequate to account for the 3 to 21% differences between the reference cross sections of Table 1 and those determined by experiment.

It should be noted from Eq. (30), that our effective production rate K' is a function of the doping ratio ρ_f/ρ_o for p-type starting

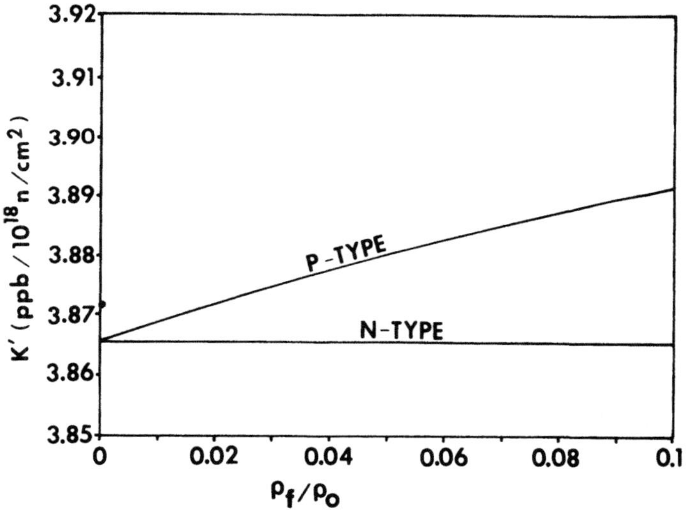

Fig. 4. Apparent phosphorus production rate when it is assumed that K_p' and K_n' are correct values.

material or for a mixture of n-type and p-type. This has the effect of causing a confusing scatter in our apparent production rate from sample to sample which will most likely be attributed either to poor probe measurements or to poor reactor performance in providing reproducible fluences. So long as the doping ratio is relatively constant, however, the above procedure will continue to produce NTD material near target.

What happens if we change the doping ratio by using a different starting resistivity ρ_o' or a new target resistivity ρ_T'? The new target fluence, Φ_T', is erroneously calculated to be

$$\Phi_T' = \frac{1}{K'}\left[\frac{K_n'}{\rho_T'} + \frac{K_p'}{\rho_o'}\right] . \tag{32}$$

What we actually will get is a new final resistivity, ρ_f', which differs from our new target, ρ_T'. As in Eq. (25),

$$\rho_f' = \frac{K_n \rho_o'}{K \Phi_T' \rho_o' - K_p} \tag{33}$$

$$= \frac{K_n}{\frac{K}{K'}\left[\frac{K_n'}{\rho_T'} + \frac{K_p'}{\rho_o'}\right] - \frac{K_p}{\rho_o'}} \tag{34}$$

after substitution of Eq. (32) into Eq. (33). We can now calculate the error between what we got, ρ_f', and what we wanted, ρ_T', as

$$\frac{\rho_f' - \rho_T'}{\rho_T'} = \frac{\left(K_n - \frac{K}{K'} K_n'\right) + \left(K_p - \frac{K}{K'} K_p'\right)\frac{\rho_T'}{\rho_o'}}{\frac{K}{K'} K_n - \left(K_p - \frac{K}{K'} K_p'\right)\frac{\rho_T'}{\rho_o'}} . \tag{35}$$

For n-type starting material, we find that

$$\frac{\rho_f' - \rho_T'}{\rho_T'} = \frac{\left(K_n - \frac{K}{K'} K_n'\right) + \left(\frac{K}{K'} K_n' - K_n\right)\frac{\rho_T'}{\rho_o'}}{\frac{K}{K'} K_n - \left(\frac{K}{K'} K_n' - K_n\right)\frac{\rho_T'}{\rho_o'}} . \tag{36}$$

from Eq. (35) by making the replacements $K_p \to K_n$ and $K_p' \to -K_n'$. Substituting Eq. (31) into (36) we find that for n-type starting material

$$\frac{\rho_f' - \rho_T'}{\rho_T'} = 0 . \tag{37}$$

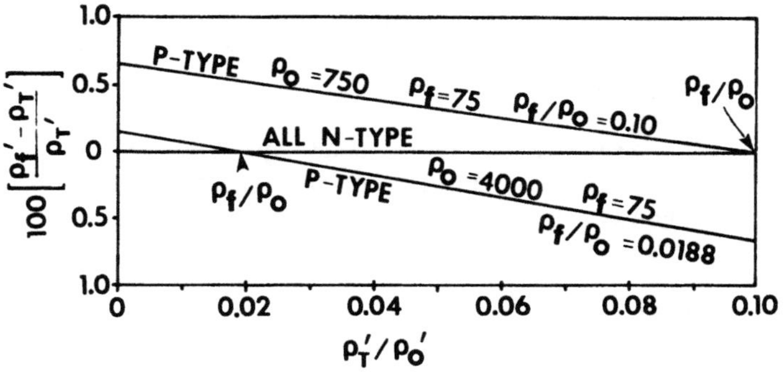

Fig. 5. Percentage error in final resistivity on second irradiation, when effective production rate K' is assumed on the basis of the first irradiation, as a function of the new target to initial resistivity ratio.

We have plotted Eqs. (35) and (37) in Fig. 5. Since the ratio of production rates K/K' is a function of our initial test resistivities ρ_f/ρ_o for p-type material, we have chosen two typical starting materials of 750 Ω-cm and 4000 Ω-cm and a final resistivity of 75 Ω-cm.

The surprising result from Fig. 5 is that the overall effect of compounding all of the above mistakes is that our final result produces only minor errors (∿ ± 0.5%) in the resultant target resistivities! We have achieved our goal by successive approximation without a precise knowledge of the mobility factors. Such a result is indeed encouraging. The above procedure will give us an apparent production rate which is in disagreement with that expected, however.

SUMMARY AND CONCLUSIONS

We have discussed a number of factors which must be taken into account in order to arrive at the production rate of phosphorus in NTD-Si. We have discussed the following:

1. There is a difference between the accepted absorption cross section for ^{30}Si which is known to about ± 2% and the effective cross section observed in NTD experiments. The NTD experiments differ by amounts ranging from a few percent to as high as 21%.

2. We have argued that epi-thermal and resonance production should yield an additional 2% phosphorus in a light water reactor and would be negligible in a heavy water reactor.

3. We have also argued, however, that epi-thermal and resonance neutrons can cause serious errors in flux wire measurements if not corrected. For gold, the error could be as high as 31% while only 4% error is expected for cobalt flux wires.

4. We have also discussed methods of phosphorus determination in Si and conclude that of the various methods, the most commonly used, i.e., resistivity, leads to perhaps the largest errors because of our poor knowledge of the mobility ratio. Unfortunately, most other techniques have relied upon electrical measurements for cross calibration.

5. In the last section we have shown that errors in the factors which convert resistivity to carrier concentration of 15 to 20% will produce similar errors in the final resistivity obtained by NTD and in the effective phosphorus production rate observed.

6. We have also shown that such errors can be overcome by successive approximation using a feedback system where new data is used to correct effective production rates. Small errors and data scatter will still occur under this system because the results are dependent on the doping ratio, however, these errors are typically lower than probing errors normally encountered if the doping ratio is relatively constant.

7. We further conclude that neutron doping coupled with a greater effort to determine phosphorus concentration (using optical methods, perhaps) can ultimately produce better data on the mobility ratio. The NTD technology can also be used to produce standard wafers for spreading resistance and four point probes to higher precision and better uniformity than has previously existed.

REFERENCES

+ Also, Departments of Physics and Electrical Engineering, University of Missouri-Columbia.

1) J. F. Mughabghab and D. I. Garber, Neutron Cross Sections vol. 1, Resonance Parameters, BNL325, 3rd ed., (Brookhaven, Upton, N.Y., 1973).
2) Table of Isotopes, 7th ed. by C. M. Lederer and V. S. Shirley. (John Wiley + Sons, Inc., New York, 1978), p. 57.
3) H. C. Schweiner, J. Appl. Phys. 30, 1125 (1959).
4) M. Tanenbaum and A. D. Mills, J. Electrochem. Soc. 108, 171 (1961).

5) J. Messier, Y. le Coroller, J. M. Flores, I.E.E.E. Trans. Nuc. Sci. NS-11, 276 (1964).
6) V. A. Kharchenko and S. P. Solov'ev, Izv. Akad. Nauk SSSR, Neorgan. Mat. 7, 2137 (1971).
7) H. A. Herman and H. Herzer, J. Electrochem. Soc. 122, 1568 (1975).
8) W. E. Hass and M. S. Schnöller, J. Electronic Mat. 5, 57 (1976).
9) H. M. Janus and O. Malmros, I.E.E.E. Trans. Electron Devices ED-23, 297 (1976).
10) H. Herzer, Semiconductor Silicon 1977, ed. by H. R. Huff and E. Sirtl, Electrochem. Soc. Proc. vol. 77-2 (Electrochem. Soc., Princeton, N.J., 1977), p. 106.
11) F. Benière and V. K. Agrawal, J. de Physique Lett. 39, L-257 (1978).
12) V. K. Agrawal and F. Benière, J. de Physique Lett. 40, L-49 (1979).
13) CRC Handbook of Chemistry and Physics, 58th edition, ed. by R. C. Weast (1977).
14) J. W. Cleland, K. Lark-Horovitz and J. C. Pigg, Phys. Rev. 78, 814 (1950).
15) D. J. Hughes, Pile Neutron Research, (Addison-Wesley, Cambridge, Mass., 1953), p. 66, 67.
16) D. J. Hughes, Ibid., p. 59.
17) D. J. Hughes, Ibid., p. 110.
18) S. L. Gunn, J. M. Meese and D. M. Alger, Neutron Transmutation Doping in Semiconductors, ed. by J. M. Meese, (Plenum Press, N.Y., 1979), p. 197.
19) R. Berliner and S. Wood, Ibid., p. 215.
20) J. E. Morrissey, T. Tillinghast, A. P. Ferro and B. J. Baligu, Ibid., p. 171.
21) J. L. Bourdon and G. Restelli, Ibid., p. 181.
22) Nuclear Engineering Handbook, 1st ed., ed. by H. Etherington (McGraw-Hill, New York, 1958).
23) Handbook on Nuclear Activation Cross-Sections, Tech. Reports Series No. 156 (I.A.E.A., Vienna, 1974), p. 17.
24) F. A. Valente, A Manual of Experiments in Reactor Physics, (Macmillan Co., N.Y., 1963).
25) J. M. Meese, J. Appl. Phys., 51, 3672 (1980).
26) A.S.T.M. Standard F84-73, 1916 Race St., Philadelphia, Pennsylvania 19103.
27) J. M. Meese, Silicon Detector Compensation by Nuclear Transmutation, Technical Report AFML-TR-77-178, (Air Force Materials Laboratory, WPAFB, Ohio, 1978).
28) M. L. Schultz, Infrared Phys. 4, 93 (1964).
29) J. Messier and J. M. Flores, J. Phys. Chem. Solid 24, 1539 (1963).
30) H. Jaskolska, L. Rowenska and L. Walis, J. Radioanal. Chem. 38, 29 (1977).

31) R. R. Hart, L. D. Albert, N. G. Skinner, M. H. Young, R. Baron, and O. J. Marsh, Neutron Transmutation Doping in Semiconductors, ed. by J. M. Meese, (Plenum, N.Y., 1979), p. 345.
32) R. B. Hammond and J. M. Meese, this volume.
33) M. Tajima and A. Yusa, this volume.
34) B. Pajot, this volume.
35) R. B. Hammond, private communication.
36) S. M. Sze, Physics of Semiconductor Devices, (Wiley, New York, 1969).

NEUTRON DOPED SILICON IN GRENOBLE REACTOR FACILITIES

P. SCIERS - J.F. VEYRAT

COMMISSARIAT A L'ENERGIE ATOMIQUE
CENG - SERVICE DES PILES
85 X - 38041 GRENOBLE CEDEX - FRANCE

INTRODUCTION

The industrial use of neutron doped silicon by means of thermal neutron irradiation has led to the development and adaptation of experimental reactors to meet the demand in this field.

Since 1976, the Reactor Department of the Grenoble Nuclear Research Centre (CEN-G) has been producing several tons of irradiated silicon each year in the MELUSINE reactor. To ensure product quality and satisfactory delivery schedules, suitable equipment procedures and controls have been set up that, in conjunction with the other C.E.A. experimental reactors, will authorize an increase in the number of irradiations in the future.

EXPERIMENTAL FACILITIES FOR THE NEUTRON DOPING OF SILICON AT THE CENG

Mélusine Reactor (Fig. 1)

The Reactor Department carries out silicon irradiations in the Mélusine reactor at the Grenoble Nuclear Research Centre. This 8 MW pool-type reactor has the following characteristics :

- Standard flat-plate MTR-type fuel elements (Uranium enriched to 93 %)
- Heavy water tank

Fig. 1. Melusine Reactor

- Hot cell
- 4-weeks operating cycle with 5-days shutdown after each cycle and one-month summer shutdown.

Irradiation Locations and Available Neutron Flux

The irradiation locations are set out around the core block, in the light water reflector or behind the heavy water tank. The location dimensions are governed by the spacing of the grid, holding the fuel elements (77 x 81 mm).

The characteristics of the silicon irradiation locations presently available are given in Figure 2 and Table 1.

Table 1. Characteristics of the silicon Irradiation Locations

Total number of facilities in the reactor	∿ 15	
Location of these facilities	Behind D_2O Tank	H_2O Pool
Number of 2-inches facilities	2	10
Maximum ingot diameter (mm)	55	55
Maximum ingot length (mm)	500	500
Number of 3-inches facilities	2	4
Maximum ingot diameter (mm)	85	85
Maximum ingot length (mm)	500	500
Number of > 3-inches facilities (possible)	0	1
$\frac{\phi th}{\phi r}$ (ϕr fast flux > 1 MeV)	∿ 600	∿ 5
Gamma heating (W/g)	0.1	0.5

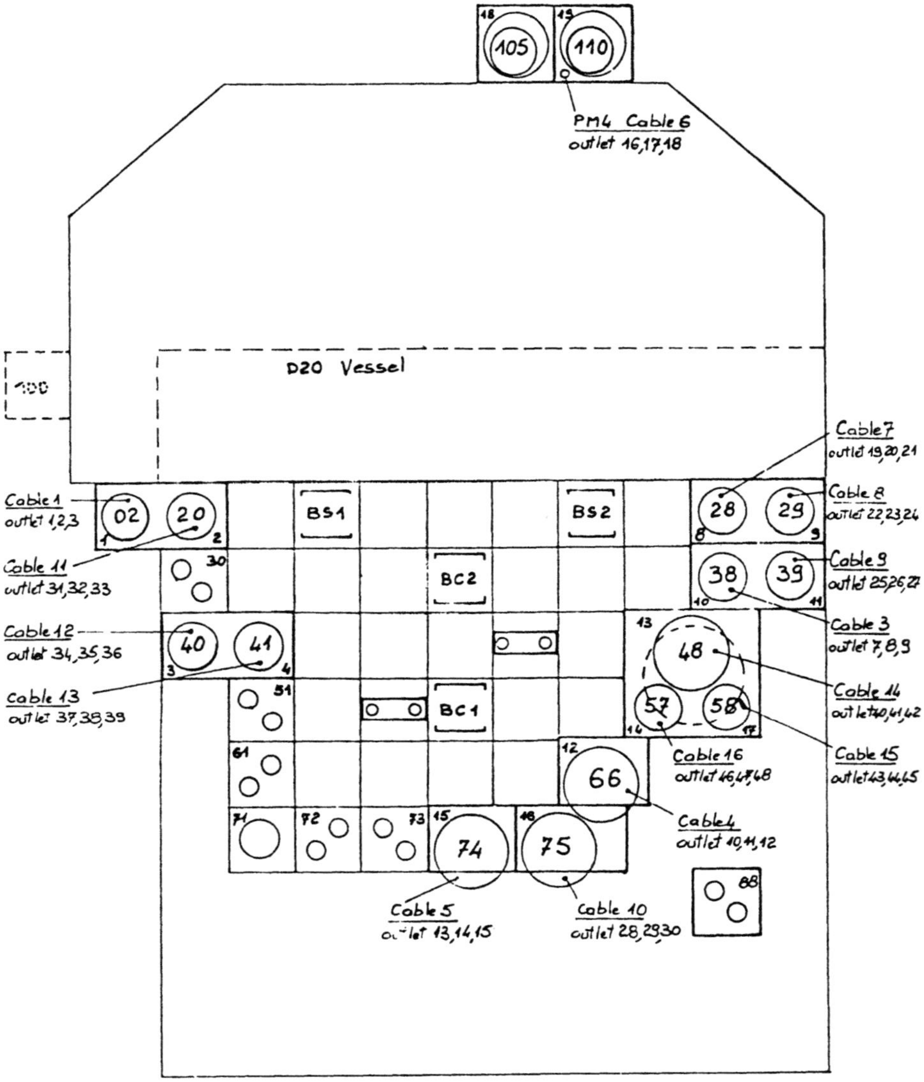

Fig. 2. Melusine core configuration

With respect to the neutron flux, the ration Øth/Ør is between 5 and 10 for all the locations on the periphery of the core and approximately 600 for the locations behind the heavy water tank.

On the periphery the thermal neutron flux is between 1.5 and 4.5×10^{13} n/cm^2.s, depending on the location ; behind the heavy water tank the flux is about 3.5×10^{12} n/cm^2.s.

The princip of a constant environment is respected to ensure a more reproducible flux and to avoid interferences between locations. For this reason, each location contains either silicon for irradiation or aluminium filling blocks (see fig. 3).

In the same way, in the vertical direction, the charging is completed by aluminium blocks above and below the 370 mm silicon ingot.

In-Pile Support Equipment (see fig. 4)

The in-pile support equipment is basically made up of :

- a tube mounted on a base plugged in the grid containing the irradiation basket.

For the dosimetry measurements, 3 rhodium self-power neutron detectors are wound around the tube at mid-height and 150 mm above and below mid-height, covering the entire circumference of the tube. A nickel chromium screen of varying thickness is used to ensure a uniform axial distribution of resistivity to within ± 5 %.

Useful dimensions for 2" ingots : 55 x 350 mm
Useful dimensions for 3" ingots : 85 x 350 mm
Useful dimensions for 4" ingots : 105 x 350 mm (construction underway).

- an irradiation basket (fig. 5 - loading the basket)

The ingots are placed in the baskets which rotate during the irradiation in such a way as to obtain a uniform radial resistivity distribution to within ± 2 %.

The baskets are centred on both sides of the reactor mid-height. They are active and are loaded and unloaded remotely under water and connected directly to the rotational drive motor by a long rod.

- a panel for the motor driving the basket, connected to a computer, on which are displayed the various signals for the start-up and completion of irradiation.

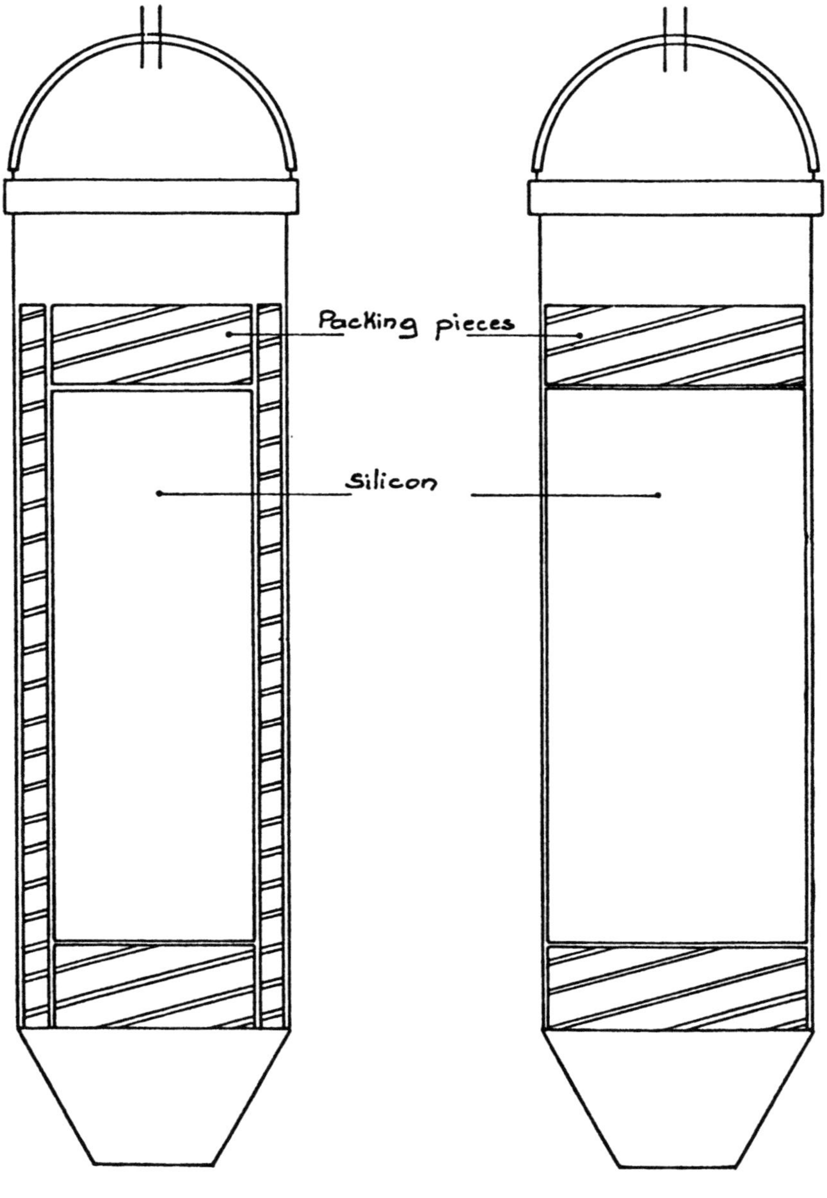

Fig. 3. SILICON LOADING WITH PACKING PIECES.

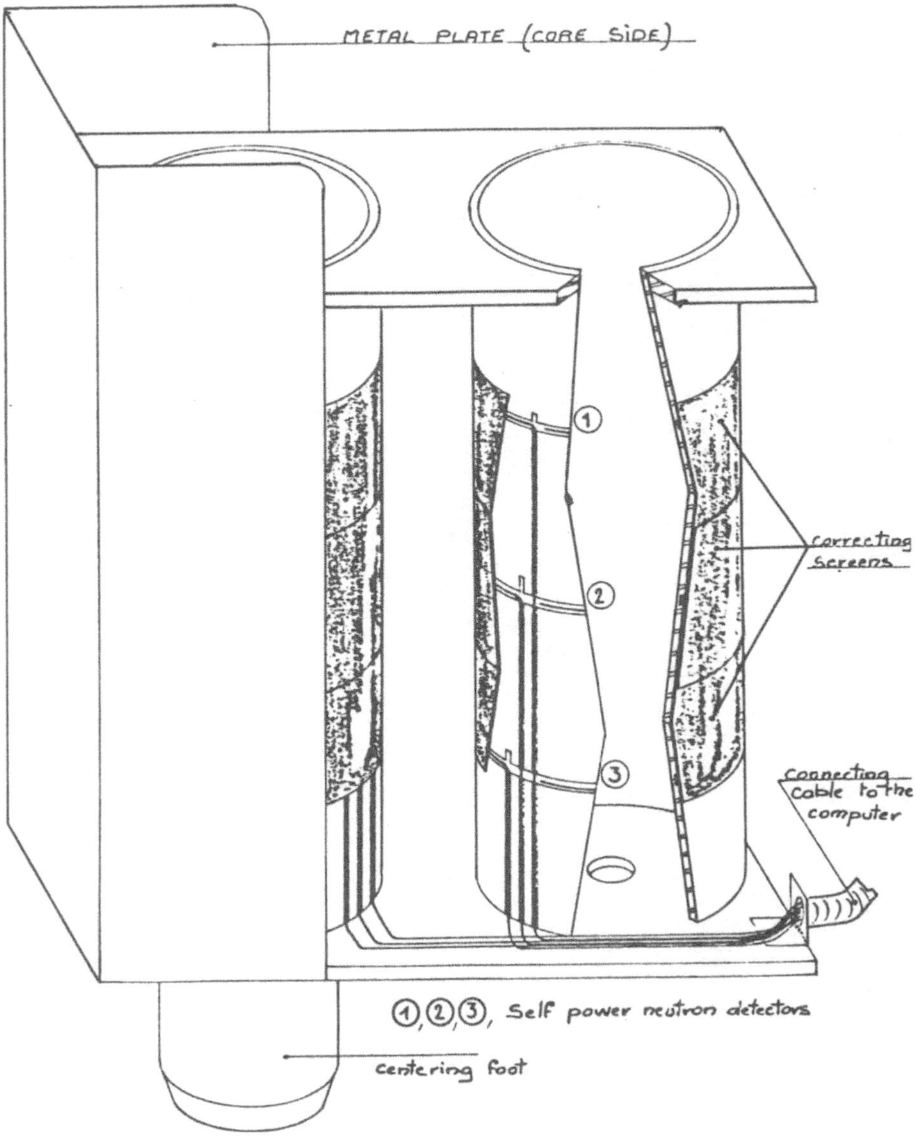

Fig. 4. Silicon in pile support

Fig. 5. Loading of the irradiation basket

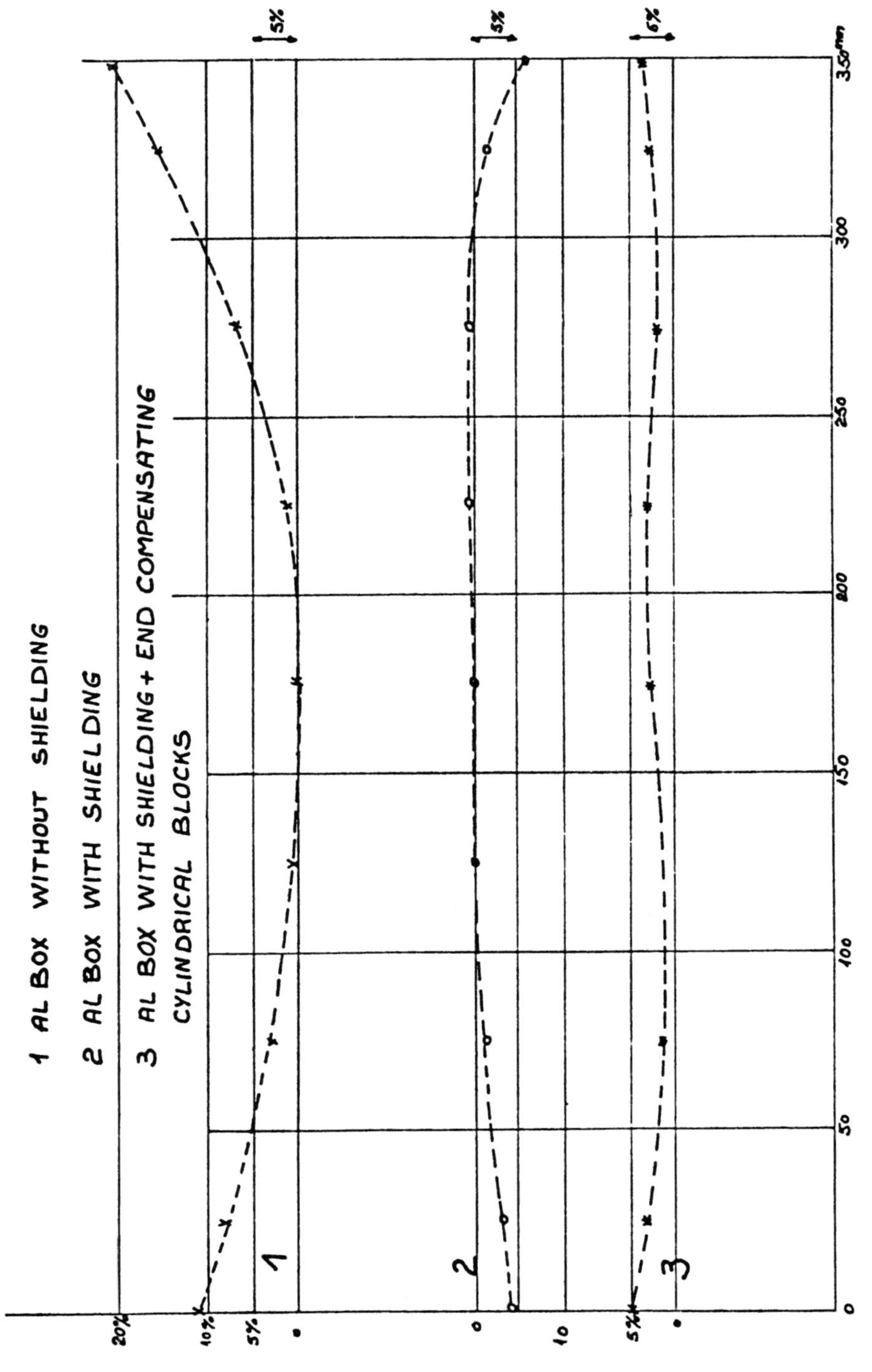

Fig. 6. RESISTIVITY VARIATION ALONG 350 mm OF SI INGOT

The diagram in figure 6 gives the axial resistivity results obtained for :

- irradiation with support and without shield
- irradiation with support and with shield
- irradiation with support, with shield and aluminium end blocks.

IRRADIATION CONTROL

Figure 7 is a general diagram of the installation showing the different components :

Digital Control Panel (Fig. 8)

The digital control bay is equipped with an INTERTECHNIQUE MULTI 20 computer with a 32 K memory, an A 43 teleprinter, a digital analog converter and a analog input rack (100 channels).

This unit is linked to the central computer of the Reactor Department. The main functions of the digital control panel are the following :

. the reading of the label for an irradiation

. recognition of an irradiation entry or exit

. irradiation start-up or stop

. calculation of the integrated current, the flux and the fluence on each self-power neutron detector

. correction for wear of the neutron detectors depending on the fluence

. calculation of the flux and the fluence received by the silicon ingots under irradiation

. storage of parameters pertaining to the irradiations to be carried out

. storage of the results pertaining to irradiations carried out

. transmission of the data report concerning each irradiation.

Control of Flux and Fluence

As already indicated, three self-powered neutron detectors are fixed at three different levels on the support holding the irradiation basket. From these detectors, the neutron detector flux, the neutron detector flux corrected for wear, the actual flux received by the silicon (calibration between the neutron detector flux, the silicon

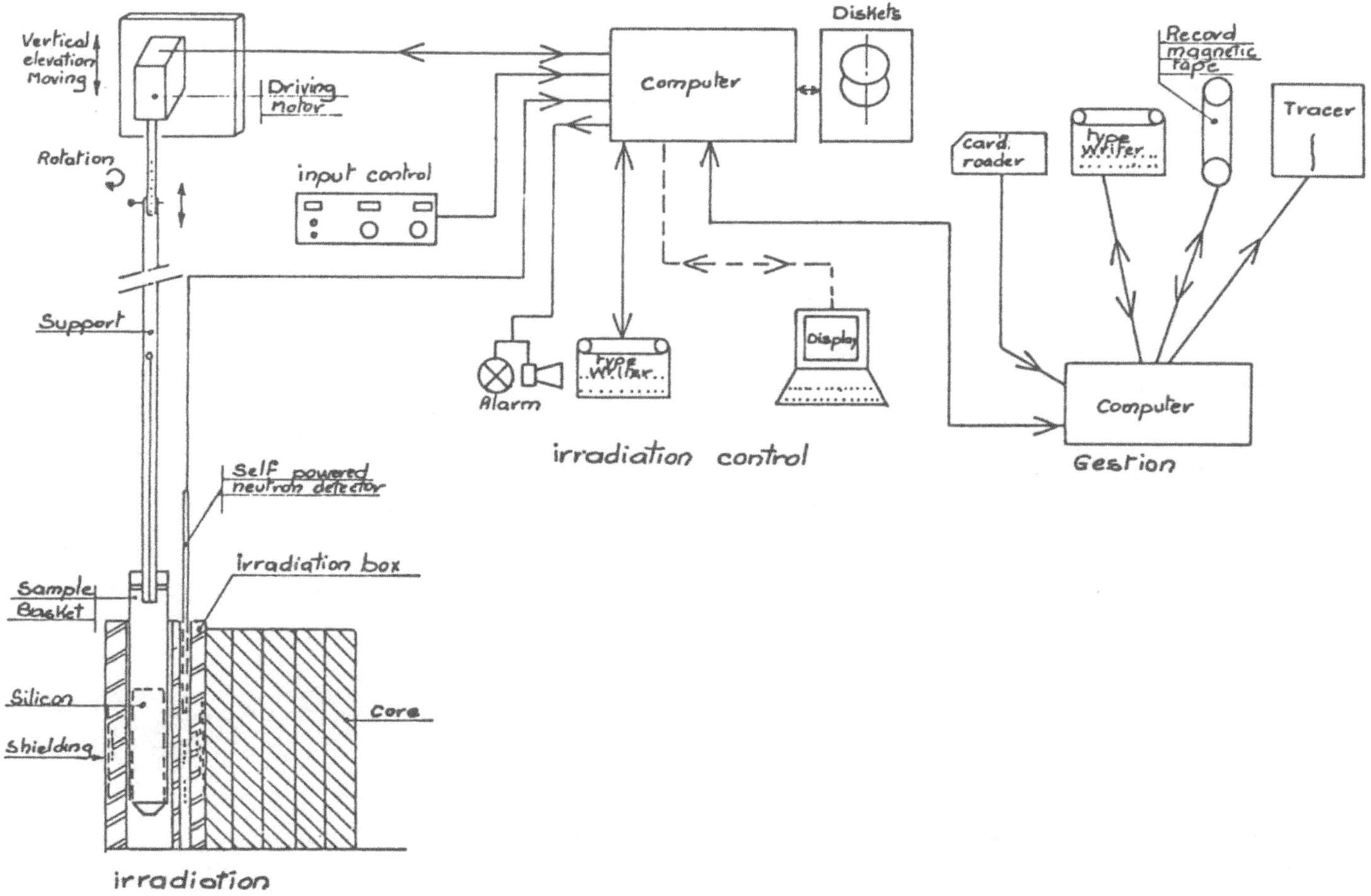

Fig. 7. GENERAL VIEW OF IRRADIATION AND CONTROL FACILITIES

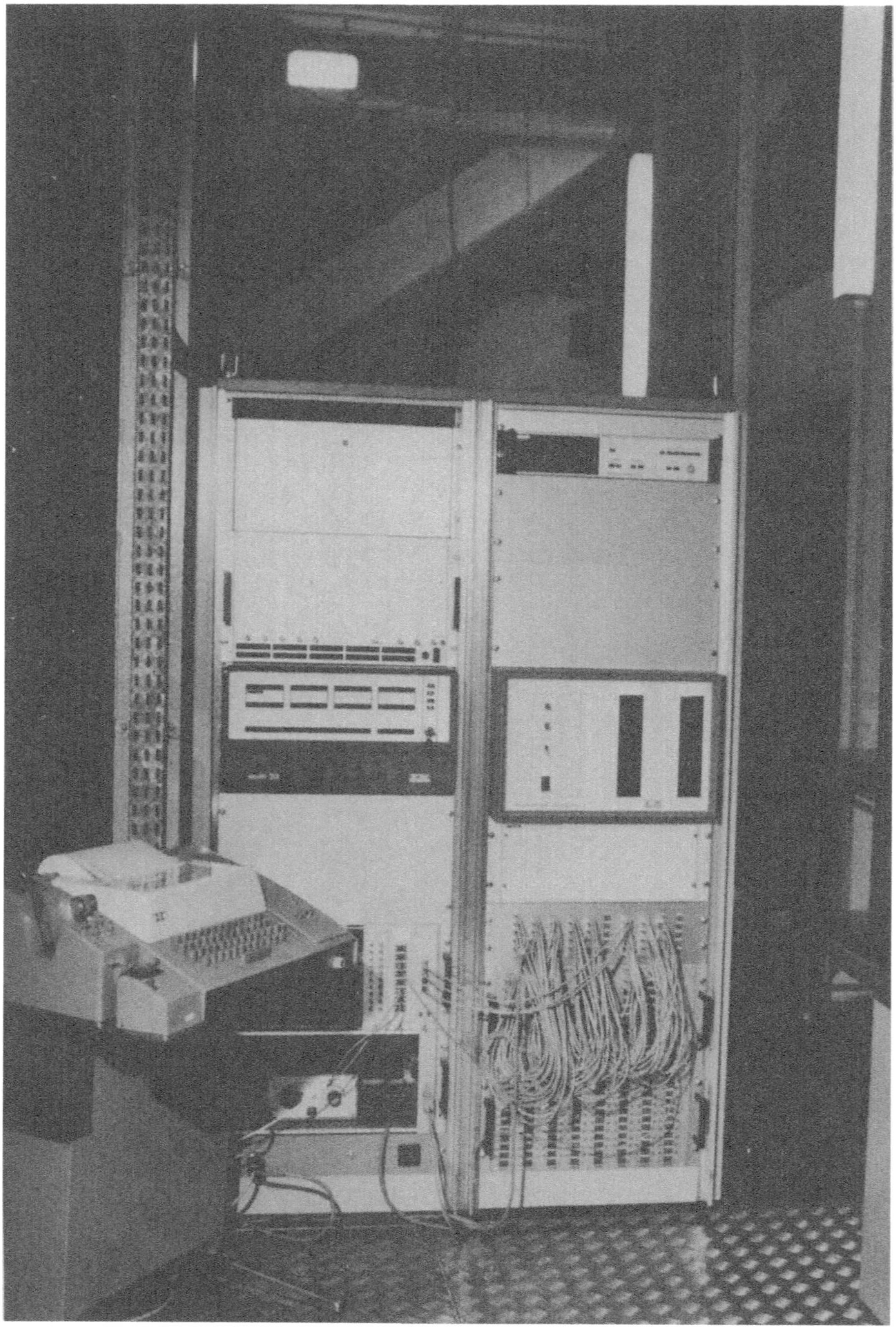

Fig. 8. Digital control panel

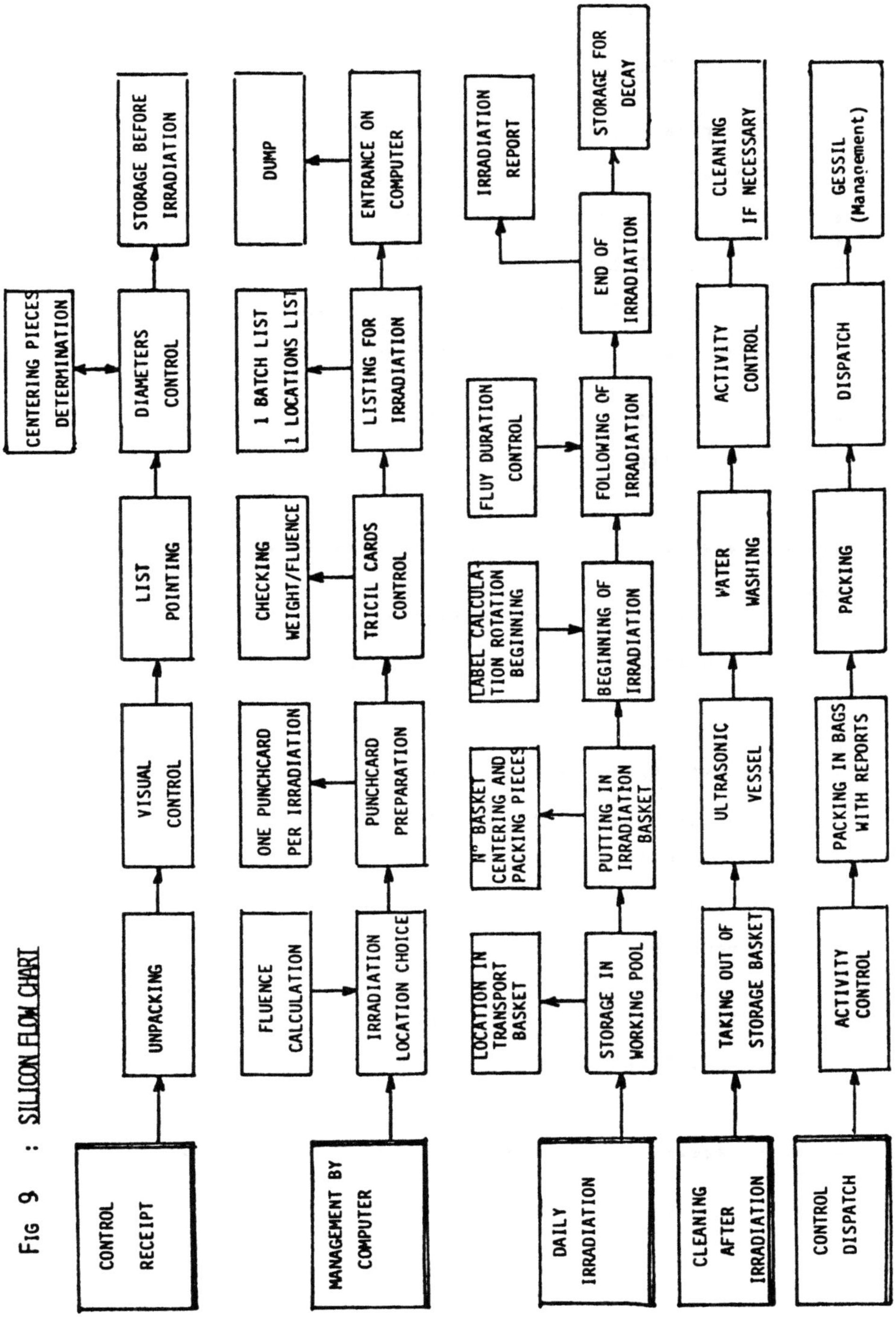

Fig 9 : SILICON FLOW CHART

```
GTM-PILES
*************************************************************
*                                                           *
*     IRRADIATION     :      1      POSITION : 074          *
*                                                           *
*     NO ECHANTILLON : ESSAI        POIDS : 3.100           *
*                                                           *
*     FLUENCE DEMANDEE :     .3600E+17                      *
*                                                           *
*     FLUENCE RECUE    :     .3630E+17                      *
*                                                           *
*     DATE D ENTREE  :    31/10/79   14H35M29S              *
*                                                           *
*     DATE DE SORTIE :    31/10/79   14H51M53S              *
*                                                           *
*     DUREE IRRAD.   :    0H 16M 24S                        *
*                                                           *
*     EC1:  -2.4%       EC3:  -2.9%       C2:    .3620E+14  *
*                                                           *
*     OBSERVATIONS :                                        *
*                                                           *
*************************************************************
```

```
GTM-PILES
*************************************************************
*                                                           *
*     IRRADIATION     :      9      POSITION : 030          *
*                                                           *
*     NO ECHANTILLON : DOS          POIDS : 1.000           *
*                                                           *
*     FLUENCE DEMANDEE :     .1000E+18                      *
*                                                           *
*     FLUENCE RECUE    :     .1006E+18                      *
*                                                           *
*     DATE D ENTREE  :    05/11/79   14H08M13S              *
*                                                           *
*     DATE DE SORTIE :    05/11/79   14H42M29S              *
*                                                           *
*     DUREE IRRAD.   :    0H 34M 16S                        *
*                                                           *
*     EC1:   1.4%       EC3:  -2.1%       C2:    .4904E+14  *
*                                                           *
*     OBSERVATIONS :                                        *
*                                                           *
*************************************************************
```

Fig. 10. Irradiation results.

flux and the silicon resistivity values) and the fluence are calculated on-line and appear in the data report for each irradiation.

Silicon Circuit

The sequence of operations guaranteeing the quality of the production is illustrated in figure 9 :

- In-coming acceptance inspection
- Transfer to computer control
- Irradiation
- Decontamination and cleaning
- Inspection and shipment

In this sequence, the manipulations carried out underwater require special attention to ensure proper positioning of the ingots.

The delivery schedules are governed by the maximum activity required on delivery i.e. 2×10^{-3} μ Ci/g. Each irradiated ingot is accompanied by an irradiation data report in its packaging (see fig. 10).

CONCLUSION

After thorough experimental and development work, and with the facilities made available on the Mélusine reactor, the Reactor Department in Grenoble is capable of fulfilling the demand for neutron doped silicon.

Results obtained to date in supplying the silicon industry have shown our capacity to supply a high quality irradiated product within acceptable delivery schedules.

CHARACTERIZATION OF NTD SILICON IRRADIATED IN GRENOBLE REACTOR FACILITIES

D. Roche, F. Blanc and M. Vilain

LETI - Commissariat à l'Energie Atomique

85X - 38041 - GRENOBLE - Cedex, FRANCE

INTRODUCTION

Since 1976, the LETI (Laboratoire d'Electronique et de Technologie de l'Informatique) studies the performances of NTD-silicon irradiated in different reactors of the CEA (in Grenoble and in Saclay)+.

The factors investigated have been :

- the irradiation characteristics : ratio of thermal to fast neutron flux between 5 and 3,000, doping precision, influence on the radial homogeneity of silicon crystal rotation around its axis, amelioration of the vertical homogeneity,
- the optimization of annealing heat treatment cycle,
- the measurements of electrical parameters after annealing : resistivity by the four-point probe method, and its microvariations by spreading resistance probe, mobility of carriers, evolution of the minority carrier lifetime.

We have also compared performances of power-devices manufactured with conventional FZ silicon and NTD silicon, and pointed out the increase in manufacture yield with NTD silicon.

Some preliminary results have been reported at the 3rd Symposium on Silicon Materials Science and Technology-Semiconductor Silicon - 1977 (1).

Table 1. Experimental conditions

Ingot number / Characteristics	115	62		67	
		Top	Bottom	Top	Bottom
Size	small bars	cylinder			
Dimensions	$S = 1.5\ cm^2$, $L = 5cm$	diameter: 54 mm, length: 40 mm			
Initial type	p	p	n	n	n
Initial resistivity ($\Omega \times cm$)	20,000 - 40,000	about 10^5	2,000-3,300	2,000	1,000-1,600
Irradiation	Water pool	Behind heavy water tank			
Thermal/fast	10	600			
Thermal neutron fluence (cm^{-2})	10^{16} to 10^{18}	about 5×10^{16}		about 3×10^{16}	
Annealing conditions	quartz tube with flowing nitrogen 740°C x 1hr slow cooling	quartz tube with flowing argon 700°C x 1hr slow cooling			

LETI is now interested in the doping with low phosphorus concentration (10^{12} to 10^{13} cm^{-3}) in order to compensate residual boron concentrations in high purity silicon.

Two experiments of NTD silicon irradiated in the 8 MW pool-type reactor Melusine of CEN-Grenoble are presented. First, a characterization of the doping precision, and then, homogeneity measurements on NTD silicon with low doping.

EXPERIMENTAL PROCEDURE

The high purity FZ silicon used for experiments was grown in LETI. The starting characteristics of the three ingots are listed in table 1.

One ingot was cut in five small bars (1.5 cm^2 in section, 5 cm in length) and irradiated in water pool with thermal neutron fluences, ranging from 10^{16} to 10^{18} cm^{-2}. At the irradiation location, the thermal neutron flux is about 1.5×10^{13} $cm^{-2} \times s^{-1}$, and the ratio of thermal to fast neutrons is 10 : 1. The other two ingots (54 mm in diameter) were cut in 40 mm length samples, chemical polished, and irradiated behind heavy water tank in location where thermal neutron flux is about 4.5×10^{12} $cm^{-2} \times s^{-1}$ and ratio of thermal to fast neutron is 600 : 1.

In each case, cobalt detectors were positionned near the samples in order to determine the thermal neutron fluence.

After irradiation, decrease of residual radioactivity and chemical etching, the samples were annealed following the conditions listed in table 1. The small bars were cut and polished in 2 mm thick samples and the final resistivity was measured on the sample taken in the middle.

In the case of the two other ingots, four point probe resistivity measurements were performed along two reference diameters (on the top and the bottom of the samples), before and after irradiation and annealing.

DOPING PRECISION

The results of thermal neutron fluence from cobalt detector dosimetry and final carrier concentration (N_D - N_A) from resistivity are listed on table 2.

Table 2. Characterization of the doping

Ingot	Initial carrier concentration (cm^{-3})	Sample	Thermal neutron fluence (cm^{-2}) [1]	Final carrier concentration $[N_D - N_A]$ (cm^{-3}) [2]	^{31}P concentration (cm^{-3}) [3]
115	$(6 \pm 2) \times 10^{11}$ p-type	1	1.01×10^{18}	$(1.65 \pm 0.01) \times 10^{14}$	$(1.66 \pm 0.1) \times 10^{14}$
		2	4.09×10^{17}	$(6.69 \pm 0.02) \times 10^{13}$	$(6.7 \pm 0.1) \times 10^{13}$
		3	1.15×10^{17}	$(1.75 \pm 0.01) \times 10^{13}$	$(1.80 \pm 0.04) \times 10^{13}$
		4	4.04×10^{16}	$(5. \pm 0.2) \times 10^{12}$	$(5.6 \pm 0.03) \times 10^{12}$
		5	1.23×10^{16}	$(3.3 \pm 0.2) \times 10^{11}$	$(9 \pm 2) \times 10^{11}$
62	about 10^{11} p-type	top	5.35×10^{16}	$(9. \pm 0.2) \times 10^{12}$	$(9.1 \pm 0.02) \times 10^{12}$
67	$(4 \pm 1) \times 10^{12}$ n-type	bottom	2.93×10^{16}	$(8.5 \pm 0.2) \times 10^{12}$	$(4.8 \pm 0.4) \times 10^{12}$

[1] from cobalt detectors dosimetry

[2] from final resistivity with $\mu_n = 1{,}360\ cm^2.V^{-1}.s^{-1}$

[3] final carrier concentration after correction for initial doping

Table 3. Homogeneity of neutron transmutation low doping of two silicon ingots.

Sample / Characteristics	62		67	
	Top	Bottom	Top	Bottom
Initial type	p	n	n	n
Initial resistivity ($\Omega \times cm$)	10^5	2,600	2,000	1,300
Initial homogeneity (%)	50	26	10	23
Final resistivity ($\Omega \times cm$)	510	440	610	520
Final homogeneity (%)	2	5	6	8
^{31}P added (cm^{-3})	about 9×10^{12}		about 5×10^{12}	

The final concentrations (from a few 10^{11} cm $^{-3}$ to 10^{14} cm^{-3} range) have been determinated with the assumption of a constant electron mobility μ_n : 1360 $cm^2.V^{-1}.s^{-1}$. This value was measured by Hall effect in LETI for NTD silicon with 10^{14} cm-3 ^{31}P doping and is in agreement with values given by Runyan (2).

The ^{31}P concentration added by irradiation has been determined after correction for initial carrier concentration and is plotted versus thermal neutron fluence on figure 1. A good agreement with theoretical linear variation is found for samples irradiated with thermal neutron fluence higher than 2.5 x 10^{16} cm^{-2}. The samples 5 and 6 of ingot 115 give lower values than theoretical ones deduced from least squares fit on the other measurements. This fact can be attributed to the poor knowledge of the starting material carrier concentration, which had been measured on an other sample. We have verified (by resistivity measurement) on a non-transmuted sample that the annealing process does not affect the initial carrier concentration.

We have determined by least squares method the doping factor K, linking ^{31}P concentration with thermal neutron fluence :

$$K = 1.65 \times 10^{-4}\ cm^{-1}$$

The doping factor K can be identify with the theoretical ^{30}Si macroscopic thermal neutron cross-section.
Theoretical value for Σ is calculated from :
σ = (105$\pm$ 5) mb. (3), (^{30}Si) = 1.54 x 10 ^{21} cm^{-3}, 3.09 % isotopic abundance, ρ = 2.33 g x cm^{-3}, Si = 28.09 a.m.u., and gives :

$$\Sigma = (1.62 \pm 0.08) \times 10^{-4}\ cm^{-1}.$$

The comparison between K and Σ shows a good agreement.

From these experiments we conclude, that the assumptions taken for the electron mobility and the microscopic thermal neutron cross-section of ^{30}Si allow us to predict with precision the phosphorus produced by neutron transmutation, and to determine the exact amount of fluence needed to exactly compensate high purity silicon.

DOPING HOMOGENEITY

For ingots 62 and 67, we have studied the low doping homogeneity and we give on table 3, the mean resistivity and its variation before and after irradiation (on top and bottom of the samples).

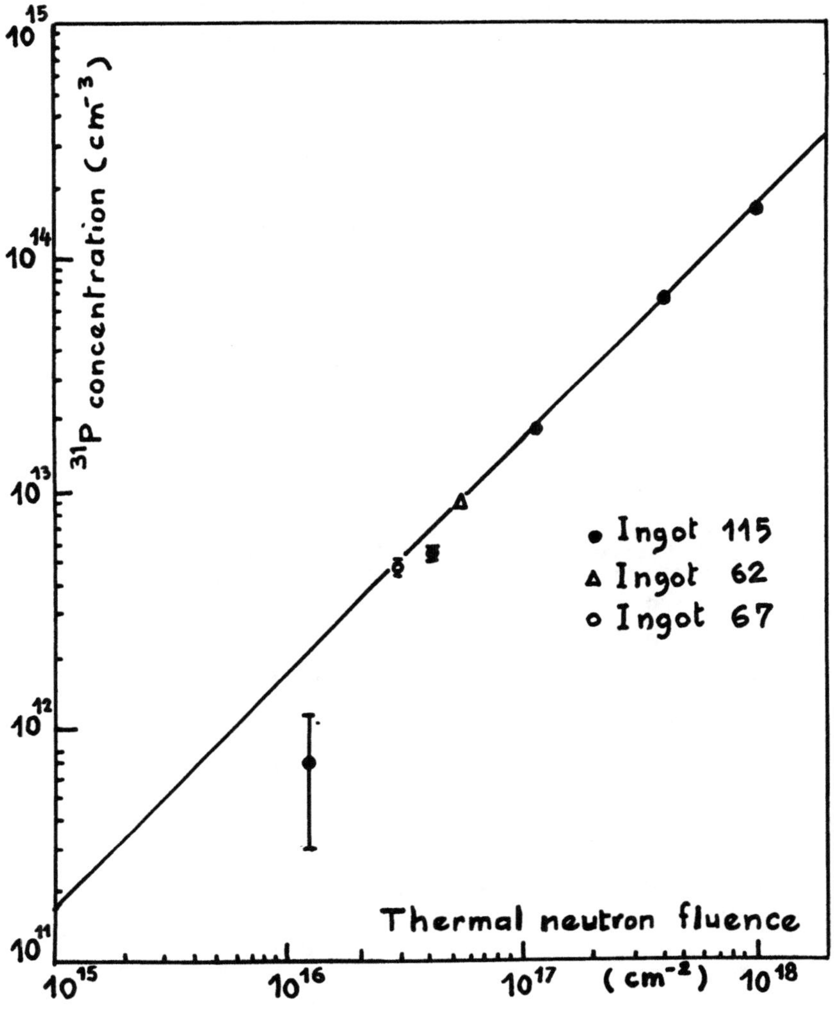

Fig. 1. Variation of neutron produced concentration (from resistivity) versus neutron thermal fluence (from dosimetry).

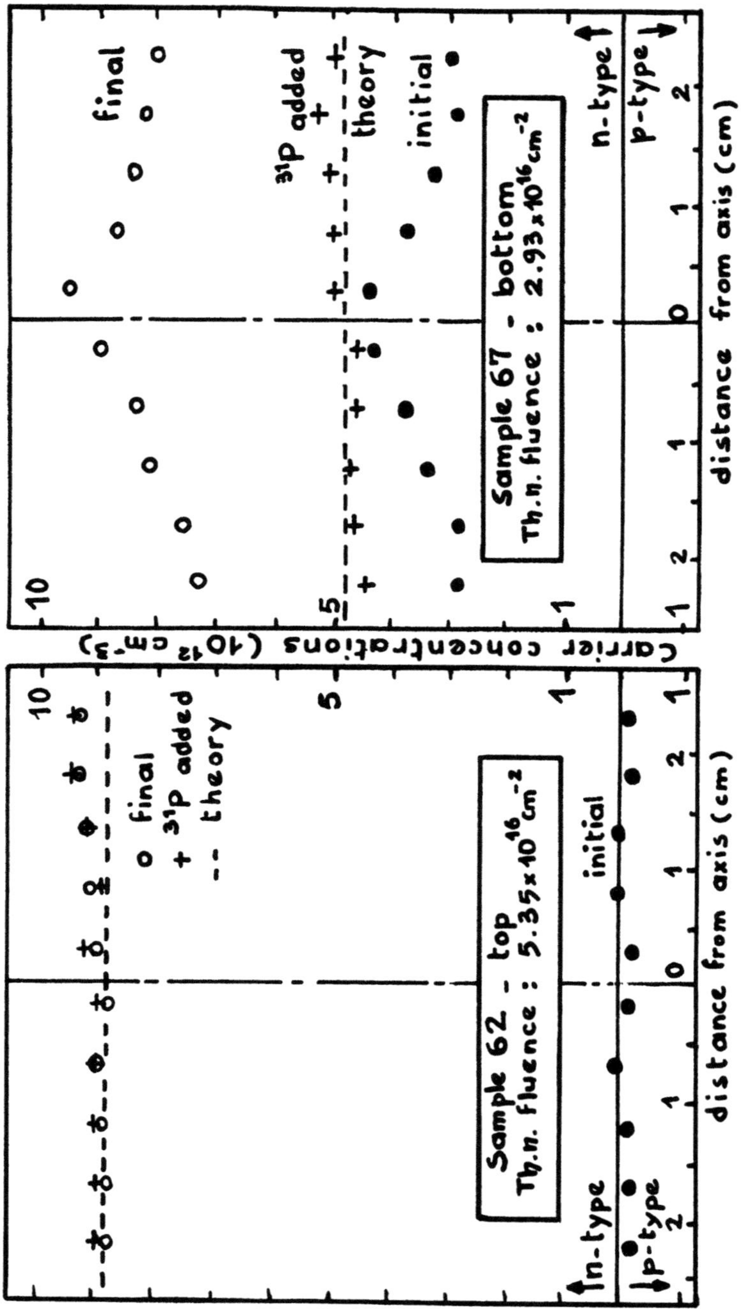

Fig. 2. Variations of carrier concentrations along a diameter before and after neutron transmutation of samples 62 and 67.

On figure 2, are plotted the variations of initial and final carrier concentrations and of ^{31}P added, along the same diameter in the case of samples 62-top and 67-bottom.

This shows that irradiation is fairly homogeneous.

The absolute variations of concentration are maintained, but, as soon as the doping due to irradiation is high enough, the relative variations of carrier concentrations become very low. This effect is evident in the case of sample 62.

CONCLUSION

These characterizations of NTD silicon irradiated in Melusine-reactor show that neutron transmutation is an adequate method for low doping of high resistivity silicon. In case of compensation of residual boron concentration, the initial characteristics of the material should be precisely known in order to predict accurately the irradiation conditions.

In any cases, inhomogeneities in the ingot should be less than the desired residual doping after compensation. This requirement remains the more difficult to fill for neutron compensation of high purity silicon.

REFERENCES

+ Research supported by the Délégation Générale à la Recherche Scientifique et Technique of the French Government in connection with the french manufacturers of power devices.

1. A. Senes and G. Sifre, M. Breant - Procedings of the Third International Symposium on Silicon Materials Science and Technology, Semiconductor Silicon (1977).

2. W.R. Runyan, Silicon Semiconductor Technology, Mc Graw Hill.

3. R. Sher, Handbook on Nuclear Activation Cross-Sections, I.A.E.A., Vienna (1974).

A PRELIMINARY STUDY ON NTD-SILICON

Lu Cuengang and Li Yaoxin

Laboratory of Fuel and Material
Institute of Atomic Energy
Beijing, Peoples Republic of China

INTRODUCTION

Some preliminary works on NTD-silicon have been done in a 3.5 MW swimming pool reactor since September 1979. In part A of this article, a brief account of the work is given, including:

1. irradiation conditions and irradiation facility;
2. the discrepancy and uniformity of resistivity of NTD-silicon;
3. methods for cleaning and identification of NTD-silicon radioactivity;
4. batch annealing results as done in the Institute of Non-ferrous Metal.

In part B, a report on separate annealing experiment is included.

PART A EXPERIMENTS AT SWIMMING POOL REACTOR

by Lu Cuengang

Neutron transmutation doping experiments with over 100 silicon samples have been done since last September. Samples used include float zone silicons crystallized under vacuum, under hydrogen or argon atomsphere. The sample diameters range from ϕ20- ϕ65 mm, with length 20-200 mm and final resistivity 30-350 Ω-cm. some preliminary results have been obtained.

IRRADIATION FACILITY AND CONDITIONS

Fig. 1 shows the horizontal section of the reactor core at which work on NTD-silicon has been done. Silicon samples were irradiated at D-11 and No.4 channels. Some of the samples were irradiated in the thermal column in order to evaluate the effect of cadmium ratio of the neutron flux. Table 1 gives the main characteristics of these channels. D-11 channel having a low irradiation temperature can accommodate samples with diameters less than 60 mm, while No.4 channel, with sample diameters less than 90 mm.

Presence of silicon samples changes neutron flux in the channel. Fig. 2 shows the axial distribution of neutron flux with and without silicon samples. Flux distribution in D-11 tends to be flattened after loading of silicon sample. In F-11 channel loading of silicon sample of Ø30 mm with length 63 mm and of Ø40 mm with length 400 mm increases the neutron flux by 5.2% and 14% respectively. Therefore, loading of silicon samples has a significant effect on the channel neutron flux, and this effect is different with different loading pattern. In view of this, cobalt-aluminium alloy foils were used to monitor the fluence, and later on self-powered detectors will be used. No axial flux flattening device has yet been used.

Fig. 3 shows a schematic diagram of a simple rotating facility used in D-11 and No.4 channels. Silicon sample and Co-Al foils wrapped with Al foil were placed in a aluminium container which was driven by a SD-09 motor at a rotating speed of 9 rpm during irradiation so as to maintain the radial uniformity.

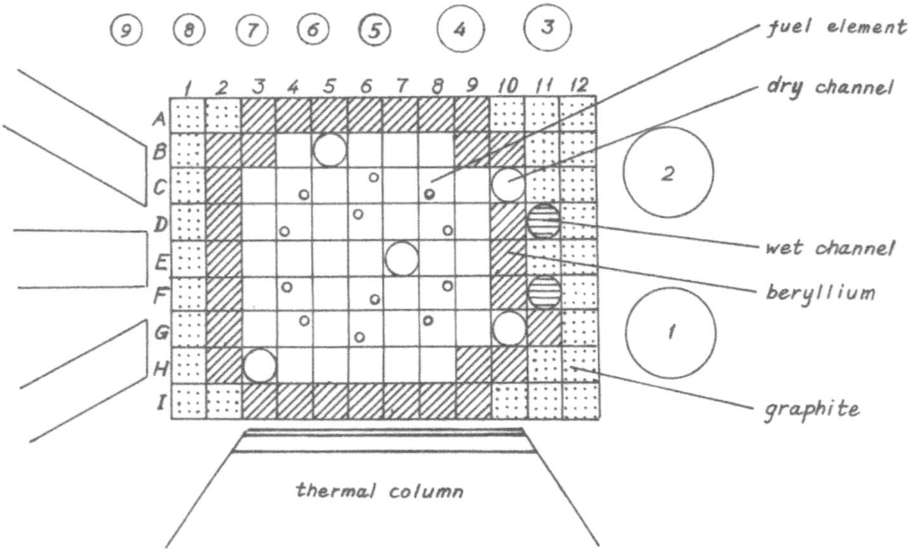

Fig. 1. Cross-sectional view of the swimming-pool type reactor

Table 1. Main data of channels for silicon sample irradiation

Channel	Location	Inner diameter (mm)	Thermal neutron flux (n/cm s/3.2MW)	Cd ratio (Au)	Irradiation temperiture (°C)	Dry or wet
D-11	in reflector	64	1.8×10^{13}	7.5	40	wet
No.4	pool place near reflector	100	3.4×10^{12}	9.3	130	dry
Thermal column	in graphite lump	150	1.1×10^{11}	320		dry

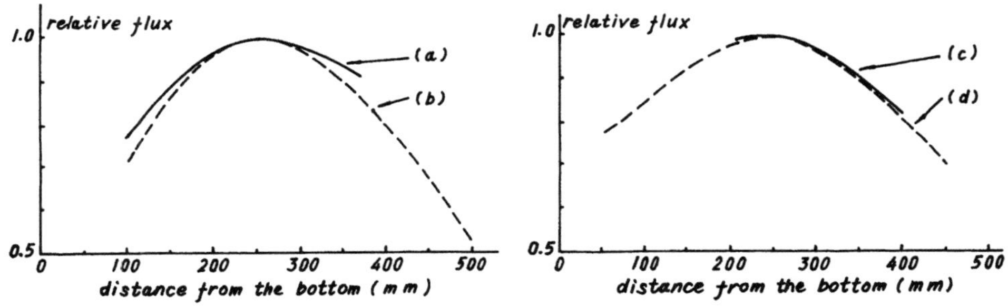

Fig. 2. Axial distribution of neutron flux in D-11 and No.4
(a) In D-11 with Ø40×260 mm² silicon sample;
(b) In D-11 without sample;
(c) In No.4 with Ø45×220 mm² silicon sample;
(d) In No.4 without sample.

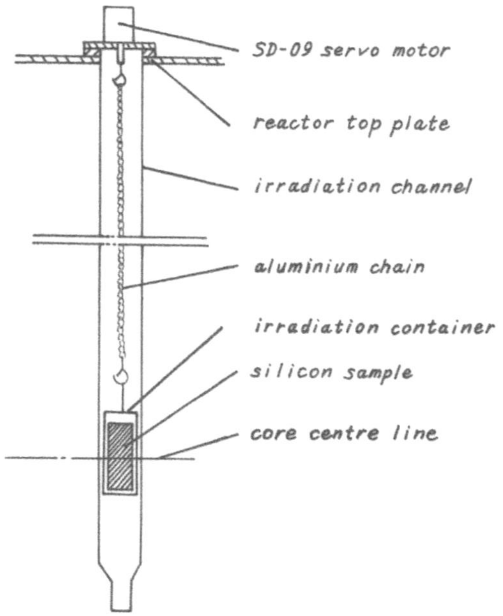

Fig. 3. Schematic diagram of the rotating irradiation facility

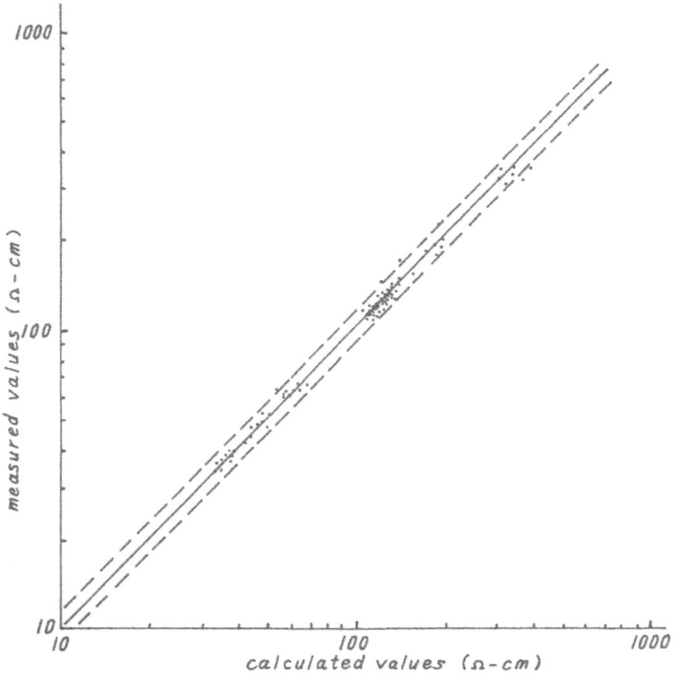

Fig. 4. Correspondence of measured and calculated values of resistivity

DISCREPANCY OF RESISTIVITY AFTER DOPING

The sample resistivity ρ_t after irradiating in a fluence of Φ can be calculated from the following formulas:[1)]
for n-type sample having initial resistivity ρ_n,

$$\rho_t = \frac{1}{e\mu_d[^{30}Si]\sigma\Phi + \frac{1}{\rho_n}} = \frac{1}{3.65\times10^{-20}\Phi + \frac{1}{\rho_n}} \qquad (1)$$

and for p-type sample having initial resistivity ρ_p,

$$\rho_t = \frac{1}{e\mu_d[^{30}Si]\sigma\Phi - \frac{1}{\rho_p}\cdot\mu_d/\mu_a} = \frac{1}{3.65\times10^{-20}\Phi - \frac{3}{\rho_p}} \qquad (2)$$

Φ values were taken from the activity of the Co-Al foils. Resistivity were measured with a four-point probe. The discrepancy between measured value ρ_m and calculated value ρ_t is given by the expression:

$$\Delta = \frac{\rho_m - \rho_t}{\rho_t} \times 100\% \qquad (3)$$

Fig. 4 gives a comparison of ρ_m and ρ_t. The solid line in the figure corresponds to the case $\Delta=0$, while dotted lines correspond to $\Delta=\pm10\%$.

The discrepancies are considered to be resulted from the quality of original materials, the exact annealing processes and the measuring accuracy. Further improvements are to be expected.

RADIAL UNIFORMITY OF NEUTRON DOPING

The resistivity nonuniformity is defined by:

$$\Delta\rho = \frac{\rho_{max} - \rho_{min}}{\rho_{min}} \times 100\% \qquad (4)$$

It is clearly seen from formulas (1) and (2) that nonuniformities of fluence Φ and initial resistivity ρ_n or ρ_p cause the nonuniformity of the final resistivity ρ_t.

Effect of Fluence Nonuniformity

If sample is irradiated at a fluence with nonuniformity

$$\Delta\Phi = \frac{\Phi_{max} - \Phi_{min}}{\Phi_{min}} \times 100\% \qquad (5)$$

then the resistivity nonuniformity $\Delta\rho_t$ in relation to $\Delta\Phi$ after doping may derived from formulas (1) and (2).
For n-type initial material,

$$\Delta\rho_t = \frac{K - 1}{\frac{K}{\Delta\Phi} + 1} \tag{6}$$

And for p-type initial material,

$$\Delta\rho_t = \frac{K + 3}{\frac{K}{\Delta\Phi} - 3} \tag{7}$$

here K is the doping ratio, defined as

$$K = \frac{\rho_{o\,min}}{\rho_{t\,min}} \tag{8}$$

where ρ_o can be either ρ_n or ρ_p.

$\Delta\rho_t = \Delta\Phi$ for large values of K.

We measured the ^{31}Si β-activity profile across the sample end surface as an index for the radial uniformity of doped phosphorus.[2] The sample used was a Ø44 mm short rod rotating during irradiation. Fig. 5 shows β-activity profile measured by a Si(Li) detector with effective diameter of 3 mm. The nonuniformity is seen to be less than 2%, including the errors in measurements.

Effect of Resistivity Nonuniformity of Initial Material

It can be easily seen that for n-type material,

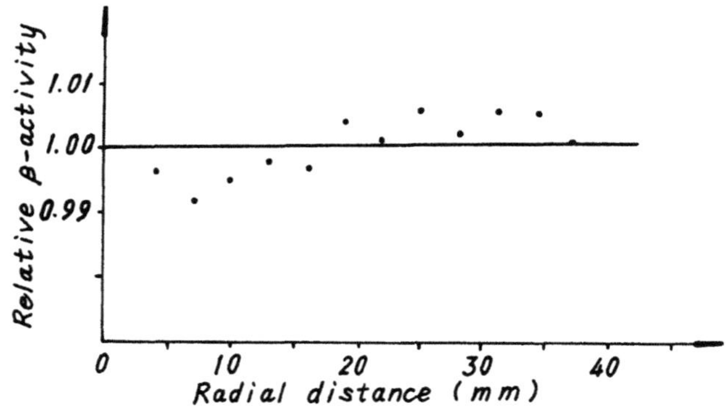

Fig. 5. ^{31}Si β-activity along the radial direction of silicon sample

$$\Delta\rho_t \doteq \frac{1}{(K-1) + \frac{K}{\Delta\rho_n}} \tag{9}$$

and for p-type material,

$$\Delta\rho_t = \frac{1}{(\frac{K}{3}-1) + \frac{K/3}{\Delta\rho_p}} \tag{10}$$

It is evident that the larger the K values, the less influence on $\Delta\rho_t$ will result from the resistivity nonuniformity of initial material.

Table 2 shows some of the results. Discrepancies between measured and calculated values are within the error of a four-point probe.

RADIOACTIVITY OF SAMPLES

The radioactivity of irradiated samples comes from both surface contamination and activation of impurities existed in the samples themselves.

A series of experiments were made to compare various cleaning method for removing the surface contamination. Table 3 lists the results. It is seen that with the methods used etching with HNO_3/HF mixed solution is most effective.[3] The activity fell below 2×10^{-5} μCi/cm^2.

Table 2. The effect of initial material on the radial uniformity

Initial resistivity $\rho_{o\,min}$ (Ω-cm)	Nonuniformity $\Delta\rho_o$ (%)	Resistivity after doping $\rho_{t\,min}$ (Ω-cm)	Doping ratio K	Nonuniformity $\Delta\rho_t$ (%)	
				calculated	measured
P-1300	15	183	7	5.9	6.0
P-1500	20	170	9	5.9	4.7
P-1750	17	183	10	4.6	1.6
P-2400	38	115	21	4.1	4.0
P-3500	6	130	27	0.6	1.9
P-3900	67	105	37	3.4	7.1
P-5500	32	111	50	1.5	7.7
P-7050	106	112	63	2.5	1.3
P-4150	59	49.5	84	1.3	5.0
N-102	34	33.5	3.0	9.0	13.0
N-160	24	35.5	4.5	4.5	8.2
N-550	118	109	5.0	12	9.2
N-913	284	119	7.7	11	8.0

Table 3. Comparison of post-irradiation cleaning methods for silicon sample

No. of sample	Decay time	Cleaning method	Results measured	Remarks
1-6-2	5 days	Supersonic cleaning 15 min. with 10% citric acid	^{51}Cr, activity 0.05 $\mu Ci/100cm^2$	Taking from same sample under same irradiation condition
1-6-3	5 days	Supersonic cleaning 15 min. with 3% chlorhydric acid + 10% citric acid solution	^{51}Cr, activity 0.05 $\mu Ci/100cm^2$	
1-6-1	5 days	Etching with HNO_3/HF mixed solution	Nil	
1-7-2	5 days	Supersonic cleaning 15 min. with 15% chromic acid + 10% citric acid solution	^{51}Cr, activity 0.05 $\mu Ci/100cm^2$	Same irradiation condition as 1-6 samples
	5 days	Etching again with HNO_3/HF mixed solution	Nil	

Table 4. Comparison of matrix impurities in two silicon samples

No. of sample	Decay time	Isotope	Half-life	Specific activity ($\mu Ci/g$)
8-11-2	4 days	^{122}Sb	2.8 days	1.7×10^{-4}
		^{124}Sb	60.1 days	1.8×10^{-5}
7-8-4	5 days	^{198}Au	2.7 days	1.9×10^{-5}

The matrix activity of samples was measured semiquantitatively by a Ge(Li) γ spectrometer with sensitivity $\sim1\times10^{-5}$ $\mu Ci/g$. The results show that the activity was undetectable after 4-5 days except few samples. Table 4 shows the highest matrix activity measured for two samples.

^{32}P activity in the case considered can be neglected according to calculation.[3]

In order to control the exempt level, a 2π gas flow low background β counting device was used. Results show that after 5 days the activity is less than 5×10^{-4} μCi/g.

ANNEALING

The annealing studies were done in the Institute of Nonferrous Metal. Samples irradiated at the reactor after annealing restored to their calculated resistivity values as shown in Fig. 4. The measured values of avalanche breakdown voltage are within the range as given by calculation from resistivity value.

After annealing the minority carrier lifetimes for our NTD-silicon having a final resistivity above 50 Ω-cm are within 100-600 μs, with most of them higher than 200 μs.

Some interesting facts are also observed. Lifetime increases with resistivity. Lifetime for sample crystallized under atomsphere is higher than that of sample crystallized under vacuum. And, lifetime is dependent on the particular channel in which sample is irradiated.

Rectifiers and Thyristers using NTD-silicon were done at Institute of Rectifier, and satisfactory results were obtained. Devices have been trial-used at customs.

Many individuals of the NTD group and the reactor group contributed to the above work. The author presents the paper on behalf of all those concerned.

PART B ANNEALING EXPERIMENT

by Li Yaoxin

In part B, preliminary experiment on the effect of different annealing temperature on minority carrier lifetime and resistivity is reported.

Samples from three different material types S_1, S_2 and S_3 were used for experiment. Each undoped sample having a diameter of 40 mm, resistivity of ~1000 Ω-cm and minority carrier lifetime of 600 μs or so was irradiated to a fluence of $1.6\text{-}4.1\times10^{17}$ n/cm^2. After irradiation samples were cooled for a week, then cleaned to remove the surface contamination. Before annealing, samples were cleaned again for removing dirt and metal ion on the surface of samples. Then they were annealed under nitrogen at a

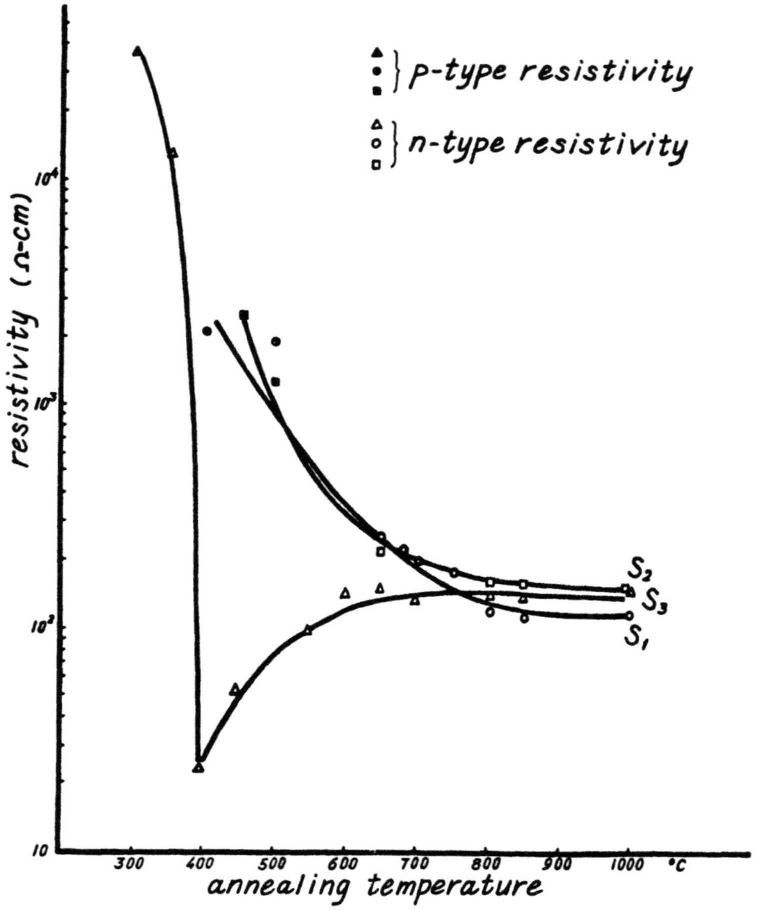

Fig. 6. Resistivity versus annealing temperature

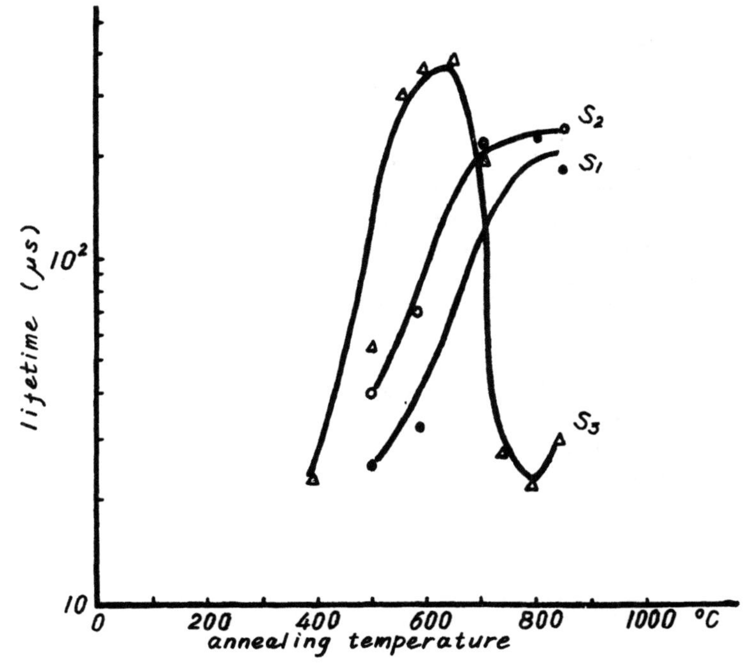

Fig. 7. Lifetime versus annealing temperature

temperature ranging from 200 to 1000 °c. Fig. 6 and 7 show that samples from the three different types of material have almost same annealing behavior. Initially they are all of p-type before annealing. Lifetimes are not measurable and resistivities are very high. As annealing temperature increases, resistivity decreases gradually.

The material converts from p-type to n-type at certain temperature and minority carrier lifetime becomes measurable. That obiously means that defects produced by irradiation begin to be recovered. The defects, and so the resistivity and lifetime are further recovered as the temperature increases.

The lifetime of S_1 and S_2 at 800-850 °c is longer. But above this temperature, lifetime measurement becomes erratic because of possible contamination. However, the resistivity only changes a little.

The lifetime of S_3 is longer at 650 °c and rapidly decreases at above that temperature. It is reported[4)] that for hydrogen float zone silicon after annealing at 1000 °c, a lot of "snowflake" defects were observed by X-ray topography. These defects result from breakdown of Si-H bonds. In our experiment the rapid decrease of lifetime at above 650 °c is probably related to the "snowflake" defects formation.

In addition, we used the positron annihilation technique to study the relation between annealing and defects. Table 5 and Fig. 8 show the results obtained for a vacuum float zone sample S_0 and S_2 sample respectively.

Table 5. Positron annihilation lifetime and intensity in sample S_0

Condition	Lifetime		Intensity		$\frac{I_2}{I_1 + I_2}$ (%)
	τ_1 (ns)	τ_2 (ns)	I_1	I_2	
Unirradiated	0.214	0.433	0.91	0.09	9
Irradiated	0.199	0.357	0.66	0.34	34
Annealed (at 850 °c)	0.215	0.435	0.92	0.08	8

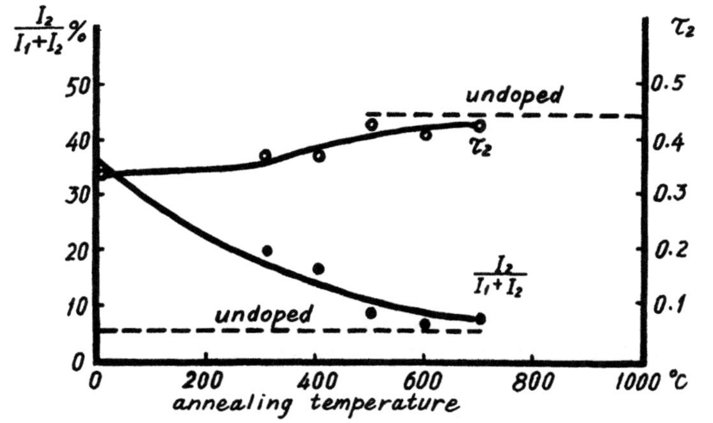

Fig. 8 $\frac{I_2}{I_1+I_2}$ and τ_2 versus annealing temperature

From table 5 it is seen that after annealing at 850 °c most of the defects in sample S_0 produced by irradiation are removed. Intensity I_2, characterizing the concentration of defects is recovered to 0.08 which is near its undoped value 0.09. The positron annithilation lifetime τ_2 characterizing the size of defects is also recovered to 0.435 ns which is near undoped value 0.433 ns too.

Fig. 8 shows the results of experiment done at a series of temperature for sample S_2. It appears that defects produced by irradiation apparently begin to be recovered at 500 °c. I_2 and τ_2 are all near undoped values, and minority carrier lifetime appears at that temperature.

From the above preliminary experiments, it seems that after annealing at 850 °c most of the defects are recovered. Further experiments are still needed to elaborate on the annealing characteristics for various NTD silicon samples.

The author wish to thank Mr. Xiong Xingmin and Miss. Cao chuan from Institute of high energy and Mr. Wang Zheng yuan from Institute of Semiconductor with help of this experiment. Many individuals in the material group participated in the investigation. The author presents the paper on behalf of all those concerned.

CONCLUSION

1. Qualified NTD-silicon can be obtained from a swimming pool reactor at a low cadmium ratio.

2. Resistivity discrepancy between measured and calculated values is within ±5% for the overwhelming samples after annealing.

3. Radial nonuniformity of doped phosphorus is less than 2% for Ø44 mm sample used in experiment; and it is within the error of a four-point probe if qualified initial material is used.

4. The defects produced by irradiation have been recovered. Minority carrier lifetimes range from 100 to 600 μs.

5. Recovery of properties of NDT-silicon depends on annealing temperature. The type convertion temperature occurs in the NTD-silicon irradiated at light water reactor. This temperature also relates to the origin of material.

6. The trial-manufacturing of device shows great advantage of NTD-silicon over conventional doping.

REFERENCES

1) J. M. Meese and P. J. Glairon, Neutron Transmutation Doping in Semiconductor, ed. by J. M. Meese (1979) P.109.

2) A . Yusa, Y. Yatsurugi and T. Takaishi, J. Electrochem. Soc. 124, No.2, 312 (1977).

3) E. W. Haas and J. A. Martin, Neutron Transmutation Doping in Semiconductor, ed. by J. M. Meese (1979) P.27.

4) Cui Shufan, Ge Peiwen, Zhao Yaqin and Wu Lansheng, ACTA Physica Sinica, 28, No.6, (1979) P.791.

DEVELOPMENT OF THE IRRADIATION FACILITIES

FOR SILICON NEUTRON DOPING IN FRANCE

Paul Bréant, François Cherruau, Jean-Pierre Genthon

Services des Piles
Centre d'Etudes Nucléaires de SACLAY
B.P. n° 2 - F 91190 Gif-sur-Yvette (FRANCE)

INTRODUCTION

Irradiation facilities for silicon irradiations in France may be classified in two categories : on the one hand the devices directly implemented in the light water of a swimming-pool reactor and on the other hand the devices implemented in a neutronic medium – heavy water here – providing a high value of the thermal to fast neutron flux ratio.

The tools used in France for silicon neutron doping are recapitulated according to this classification in table 1 and table 2, with their main characteristics ; in addition it is specified that all existing facilities are instrumented with ingot rotation, flux monitoring, automatic time integration and are associated with handling, storage, cleaning, decontamination and activity control equipment. Finally, concerning the irradiation capacities given in these tables, it is pointed out that they are realistic and present capacities, for presently running devices and according to the way they are used on an average ; that is to say that they take into account the real average dimensions of the ingot supplied and not the maximum possible loading with the maximum diameters. Further extensions of capacity are possible if need be by the installation of supplementary standard irradiation modules in the pools (n being in table 1 the number of such supplementary modules).

Particular attention will be given to the new developments :

- new irradiation facilities with a high Øth/Ør ratio being developed in the new research reactor ORPHEE.

- developments in the OSIRIS H_2O reactor.

In particular and in connection with the facilities developed in OSIRIS, neutronic problems typical of swimming-pool reactors will be contemplated and the adopted solutions given.

CLASSICAL SWIMMING-POOL REACTORS (table 1)

The reactors MELUSINE, TRITON, OSIRIS and SILOE are concerned here, with irradiation devices directly in the light water of the pool.

The Reactors MELUSINE and SILOE (Grenoble)

A summary of the irradiation characteristics in these reactors is given in table 1. Silicon irradiations are mainly performed in MELUSINE, more details about the use of which are given in another communication at this conference[1].

The Reactor TRITON (Paris - Fontenay-aux-Roses)

The "Commissariat à l'Energie Atomique" has been contemplating for several years the preparation of the shut-down of this reactor on account of its age (20 years) and limited performances on the one hand, and the full possibility to achieve its program in the other French research reactors on the other hand.

The reactor will probably be shut-down before the end of next year ; its silicon program will have been taken over by other French reactors and in particular by a development of the OSIRIS and ORPHEE capabilities. The development of OSIRIS is well advanced to day, bringing into play among others the TRITON type techniques and equipment (see below).

The Reactor OSIRIS (Paris - Saclay)

It is recalled that the reactor itself has just been transformed to use low enrichment uranium (known as "Caramel" fuel) as prototype for a possible transformation of the other research and testing reactors, in order to face a possible cut-off of highly enriched uranium supply from the United States.

The reactor has been running for more than 6 months now with the new "Caramel" fuel. It is fully operational.

Table 1. Neutron doped silicon in France irradiation facilities (July 1980) (part 1 - ratio Øth/Ør ∿5)

Reactor : name	MELUSINE	TRITON	OSIRIS	SILOE
type	←	swimming	- pool H_2O	→
location	Grenoble	Paris (Fontenay-aux-Roses)	Paris (Saclay)	Grenoble
Characteristics of the irradiation facilities :				
. maximum diameter of the Si ingot	3"	3"	3"	2"
. possible extension of the ingot diameter	yes	yes	yes	yes
. environment	H_2O	H_2O	H_2O	H_2O
. thermal flux Øth 10^{13} n cm^{-2} s^{-1}	1 to 5	1 to 5	5 to 10	5 to 10
. ratio Øth/Ør (Ør rapid flux E > 1 MeV)	∼ 5	∼ 5	∼ 5	∼ 5
. practical and present capacity (ton/year for 5.10^{17} $n.cm^{-2}$)	7	7	4	no
. possible extension of capacity (delay < 6 months)	15	-	n x 4	n' x 1

The silicon irradiation module baptized DIODON[2] is a TRITON type module[5] improved by a variable neutron screen to flatten fluxes.

Figures 1 and 2 show vertical and horizontal diagrams of the DIODON module, beside the core of OSIRIS, with :

- the two sample holders (2 and 3 inches). The neutron screen is not pictured here and will appear in the following paragraph (figure 3) with other details concerning the holders.

- the gear-wheels rotating the holders, driven by an axis and a rotation motor located above the pool.

- the two possible locations of the holders in the pool : the irradiation position, and the out of irradiation position. The unloading from the former to the later is automatically driven by the unloading motors which are computer-controlled.

- the removable plate bearing the unloading motors. It is removable in order that the environment of the reactor may be cleared when necessary to move other experiments around the reactor.

- the bridge for the technician to handle the holders.

- the self-powered neutron detector connections to perform the continuous neutron dose measurements.

Classical Swimming-pool Neutronic Problems - Case of the OSIRIS Irradiation Module

The general neutronic problems to be settled in silicon irradiations and in particular in the DIODON module in OSIRIS concern spatial and temporal flux variations possibly induced by low core height, motion of the control rods, loading and unloading of neighbouring samples and light water environment.

Of course a solution to some of these problems could consist in having a taller core and another environment. For example this is the case for the reactor ORPHEE (see paragraph below "The Reactor ORPHEE") which has been specially designed for a special purpose, and consequently happens to have a rather tall core (900 mm high) and to provide a heavy water environment.

We will keep to the neutronic problems of the classical pool reactor, and in particular the DIODON irradiation module in OSIRIS[3,4].

Figure 3 shows a vertical cross-section of the in-pile part of the device.

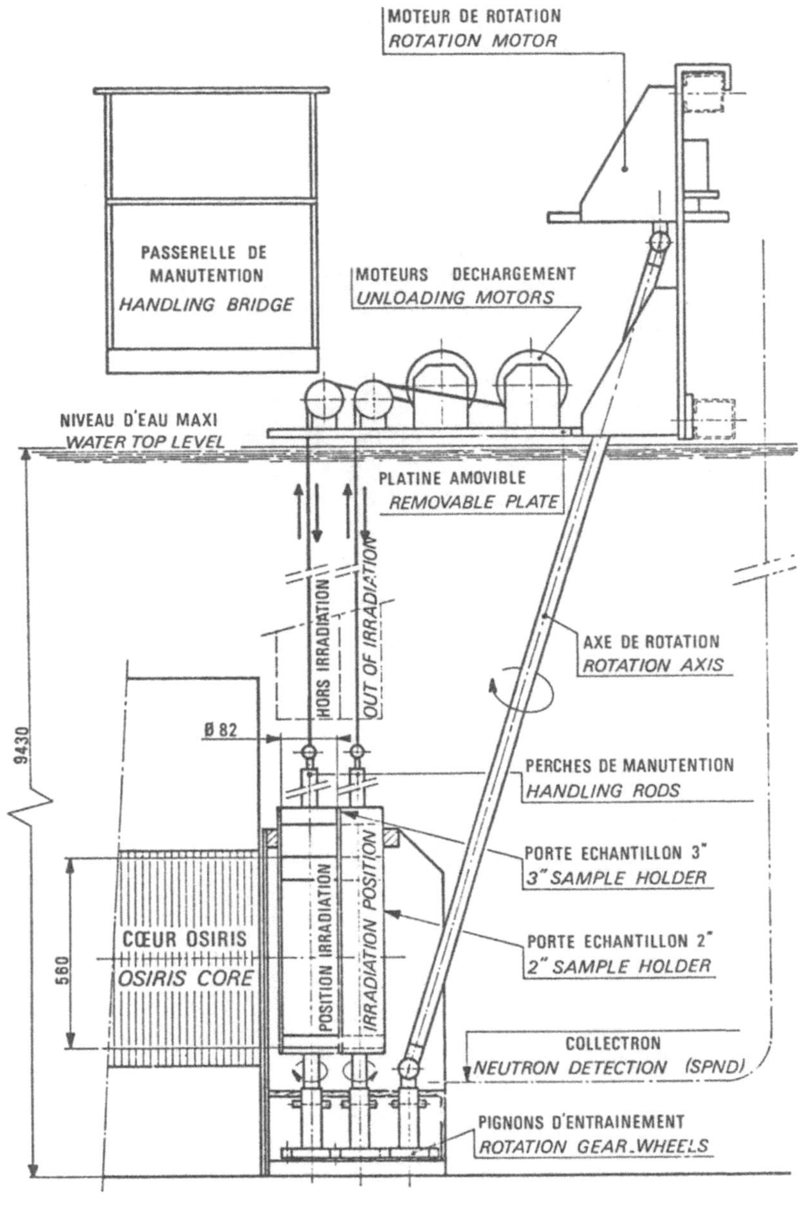

Fig. 1. Diagram of the DIODON module in OSIRIS - Vertical section

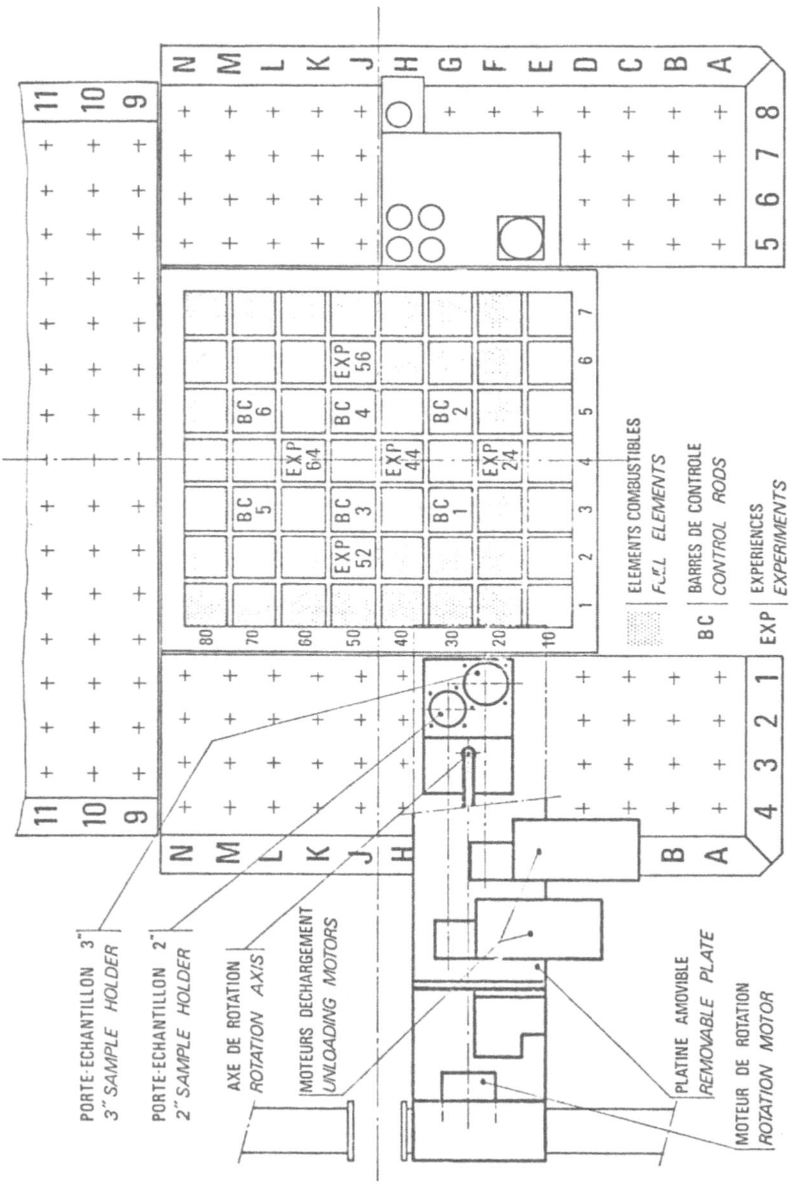

Fig. 2. Diagram of the DIODON module in OSIRIS - Horizontal projection

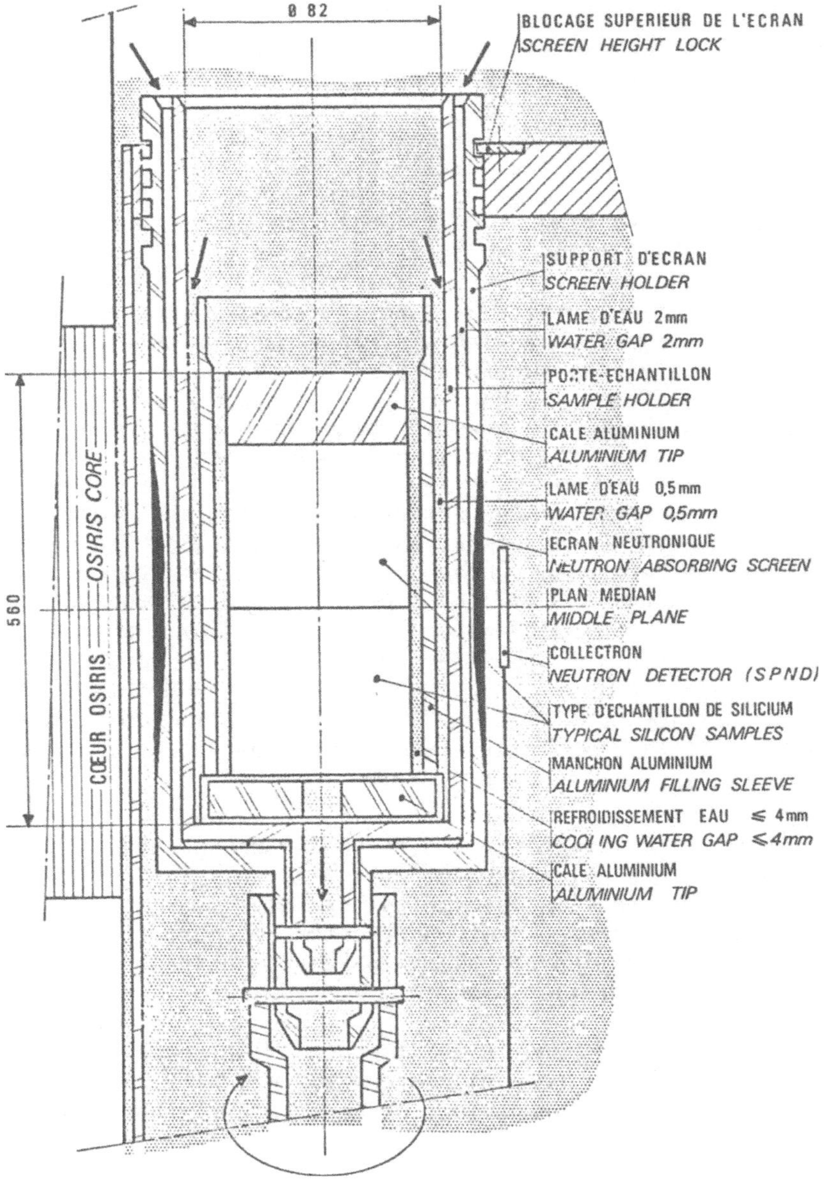

Fig. 3. Schematic axial section of the irradiated part of the DIODON module and neutronic implications

Axial Flux Distribution and Screen. A screen with a thickness variable along the samples flattens the fluxes.

Figure 4 shows the flux shape measured without the screen effect, and the flux shapes measured with a screen optimized for the DIODON module under OSIRIS conditions.

These curves refers to the main part of the OSIRIS reactor cycle where the maximum flux level in DIODON keeps 10 mm below the middle core plane, hence to a screen locked so that its middle plane remains at the same - 10 mm level (see paragraph below "Temporal Flux Variations and Variable Screen Lock").

It is seen from curves 2 and 3 that fluxes are axially constant along some 500 mm within a range much better than $\pm$ 5 % owing to the use of the screen (N.B. : the resistivity changes measured in irradiated silicon lead to the same conclusions). The original flux without screen would change by a factor two along the same length since the core is only 600 mm high.

It is also seen that the effect of a screen is clearly a function of the thickness of any water inside the screen which thermalizes new neutrons. Curve 4 shows the effect of the removal of the aluminium filling sleeve and the corresponding replacement of aluminium by water : the axial flux change is increased. Speaking of the screen efficiency, the amount of water must not change from one irradiation to another and this is the reason for calling on aluminium filling sleeves fitted to the diameter of the silicon ingots (see also below, paragraph "Environment Effects").

Temporal Flux Variations and Variable Screen Lock. The temporal flux variations are not a problem as far as only (slight) flux level changes are concerned i.e. as far as the axial flux shape does not change. The self-powered neutron detectors measure the flux level and give all information necessary to keep a constant dose by correcting the integration time.

Our main concern here is the temporal flux shape variations induced by the control rods displacements, which is not proper to the pool reactor and happens more or less in any other kind of reactor.

In OSIRIS, generally speaking, the 4 weeks cycle is controlled by the displacement of control rods BC 3 and BC 4 (see figure 2) according to the typical schedule of figure 5 :

Control rod BC 4 is far from DIODON and has no shape effect in DIODON as located in figure 2.

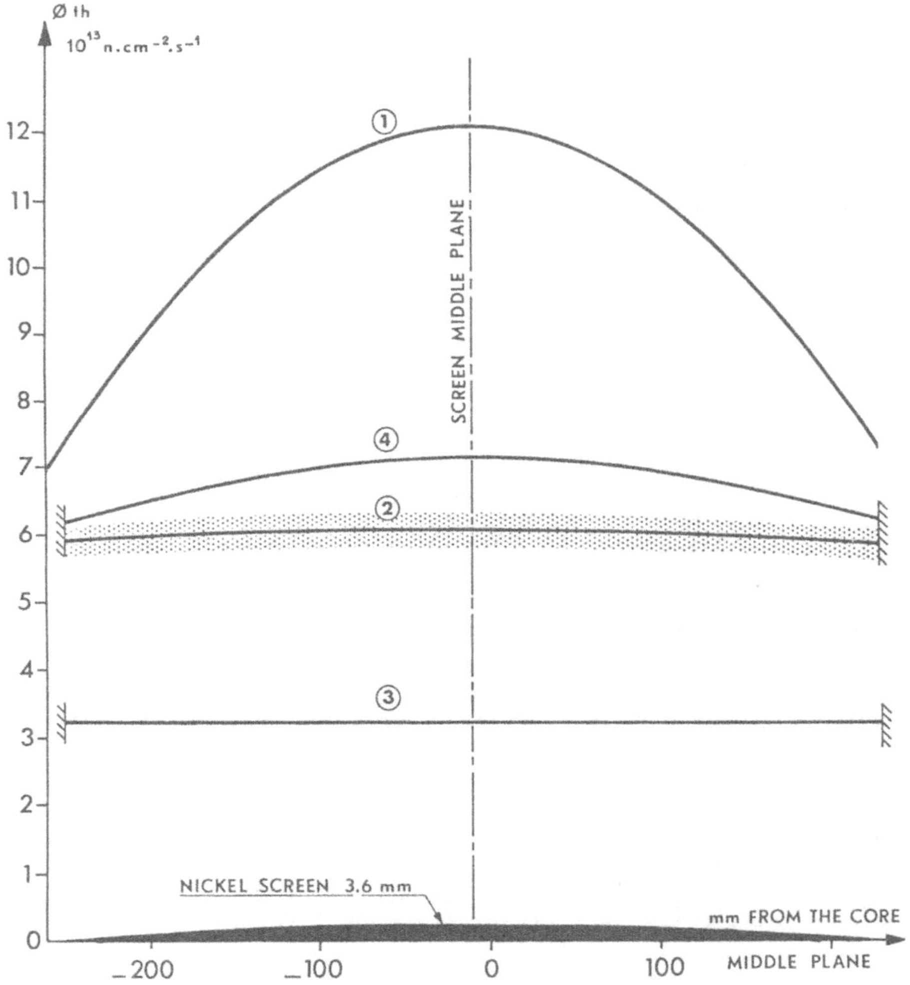

Fig. 4. Axial flux distribution - Screen effect

1 Without screen, without device ; H_2O - 3" holder site
2 DIODON with screen - 3" holder site
3 DIODON with screen - 2" holder site
4 DIODON with screen and <u>without</u> aluminium filling - 3" holder site

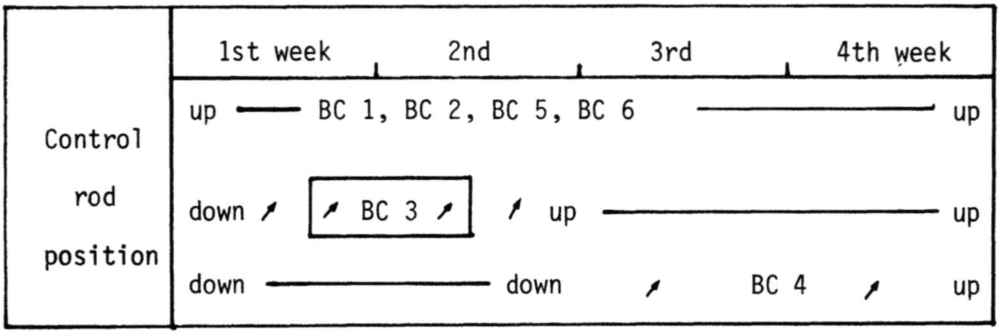

Fig. 5. Typical control rod displacement during the OSIRIS cycle

Control rod BC 3 induces a small shape effect, which is maximum when BC 3 is about half-way up, i.e. between the 4th and 11th day of the cycle (typical values).

It is clear that for another location of DIODON, another control rod could be responsible for the effect on the flux shape. The correction for this effect is based on the same principle as that developed below for this DIODON location and the adjustment is easy.

Figure 6 gives some results of a systematic set of measurements performed with a dummy loading to study the problem of flux shape variations and its solutions.

On the upper part of figure 6 appears the effect of the control rod displacement. A half-way level of BC 3 induces a change in the flux shape (curve 1) which in any case keeps constant within a range better than $\pm$ 10 %. Any other control rod displacements leave the flux shapes suitably flat, as shown by the two examples of curves 4 and 5 *.

On the lower part of figure 6 appears the effect of a screen translation. A translation of the screen middle plane from - 10 mm to - 30 mm reverses the flux shape assymetry induced by the half-way level of BC 3, and a translation to - 20 mm totally smooths down this assymetry.

*The slight difference between these curves and curve 2 of figure 4 is induced by a slight structural difference in the dummy loading (less water).

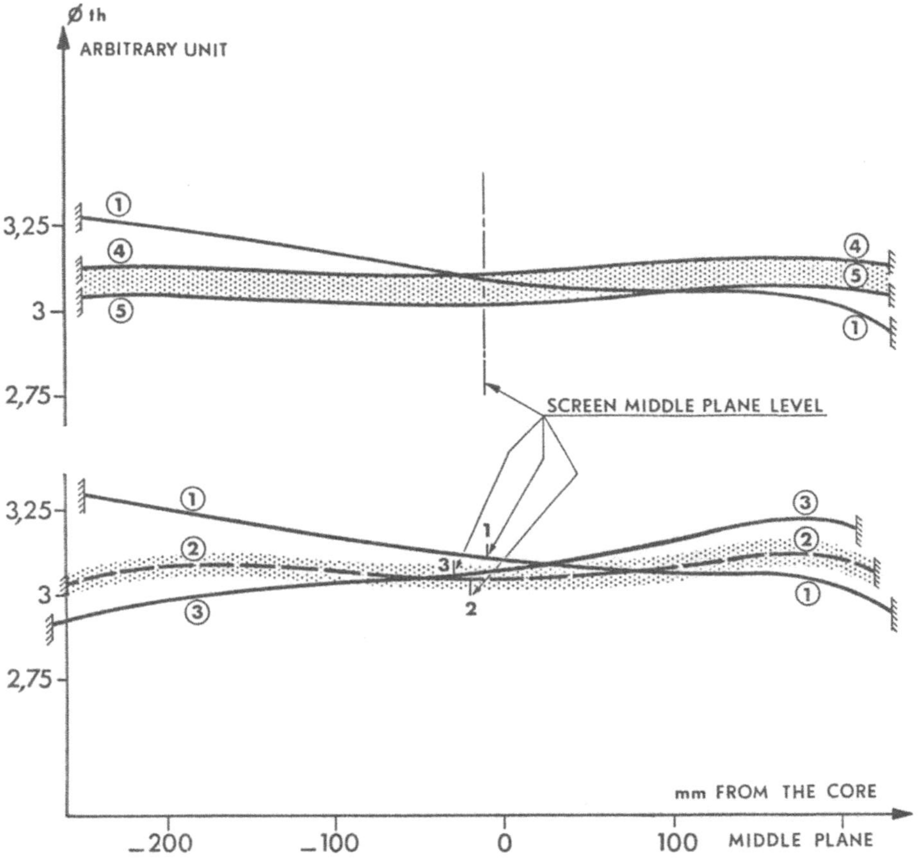

Fig. 6. Flux shape variation

- BC 3 half-way up with various screen adjustments :
  curve 1 - 10 mm
  " 2 - 20 mm
  " 3 - 30 mm
- Examples of other control rod positions with screen middle plane at - 10 mm :
  curve 4 BC 3 up
  " 5 BC 4 half-way up

In conclusion, the screen needs to be locked at the - 10 mm level (screen middle plane level) for a flux flatness better than ± 10 % ; if a flatness better than ± 5 % is needed the screen must be changed and locked at the - 20 mm level during the only week per cycle when BC 3 is close to half-way up (N.B. : as a rule, the results are better than these overestimated limits ± 5 % and ± 10 %).

Environment Effect. Light water is the most efficient moderator and a change from silicon to light water might induce flux changes. This can be partly (and only partly) characterized by calling on the neutron diffusion cross-sections Σs which also show how aluminium may be a good neutronic substitute of silicon :

	H_2O	Si	Al
10^3 Σ s cm^{-1}	3 450	84	89
10^3 Σ a cm^{-1}	22	8	15

The absorption cross-sections Σa are not significantly different here.

(a) Aluminium sleeves and tips (figure 3). These are used to prevent silicon from being influenced by direct contact with a significantly thick gap of water and to suffer from the possible following flux change in the external layer of the silicon sample.

The peripheral sleeve, upper and lower tips are provided to bear this change.

A thin water gap is left deliberately to allow for the deformation of an unmachined sample and to standardize the sleeves. This gap must keep below an average 4 mm along the samples and 8 mm locally (unmachining effect).

(b) Loading and unloading in the neighbouring location. No shape effects are measured but only an effect of flux level, well measured and taken into account (see 1st paragraph of "Temporal Flux Variations and Variable Screen Lock").

Anisotropy Effect - Diametrical Flux Shape. The rotation of the ingot gives a homogeneous irradiation even with flux variations along the diameter of the sample if this variation is linear.

The measurements show that the flux variation is linear along a diameter in the silicon. Typically, the flux decreases linearly by 5 % of the maximum per cm when going away from the reactor core inside the silicon.

In the water the flux would decrease exponantially by 20 % each cm, and the rotation would not help.

Fortunately here, even if a deformed unmachined sample is not well centered in the holder, fluxes are to be considered mainly in silicon and partly in a mixture of aluminium and some water ; actually they do vary linearly, and no anisotropy has been found in the irradiated samples.

FACILITIES WITH HIGH THERMAL TO FAST FLUX RATIO VALUE

The reactors ORPHEE, MELUSINE with its D_2O tank, and EL3 (for memory) are concerned here (table 2).

The Reactor MELUSINE

The reactor MELUSINE is considered here with its contiguous heavy water tank. The corresponding characteristics are summarized in table 2 and are presented in more detail in an other communication at this conference[1].

The Reactor EL3

The reactor EL3 is recalled for memory. It was shut-down in March 1979. Silicon irradiations, among others, are going to be transfered to the reactor ORPHEE which will be a great improvement, in particular where the flux level is concerned (factor 10).

The Reactor ORPHEE (figures 7 and 8)

Due to start during the second semester of 1980, ORPHEE is a pool-type research reactor specially designed for generating high intensity neutron beams. Its thermal power is approximately 14 MW. The compact-design core uses light water as coolant ; to allow for production of high intensity thermal neutron beams the reflector of the reactor is a heavy water reflector where the thermal flux reaches 3 10^{14} n cm^{-2} s^{-1}.

Two vertical channels R 1 and R 2 are going to be devoted to silicon irradiations from the middle of 1981.

Table 2. Neutron doped silicon in France irradiation facilities (July 1980) (part 2 - high values of the ratio Øth/Ør)

Reactor : name	ORPHEE	MELUSINE	EL3
	to start in 1980		for memory closed in 1979
type	D_2O	swimming-pool	D_2O
	H_2O cooled	D_2O tank	
location	Paris (Saclay)	Grenoble	Paris (Saclay)
Characteristics of the irradiation	to start in 1981	D_2O tank	for memory
facilities :			
. maximum diameter of the Si ingot	4"5	3"	3"
. possible extension of the ingot diameter	no	no	no
. environment	He or CO_2	H_2O	CO_2
. thermal flux Øth 10^{13} n cm^{-2} s^{-1}	3 (to be confirmed)	0.3	0.3
. ratio Øth/Ør (Ør rapid flux E $\geq$ 1 MeV)	$>$ 10.000	600	4.000
. practical capacity (ton/year for 5. 10^{17} $n.cm^{-2}$)	($\sim$10) to be specified	0.4	(1)
. possible extension of capacity	no	no	-

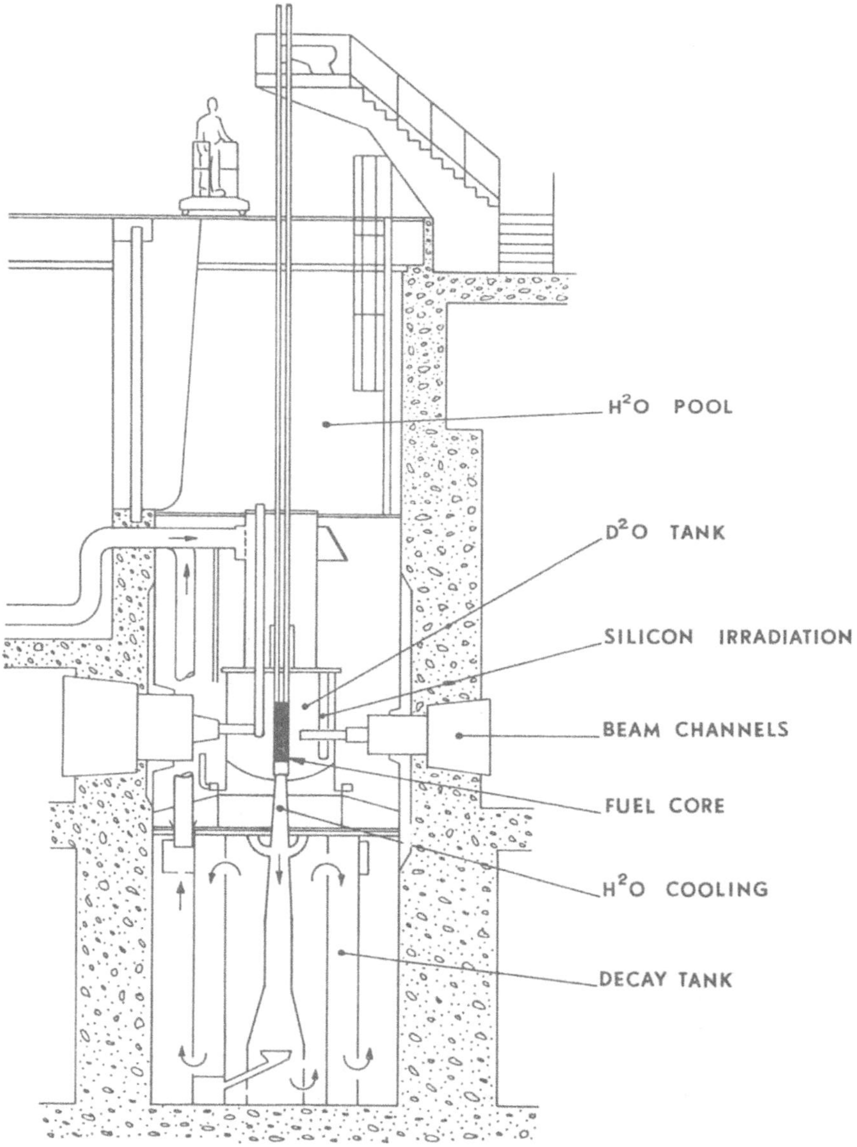

Fig. 7. ORPHEE - Vertical section

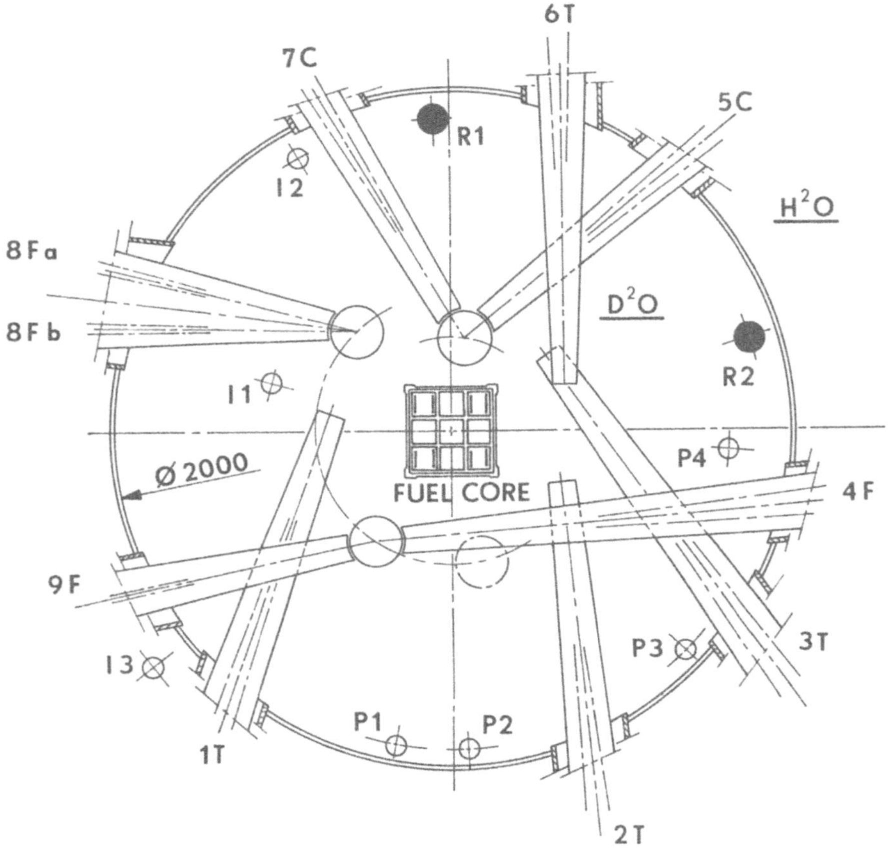

Fig. 8. ORPHEE - Horizontal cross-section of the core

- Horizontal double beam channels :
  - IT 2T 3T 6T : thermal neutrons
  - 4F 9F 8Fa 8Fb : cold neutrons
  - 5C 7C : hot neutrons
- Vertical irradiation channels :
  - P1 P2 P3 P4 : pneumatic rabbits
  - I1 I2 I3 : radioisotope production
  - R1 R2 : silicon irradiation

These channels are located inside the D_2O reflector in the peripheral zone of this reflector where the flux conditions given in table 2 may be expected.

Owing to the height of the ORPHEE core (900 mm) and to the use of an adapted screen the flux is expected to be flat over 500 mm at least.

The device will rotate the samples which will be allowed to have a diameter of up to 4.5 inches.

The ORPHEE capacity will have to be specified, in particular as a function of the actual flux. Roughly 10 ton/year are expected.

CONCLUSION

The set of reactors and silicon irradiation devices mentionned here follows a continuous renewal towards better performances and is going to consist of four pool reactors equipped with well suited irradiation devices and showing the following advantages :

- large thermal to fast flux ratio range (10 000 to 5).

- permanent availability owing to a convenient synchronization of the reactor time-schedules.

- advantages of the pool, i.e., easy handling in and out of irradiation with possible visual control if need be, easy handling in the pool for storage, deactivation and decontamination.

REFERENCES

1. P. Sciers, J.F. Veyrat, D. Roche, M. Villain and F. Blanc, "Neutron doped silicon in Grenoble", communication at this conference.
2. R. Maillot, "Le dispositif expérimental DIODON", personnal communication, Saclay, April 1980.
3. C. Morin, "Essais d'écran nickel pour irradiations silicium", personnal communication, Saclay, January 1980.
4. C. Morin, "Dosimétrie du dispositif DIODON", personnal communication, Saclay, July 1980.
5. J. Berthon, R. Bombaglia, J.L. Poinssot et G. Uzureau, "Dispositif d'irradiation d'échantillons dans un réacteur de type piscine", patent 77 21876, July 1977.

NEUTRON TRANSMUTATION DOPING OF SILICON SLICES

Bobbie D. Stone[+], Aliene D. Henry[+], and Steve L. Gunn[#]

[+]Monsanto Company St. Peters, MO

[#]University of Missouri, Columbia, MO

ABSTRACT

A process for irradiating float zone crystal in slice form to reduce irradiation charges is described. Results indicate that breakage losses are a maximum of 8% and may be as low as 3%. In this process, wherein ingot identity is maintained, targeting precision and resistivity uniformity are equivalent to rod doped NTD crystals. Lifetime measured on 6mm slugs handled along with the slices are significantly lower than those observed on rods, but a 100 μ second minimum specification is still possible at resistivities above 25 ohm cm. Cosmetic stains on as-cut slices were observed, but techniques for avoiding them were devised. A qualitative cost comparison of the two doping methods is made.

INTRODUCTION

In the past few years NTD float zone silicon has been universally accepted in the power device field as a material with resistivity properties vastly superior to crystals produced using processes wherein the dopant is introduced either in the poly deposition or crystal growth step. However, the advantages of NTD to the device maker depends a great deal on the type of device he produces--it is almost essential for high voltage rectifiers and thyristors and of somewhat less value for low voltage transistor manufacture. In all cases, cost of the crystal is an important parameter to the user and the crystal producer must tailor his processes to the user's specifications needs consistent with the lowest production costs.

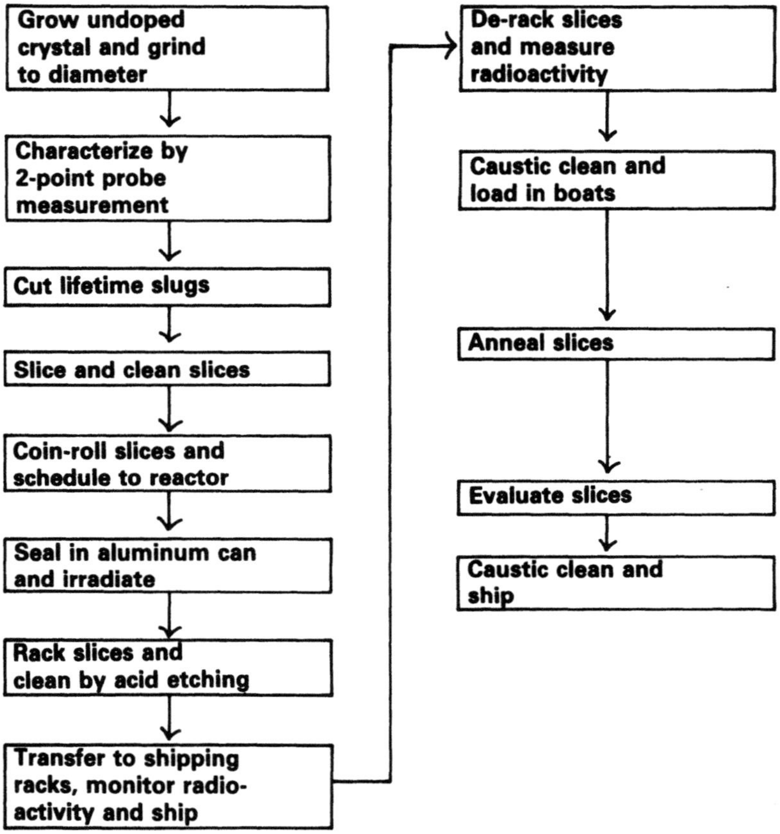

Figure 1. Schematic Diagram of Float Zone Slice Doping Process

The cost of neutrons is an added expense item in the production of NTD silicon, and, due to the inverse carrier/resistivity relationship, increases as the reciprocal of resistivity. Although pricing schedules of the various reactor facilities vary in detail, all are based on the product of thermal neutron fluence and volume occupied in the reactor. Since the necessary fluence is fixed by the resistivity desired, the only method left for reducing irradiation costs for a given quantity of product is to reduce the volume introduced into the reactor. Since all crystal is sliced before use, one obvious method is to slice the material before irradiation to save the volume occupied by kerf and rejected slices. This paper will describe a process that has been used successfully to dope specific ingots in slice form, i.e., ingot identity was maintained to allow measurement of lifetime. A procedure will be proposed that would be even more efficient and provide closer resistivity control if the need for ingot identity can be sacrificed. Agreement on a method of estimating minority carrier lifetime in slices would facilitate such a process.

PROCESS DETAILS

The stepwise procedure is shown schematically in Figure 1. It is interesting to note how many of these steps are required as a consequence of the need to maintain ingot identity. A discussion of the details of the various steps follows:

The crystal growing and grinding steps are identical to that for a rod irradiation process. An undoped crystal is ground to the desired diameter with appropriate flats, etc.

The ground rod is characterized by means of a 2-point probe resistivity measurement. It has been found that this method is best for characterizing the bulk base resistivity and allows one to calculate a single value which best describes the crystal. For this experiment a doping ratio (ratio of beginning-to-final resistivity) of at least 10 was used and this dictates that the rod be all of one type. Consequently, the p-type section that often appears near the seed end of an n-type rod due to rapid impurity segregation was cropped off. The average of the two-point measurements was used to arrive at a single resistivity value representing the entire rod.

Minority carrier lifetime cannot be determined on a thin slice by the photoconductivity method used for rods without using complicated correction factor techniques.[1] Consequently, slugs about 6mm thick are cut to be irradiated, cleaned and annealed along with the slices to provide a method of estimating lifetime of the slices. For convenience, the ingot identification number is inscribed on these with a tungsten carbide scribe.

TABLE I. Results of Gamma Spectroscopic Analysis of Wafers Irradiated to ~2 Ohm Cm Before an After Acid Etchings. (All results in disintegration/second per wafer).

Sample	^{122}Sb			^{124}Sb			^{198}Au			^{110}Ag		
1	250	9	–	–	–	< 1	214	16	–	< 20	< 5	–
2	448	8	–	–	–	< 1	462	61	–	< 20	< 5	–
3	1186	13	–	–	–	< 1	527	17	–	< 22	–	–
4	381	12	–	–	–	< 1	272	11	–	< 29	–	–
5	825	< 5	–	–	–	< 1	1446	83	–	111	12	–
6	631	< 5	–	–	–	< 1	1339	47	–	35	8	–
7	371	15	–	–	–	< 1	279	13	–	23	–	–
8	536	37	–	–	–	2.6	432	36	–			
9	749	25	–	–	–	< 1	756	35	–	2137	52	–
10	1104	143	–	–	10	11	685	29	–	4972	5	–
11	756	78	–	–	–	4.6	574	23	–	41	8	–
12	675	264	–	–	25	18	195	7	–	< 17	–	–
13	–	–	–	–	–	–	–	–	–	–	–	–
14	1385	705	–	–	61	52	242	22	–	29	–	–
15	765	201	–	–	24	14	312	20	–	23	–	–
16	626	281	–	–	32	21	259	26	–	–	–	–
17	659	10	–	–	–	< 1	1554	55	–	32	–	–
18	886	–	–	–	–	< 1	695	88	–	665	9	–
19	509	–	–	–	–	< 1	356	24	–	35	5	–
20	313	–	–	–	–	< 1	570	62	–	41	–	–

Sample	^{24}Na			^{82}br			^{51}Cr			^{65}Zn		
1	3798	–	–	–	–	–	–	–	–	68	–	–
2	3397	–	–	50	–	–	238	–	–	116	–	–
3	3516	–	–	55	–	–	–	–	–	–	–	–
4	3882	–	–	40	–	–	–	–	–	119	–	–
5	4374	–	–	154	–	–	991	–	–	228	–	–
6	3504	–	–	78	–	–	1031	–	–	133	–	–
7	3945	–	–	–	–	–	160	–	–	57	–	–
8	4701	–	–	125	–	–	208	–	–	59	–	–
9	4615	–	–	354	–	–	282	–	–	262	–	–
10	3767	–	–	472	–	–	–	–	–	142	–	–
11	4127	–	–	67	–	–	145	–	–	162	–	–
12	3559	–	–	49	–	–	109	–	–	–	–	–
13	–	–	–	–	–	–	–	–	–	–	–	–
14	4053	–	–	32	–	–	135	–	–	–	–	–
15	3447	–	–	53	–	–	230	–	–	–	–	–
16	4294	–	–	75	–	–	126	–	–	101	–	–
17	3121	–	–	71	–	–	177	–	–	145	–	–
18	3726	–	–	66	–	–	236	–	–	88	–	–
19	3725	–	–	–	–	–	109	–	–	–	–	–
20	3263	–	–	50	–	–	106	–	–	–	–	–

Under each nuclide:

1st column - After 4 days and no etching
2nd column - After 4 days and 1 etch
3rd column - After 20 days and 2 etches

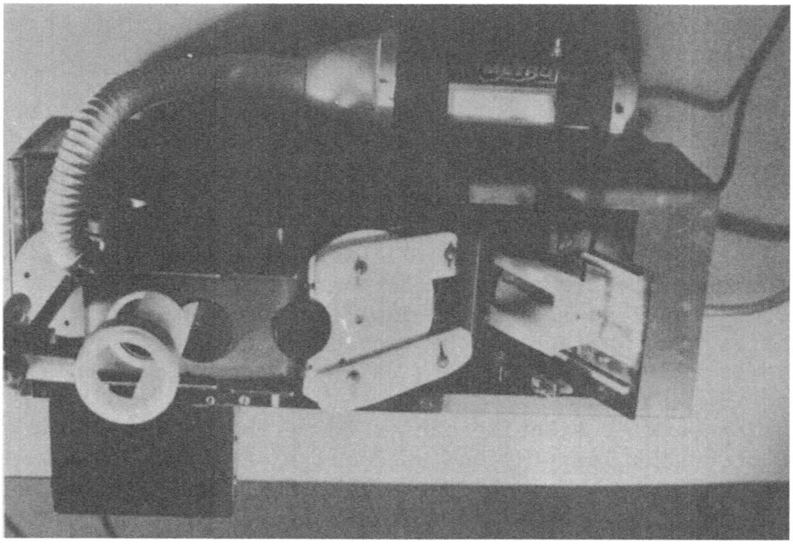

Figure 2. I.I.I. Company Model CC-2 Wafer Cartridge Loader. Modified for Automatic Feeding from Stack of Slices.

Slicing is carried out in the usual way except that the target thickness is increased by 5μm to compensate for removal in the cleaning steps. A caustic cleaning train with ultrasonic agitation is used for cleaning the slices. They are then given a cursory evaluation to remove broken or severely chipped slices.

The undoped slices are stacked and rolled together in coin roll fashion with ingot identity being maintained by placing the scribed lifetime slugs between ingots. An irradiation batch is about 10 inches long, so slices from several ingots are packaged together and it is necessary to match base levels so they all require the same neutron dose. The fluence is calculated from the target and base level resistivity levels, and the reactor facility instructed to give this exposure. The entire 10-inch length is rolled in aluminum foil in a jig built for that purpose.

Since the as-cut or lapped surface is easily stained, the slices must be in a carefully sealed container to prevent contact with the reactor pool water. The aluminum foil roll is inserted in an aluminum can, a lid welded on, and the sealed can leaked checked under a vacuum in a pool of water. The batch is then irradiated to the required fluence by placing the can at the appropriate spot in the reactor. As in the case of rods, the fluence is measured by an instrument that measures accumulated counts.

Probably the biggest drawback of slice doping is that in spite of rigorous pre-cleaning, the slice surface is contaminated with several metals that, after activation by neutron irradiation, result in an unacceptably high level of radioactivity. In a paper presented at the Second International Conference[2], we showed that several radioactive species are present on the surface and are only removed by an etching technique that removes some silicon. A similar conclusion was reached by Haas and Martin[3]. The data from our paper cited are shown in the Table I. To remove these contaminants a 10 second dip in a 10-20-70 Hydrofluoric, Nitric, Acetic acid mixture at 70°C $\pm$ 10°C is followed by several rinses in flowing DI water. This light etching treatment is very effective in removing surface radioactivity but removes less than 2.5μm of silicon. This process requires that the coin rolled slices be placed in racks, an operation best done with an automatic racker. One such apparatus, the I.I.I. Company Model CC-2 Wafer Cartridge Loader is shown in Figure 2. This model has been adapted to accept a belt loading mechanism which automatically removes slices from a stack and feeds then to the racker. This modification adds some convenience, but is probably not any faster than a trained operator feeding by hand. After rinsing, the slices are dried in their racks by a heated centrifuge dryer. High volume systems for carrying out operations such as this are marketed by Fluoroware, Inc. and Fluorocarbon Corporation and others such and automated system would be necessary for high volume operation.

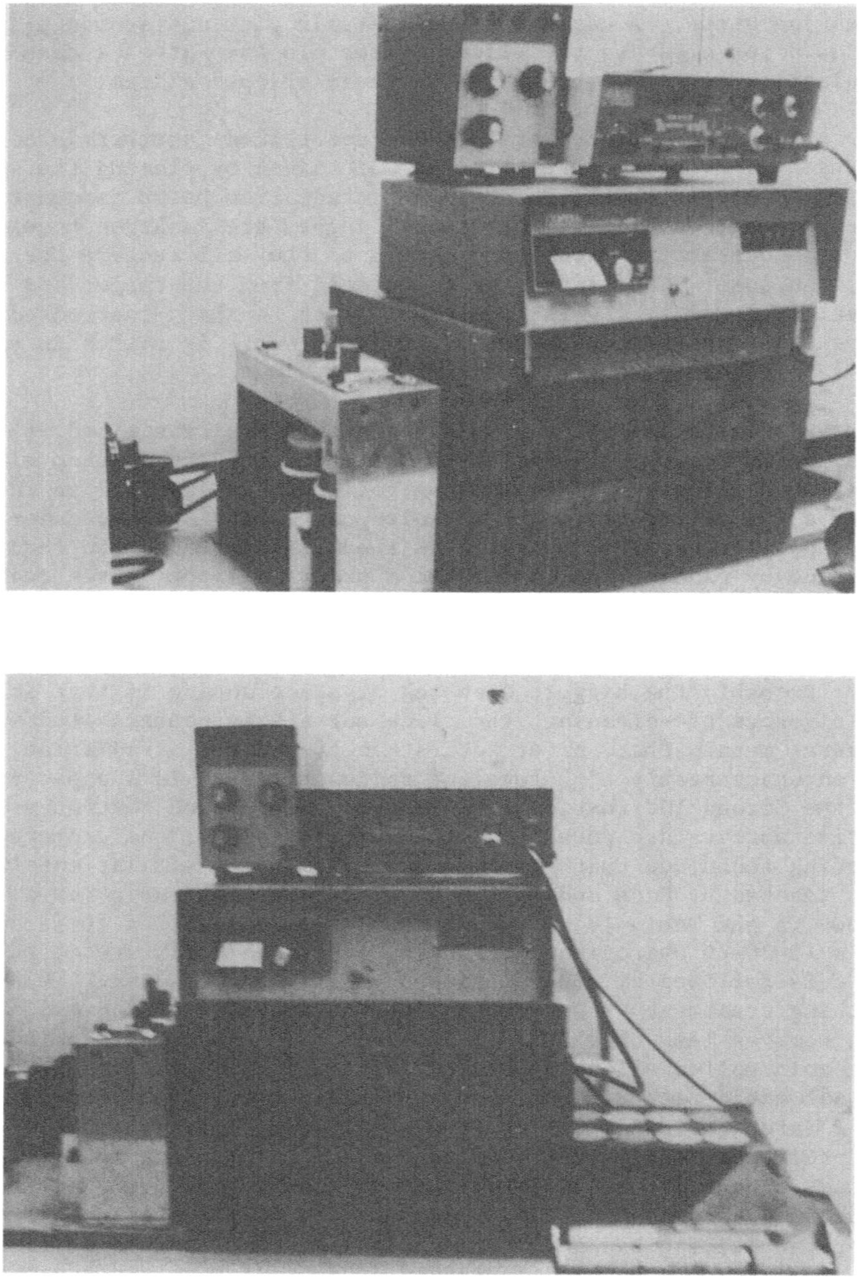

Figure 3. Apparatus for Automatically Measuring and Recording Radioactivity on Slices and Slice Annealing Boat.

It is most convenient to transfer the cleaned slices from the reactor facility in rack boxes fitted with a cover. Since these can be made from inexpensive plastics such as polyethylene, it is cheaper to transfer from the teflon racks required for the acid etching by a flip-transfer technique. The racks are designed to facilitate this and it has become a standard slice handling technique. Before shipping, the slices are measured in the shipping box with a flow proportional rate meter such as Eberline Instruments PAC-4G to confirm that the radioactivity level does not exceed that permitted to be released by the reactor. The box racks are then placed in cardboard cartons and shipped from the reactor facility.

In order to satisfy requirements of the U.S. Nuclear Regulatory Commission that all silicon be shown to contain less than the exempt concentration of a particular by-product nuclide, provisions were made for measuring the radioactivity on each slice. The device that was developed for doing this is shown in Figure 3. It consists of a large area gas flow proprotional probe (6" x 9") with a thin Mylar window (Eberline Instrument Co. Model AC-23) connected to a conventional counter (Eberline Miniscale Counter Model MS-2). The output of the counter is fed into a counter printer. To reduce background counts, the probe is mounted in a "cave" made from 5 cm thick lead bricks. An endless belt passes through the "cave" at a fixed distance from the probe face and the belt motor is tied to a timing device that moves the belt the distance required to bring a load of slices under the probe face. Slices are fed on the belt from a stacking tube which rests on the belt with enough clearance for one slice at a time to be picked up by the belt. The timer turns the counter on as soon as the batch is in place, times out a 1 minute count and the printer prints the counter result. The counter is set to alarm if the count exceeds that calculated to represent the NRC exempt concentration of P-32. In the figure, the stack loader is at the left, the probe is at the top of the "cave", and a load of slices can be seen moving out of the counting area at the right of the picture.

After counting, the slices are again cleaned in the ultrasonically-agitated caustic bath. After drying, they are loaded in silicon boats with the ingots being separated by the slugs cut for lifetime measurement. A loaded boat is shown in the forefront of Figure 3. The slices are annealed in conventional fashion in a nitrogen atmosphere with slow cooling to maximize lifetime. After annealing, the slices are separated into ingot lots and evaluated by standard procedures for resistivity, resistivity uniformity, physical parameters and cosmetic appearance. Minority carrier lifetime is measured on the slugs by the standard photoconductivity decay method. It is necessary to grind slight flats on the wafer edge to make good contact with electrodes so the illuminator can be focused on the slug surface.

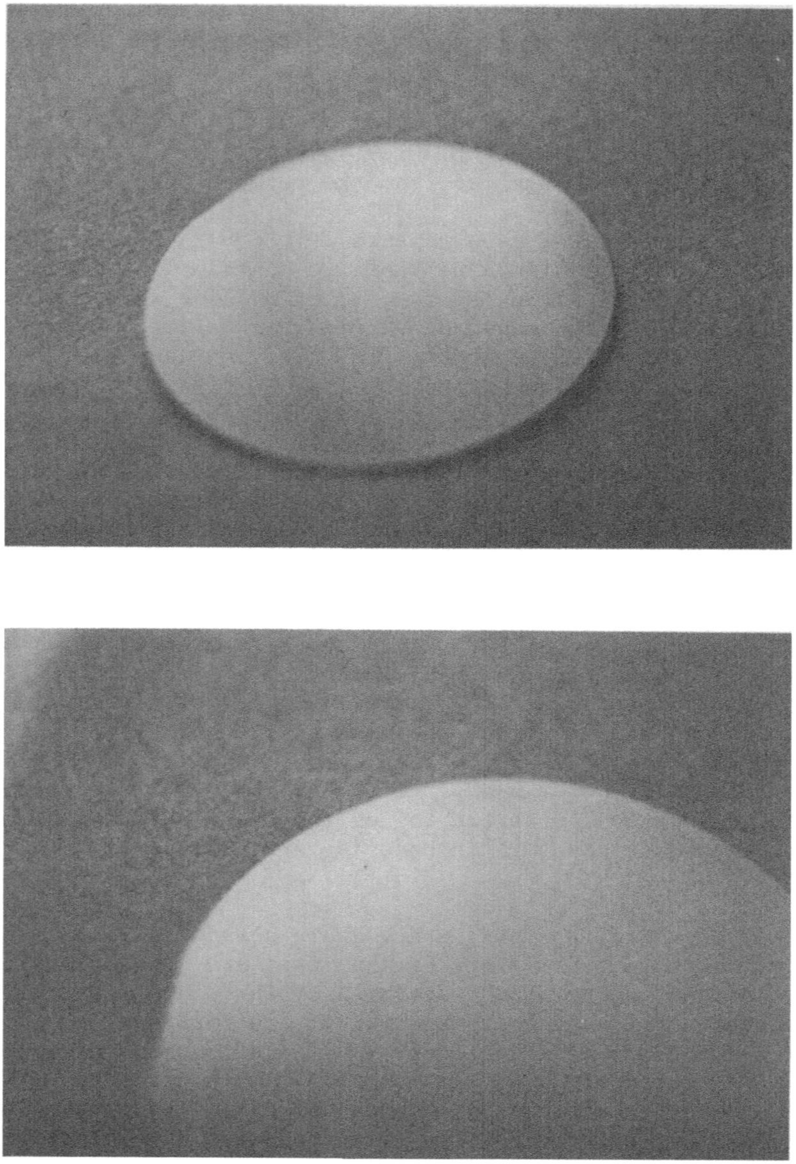

Figure 4. Photographs of Irradiated Slices Showing Stains on Periphery Decorated by Alkaline Etch Cleaning.

RESULTS

Resistivity

When ingot identity is maintained with base resistivity determined on the rod before slicing, there is no reason to expect either the resistivity distribution or radial uniformity of the slices to be any different for slice doping than for rod doping. This proved to be the case with the same excellent targeting capability and extremely low radial graidents being observed that are characteristic of rod irradiation processes. The results of several batches of slices are summarized in Table II. From these data, it can be seen that doping precision is approximately $\pm$ 7% with radial gradients measured in the standard fashion all under 5%. These are similar to values obtained routinely in rod doping at these resistivity levels.

Minority Carrier Lifetime

Lifetimes measured on the 6mm thick slice consistently gave a lower distribution of values than standard rod-doped material. The reasons for this were not identified. However, the effect does not appear to change the minimum specification attainable at ordinary resistivities. For material above 20 ohm cm resistivity, 90% of the slugs met or exceeded the standard minimum level of 100 microseconds.

Slice Breakage

Clearly the most crucial factor in the economics of the slice doping process is the number of slices broken, chipped or scratched in the several operations involved. In three different tests, for slices in the 250-300 micron thickness range overall losses were 7.5, 8.0 and 3% respectively. The third value, which is based on 12,000 slices, gives encouragement that in a large scale operation breakage can be held in the 2-3% range.

Slice Surface Appearance

Another key factor in the success of this approach is the capability of maintaining the slice surface cosmetically appealing. This is, of course, important only in the case of slices sold "as cut" as any lapping or etching steps can be carried out at the end of the process. In the present work using the sequence of steps outlined, a persistent problem occured in the appearance of a faint stain near the periphery of the slice. This stain which is shown in the pictures in Figure 4, developed only during the final alkaline cleaning step which is done to remove dust generated during the slice evaluation steps. Numerous visual inspections confirm that the stain was developed in the final caustic treatment and was not present even

TABLE II. Summary of Resistivity Trageting Precision and Uniformity in Irradiated Slices.

Resistivity Specification Range (Ohm cm)	30-40	10.6-13.0	16-20	30-36
Center Resistivity (Ohm cm)				
Median ($\bar{x}$)	36.8	12.0	18.3	34.0
Std. Dev. (s) (%)				
within rod	3.0	2.0	1.8	2.3
between rods	2.5	2.0	1.1	5.7
overall	3.9	2.8	2.1	6.2
Radial Gradient (%)				
Center - R/2				
Median ($\bar{x}$)	1.7	1.1	1.2	1.3
Std. Dev. (s)	1.6	1.0	1.1	1.2
Center - 1/8" from edge				
Median ($\bar{x}$)	2.1	1.7	1.7	2.2
Std. Dev. (s)	1.6	1.4	1.3	1.7

TABLE III. Minority Carrier Lifetime on Slugs of Various Thicknesses Compared to That Measured on NTD Rod.

Slug Thickness (mm)	Rod	10	5	2.5	1.25
Lifetime Values	200	250	250	250	150
(μ sec.)	500	500	500	450	200
	300	500	500	500	200
	800	1500	1000	500	200
	300	500	500	500	200
	400	1000	1000	500	200
	200	400	400	400	180
	200	400	450	300	150
	400	800	500	400	180

after the annealing step. The cause was not established conclusively, but it was theorized that pyrolysis of dust picked up in processing combined with a trace of oxygen gave rise to a film on the slice periphery that decorated on exposure to the alkaline etch. It was eliminated by instituting another alkaline cleaning step before annealing and carrying out this step in a nitrogen atmosphere rather than in argon.

DISCUSSION

The only unexpected technical issues that arose during this investigation were the low lifetime values and the peripheral stains. The reproducible measurement of lifetime is difficult at best and adapting the photoconductivity decay method designed for measuring rods to the measurement of slug surfaces can be expected to introduce more uncertainty. Good agreement was obtained, however, between rod and slug measurements in an experiment to determine the thickness of the slug necessary to give an accurate measurement. The experiment consisted of measuring lifetime on a series of slugs of various thicknesses cut from adjacent positions where the lifetime had been measured on the NTD rod. The results are shown in Table III. It can be seen that in all cases, measurement on slugs 2.5mm or thicker gave values at least as high as those made on the rod. Since the photoconductivity decay method measures only a thin "skin" on the surface[4], the fact that slug measurements are higher than the rod indicates that the center of the rod may have a higher lifetime than the surface. It can be concluded from these data that slug thicknesses of 2.5mm are sufficient for accurate measurement at the levels in question, and the 6mm thickness used in these tests were more than adequate. These results are in good agreement with statements by Chu[5] and others[6] that thicknesses of at least 2.5 diffusion lengths are necessary to obtain meaningful lifetime measurements. Since diffusion length and lifetime are related by the expression:[5]

$$\tau = \frac{L^2}{D}$$

where L = Diffusion length and
D = Diffusion coefficient
$D_{electron} = 35$ and $D_{hole} = 12.5\ cm^2/sec$ at room temperature,

it can be calculated that the diffusion length in N-type material with a minority carrier lifetime of 300 microseconds is about 625 μM.

Since there is no reason to doubt the reliability of the measurement, the low lifetime readings can probably be explained as due to impurities diffused into the crystal during the annealing step, even though all slugs were carefully etched before annealing.

The effect on total product cost of slice irradiation compared to the conventional rod irradiation is dependent on the neutron

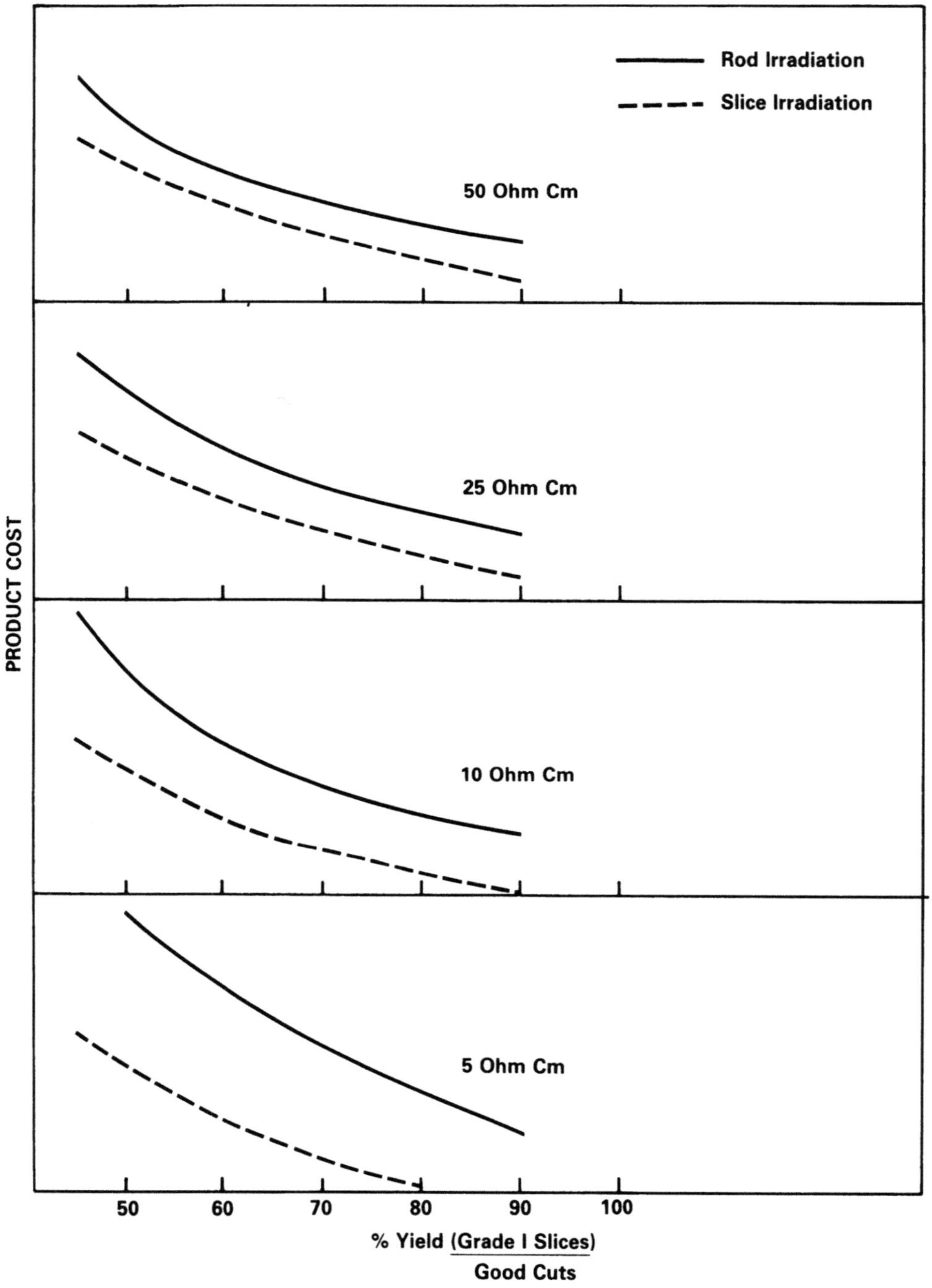

Figure 5. Relative Costs for Slice and Rod-Doping as Function of Yield at Various Resistivities.

charges, handling charges and on the cost of the undoped crystal. It is obviously not possible to make specific comparisons, but some generalizations are possible which indicate where the technique may be advantageous. The cost comparisons involve the trade off of neutron cost savings against increased handling costs and yield losses. (It is apparent that slice doping yields can never exceed those for rod doping since the added steps can only result in increased breakage). In the series of curves shown in Figure 5, product costs for the two doping methods are plotted in arbitrary units on the ordinate with yield plotted on the abscissa for different resistivity levels. For a given yield, slice doping gives the lower cost in all cases, but the curves become farther apart with decreasing resistivity due to the increased neutron costs. By connecting the two curves with a horizaontal line at any point, the yield loss that can be tolerated in going from rod to slice-doping can be identified. For example, assume that rod doping will give a Grade I yield of 90%, it can be deduced that a yield of 73% in slice doping would give equal costs at 50 ohm cm while a 62% yield would give equal costs at 5 ohm cm. From these comparisons, it is apparent that even with an 8% breakage, slice doping has significant cost savings potential at resistivities below about 25 ohm cm.

The process described here is one wherein slice doping is used in the usual specification relationship wherein ingot identity is maintained and a representative lifetime value is assigned to each batch. A much more efficient method would be to sort the slices only on the basis of undoped resistivity. (This can be done economically with an automatic sorting machine). Doping precision could be improved by irradiating slices of similar starting resistivity together. After annealing they could be sorted again if necessary. This would result in the lowest cost and the most accurate resistivity control. No control over lifetime would be possible, however, unless a mutually acceptable method of estimating lifetime on the slice could be devised.

REFERENCES

1. P. D. Blais and C. F. Seiler, "The Measurement and the Effects of Processing on Recombination Lifetime" ASTM Committee F-1 on Electronics Symposium on Lifetime in Silicon, San Diego, CA. February 15-16, 1979. Paper 12.
2. B. D. Stone, D. B. Hines, S. L. Gunn and D. McKown, "Detection and Identification of Potential Impurities Activated by Neutron Irradiation of Czochralski Silicon" Neutron Transmutation Doping in Semiconductors, J. M. Meese, Editor, Plenum Press, New York 1978 Page 11.
3. E. W. Haas and J. A. Martin, "Nuclear Transmutation Doping from the Viewpoint of Radioactivity Formation" Ibid p. 27.

4. F. G. Vieweg-Gutberlet, "Minority Carrier Lifetime Measurement on Silicon - An Overview" ASTM Symposium loc. cit. Paper 1.
5. T. L. Chu, Southern Methodist University, Personal Communication.
6. H. Reichl, I. Ruge, P. Eichinger and J. Muller, "Minority Carrier Lifetime Profile MEasurement by Use of the Photocurrent Technique", ASTM Symposium, loc cit Paper 14.
7. D. T. Chu, Solar Semiconductor Instruments, Bulletin 278 "Diffusion Length Measurement Service".

CHARACTERIZATION OF UNIJUNCTION TRANSISTORS FABRICATED ON NTD-SILICON

James R. Washburn

Motorola Semiconductor Group Inc.
Thyristor Power Device Engineering
Phoenix, AZ 85008

ABSTRACT

The vertical planar diffused unijunction transistor (UJT) has two commonly referenced D.C. electrical parameters which can be employed to indicate local homogenity of impurity doping in silicon substrates. Specifically these parameters are: interbase resistance, which is a spreading resistance from a small diffused layer, and intrinisic standoff ratio. The first parameter, interbase resistance depends primarily on local silicon resistivity in the finished wafer, and diffusion geometry. The second parameter, intrinsic standoff ratio, depends primarily on diffusion geometry, so that local resistivity is easily obtainable from measurements of interbase resistance and intrinsic standoff ratio. Results of these measurements are presented, and are shown to point out fundamental differences between neutron transmutation doped (NTD) and float zone processed (FZ) silicon in the resistivity range of 50 to 70 ohm·cm. It is also presented how measurements of these parameters have, in turn, been used to improve the design and electrical yield of these devices, in particular, with the adaptation of the process to NTD silicon.

A review of various electrostatic models for spreading resistance is presented. The results are compared to results obtained using a three-dimensional electrolytic tank apparatus with a surface voltage probe to simulate the electric field in a chip geometry.

Devices fabricated from NTD silicon and FZ silicon are compared on the basis of local impurity doping homogeneity by the method given in this paper. The results indicate unmistakable advantages in control of parameters are obtainable from NTD silicon. These results show substantial agreement between the parametric determinations of local resistivity on finished UJT wafers and local resistivity measurements made with a spreading resistance probe. This strong correlation suggests the effects of doping inhomogeneity on some primary electrical parameters are still present after wafer processing.

INTRODUCTION

Various structural geometries have been employed in fabricating unijunction transistors (UJT's) since the inception of the device. These structures, beginning with the alloyed germanium bar structure originally discussed by Shockley et al. [1] in 1949 (then called filamentary transistors), have evolved through the cube structure into a variety of diffused planar and epitaxial planar structures. The vertical planar structure has found wide acceptance, mainly due to simplicity in packaging, cost savings realizable in wafer fabrication without epitaxial substrates, and the fact that the surface geometry lends itself to providing space for two base one B_1 diffusions with contact openings and bonding pads for optional parameters.

Despite the advantages, the vertical planar structure depends far more on local substrate resistivity and certain aspects of diffusion geometry for control of parameters than the other structures do. The B_1 diffusions appear as small round dots which are contacted by much larger metalization pads, which in turn overlay the surrounding oxide layer. Through one of these openings all of the device current must flow, resulting in a high electric field in that vicinity. Local variations in substrate resistivities underlying the B_1 regions within a wafer result in variations in device parameters. This variation is compounded with variations that arise from dimensional tolerances in structural geometry. From these considerations it seemed imparative in this work that new mathematical models be investigated or old ones further developed to better explain and demonstrate the more critical interdependence of certain parameters.

Two parameters that vary almost exclusively with local resistivity and dimensional control are interbase resistance R_{BB} and intrinsic stand-off ratio . R_{BB} is the combined resistance of base one (B_1) and base two (B_2). R_{BB} depends on both local resistivity and dimensional geometries while depends on only dimensional geometries, providing no appreciable resistivity gradients exist in the immediate vicinity of B_1. If large gradients in resistivity exist in the immediate vicinity of B_1 it is expected that measurements of both R_{BB} and will depart substantially from the norm. The spreading resistance profiles of both FZ and NTD wafers, within the usable area, show that variations of more than 15% within a radius equal to one base-emitter separation distance (40 m) are rare and some of this may be due to probe contact conditions. Since represents the division of resistance between one part of the chip and another, local resistivity variation in the region near B_1 should, at worst produce no more than approximately the same percent variation in Variation in η exceeding 15% can best be explained by defective diffusion masking patterns.

THEORY OF OPERATION

UJT's may be distinguished from other bipolar transistors by the characteristic curveof emitter current I_E as a function of emitter voltage V_E as is schematically illustrated in Fig. 1. Although the precise relationships between electrical parameters may change with various structures, for example the bar structure in Fig. 1 (a) and the vertical planar structure in Fig. 1 (b), the emitter voltage-current characteristics are fundamentally the same.

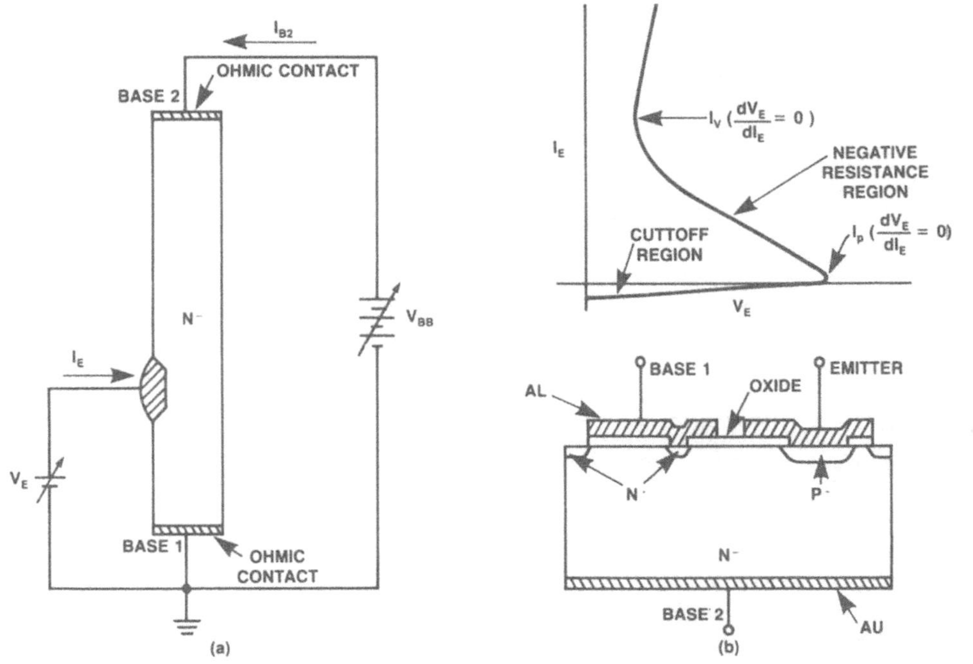

Figure 1. Schematic Voltage-Current Characteristic and (A) Alloyed Bar Structure. (B) Vertical Planar Structure.

Thus for circuit applications and preliminary device design work, UJT emitter voltage-current characteristics are adequately modeled mathematically through equivalent circuit analysis [2] , [3] . In these no quantitative account is taken of structure except as inferred by the variables used. Sze [4] gives valley current I_V and peak current I_P as

$$I_V = \frac{1}{\left(\frac{\mu_n}{\mu_p} + 1 - \frac{1}{\gamma}\right)} \cdot \frac{V_{BB}}{R_{B2}} \qquad (1)$$

$$I_P = \frac{kT}{q} \cdot \frac{1}{\left(\frac{\mu_n}{\mu_p} + 1 - \frac{1}{\gamma}\right)} \cdot \left(\frac{1}{R_{B1}} + \frac{1}{R_{B2}}\right) \qquad (2)$$

where

R_{B1} = base 1 resistance
R_{B2} = base 2 resistance
k = Boltzmanns constant
T = temperature (^oK)
μ_n = mobility of electrons
μ_p = mobility of holes
γ = emitter efficiency
V_{BB} = interbase voltage
q = electron charge

In the absence of empirical data, R_{B1} and R_{B2} are determined by solving electrostatic field equations according to structure.

To further refine the equivalent circuit approach used by [2] [3] and others consider the equivalent circuit shown in Fig. 2. Intrinsic standoff ratio η is defined by [4] as

$$\eta = \frac{R_{B1}}{R_{B1} + R_{B2}} = \frac{R_1}{R_{BB}} . \qquad (3)$$

where R_{BB} is the interbase resistance
V_E is given by

$$V_E = V_D + \eta V_{BB} \qquad (4)$$

V_D, the diode forward voltage, is obtainable [4] from ideal diode equation in the pure diffusion current range [5], [6] as

$$V_D = \frac{kT}{q} \ln \frac{I_E}{I_S} \qquad (5)$$

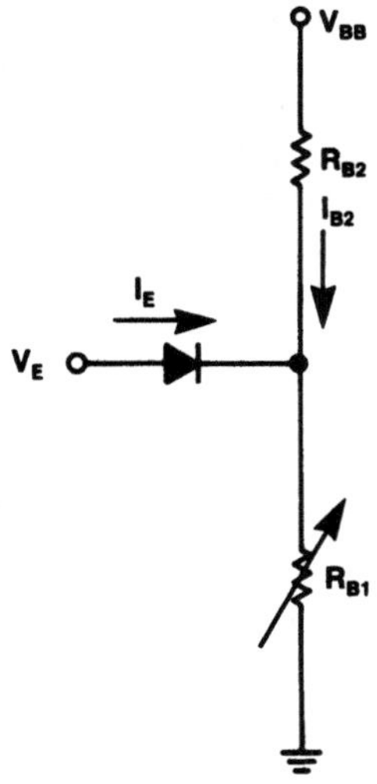

Figure 2. Equivalent Circuit of UJT.

where

I_S = diode saturation current

For convenience R_{B1} can be defined in terms of its conductivity modulated values and geometry as

$$R_1 = \frac{1}{\sigma + \sigma_{mod}} \cdot g(u_1, u_2, u_3) \qquad (6)$$

where

σ = conductivity of substrate

σ_{mod} = modulated conductivity of substrate

$g(u_1, u_2, u_3)$ = geometrical function in arbitrary coordinate space

Conductivity for a semiconductor material is given by [6] as

$$= q(\mu_n n + \mu_p p) \qquad (7)$$

where

n = electron concentration

p = hole concentration

In the diffusion current range for ideal diodes the excess free hole charge concentration p_n varies linearly with I_E. If $L_p \gg W_b$, where L_p is diffusion length of the minority carrier holes, and W_b is the effective base width, the hole concentration p_n may be assumed to fall off linearly with distance from the junction. The effective free charge in the region is therefore directly proportional to $(1+\beta)/2$. At higher levels of I_E, n_n is proportional to p_n. The conductance of the region is shown by [3], [4] to vary in direct proportion to $1+\mu_n/\mu_p$. Combining terms we have the following expression for the modulated contribution:

$$\sigma \text{mod} = C \cdot I_E \tag{8}$$

where

$$C \propto \left(\frac{\mu n}{\mu p}+1\right) \cdot \gamma \cdot \left(\frac{1+\beta}{2}\right)$$

Upon substituting (3), (5), (6) and (8) into (4) the following expression is obtained:

$$V_E = \frac{kT}{g} \ln \frac{I_E}{I_S} + \frac{V_{BB} \cdot g(u_1, u_2, u_3)}{g(u_1, u_2, u_3) + R_2(\quad + CI_E)} \tag{9}$$

At I_V and I_p, $dV_E/dI_E = 0$ [4]. Applied to (10) a quadratic algebraic expression results which has the form

$$I_E = \frac{-b \pm \sqrt{b^2 - 4ac}}{2a} \tag{10}$$

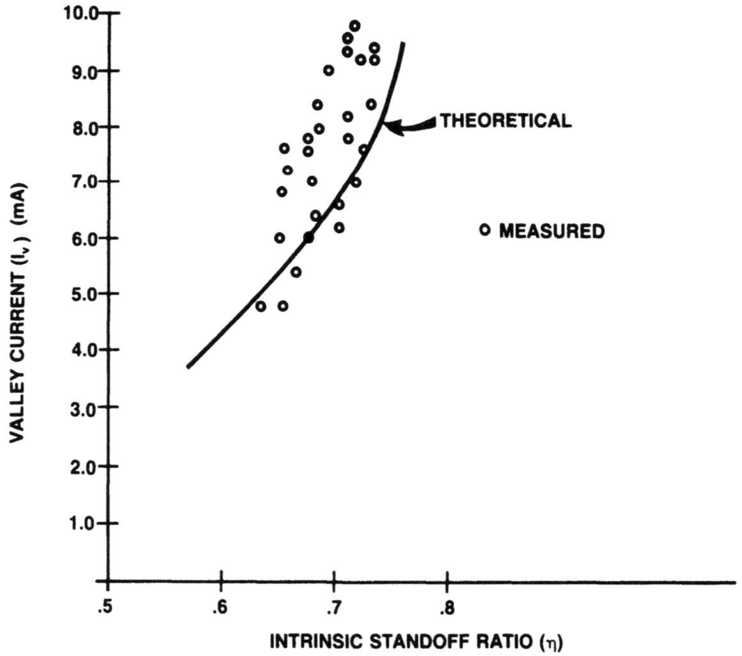

Figure 3. Valley Current, Measured Compared to Theoretical for 28 UJT Devices with $6K\Omega < R_{BB} < 6.5K\Omega$.

where

$$a = R_2^2 C^2$$
$$b = 2R_2C\left[g(u_1,u_2,u_3) + \sigma R_2\right] - \frac{q}{kT} \cdot V_{BB}R_2 g(u_1,u_2,u_3)$$
$$c = \left[R_2\sigma + g(u_1,u_2,u_3)\right]^2$$

The two roots are I_V and I_P. The term 4ac is insignificant in the I_V root and may be disregarded, whence the following expression for I_V is obtained from (10):

$$I_V = \frac{V_{BB} \cdot g(u_1,u_2,u_3)}{R_2C} \cdot \frac{q}{kT} \qquad (11)$$

The other root from (10) is I_P and has the form

$$I_P = \frac{-b + \sqrt{b^2 - 4ac}}{2a} \qquad (12)$$

The term C in (8),(9),(10),(11) and (12) is a factor for gain and will fall off somewhat at lower levels of I_E near I_P. A discussion of γ and β as functions of current density are given by [7] and [8]

In Fig. 3 a comparison is made between I_V as a function of η as measured on actual devices fabricated with the vertical planar structure and the theoretical curve from (11) V_{BB} was set at and C was calculated to be $\sim 4 \cdot 4 \times 10^3$ $(\text{ohm}\cdot\text{cm})^{-1}$ amp^{-1} from considerations given in [7]. The devices were selcted to have R_{BB} values between 6.0 K and 6.1 K . The value for $g(u_1,u_2,u_3)$ was set at ~ 60 from the embedded hemishpere model (14), discussed by [9].

The range of interest for η is typically 0.58 to 0.83 for most circuit applications. As is shown in Fig. 3, in this range small increments in η produce large increments in I_V. It is therefore essential to excercise tight control over η in the wafer fabrication process.

ELECTROSTATIC MODELS FOR VERTICAL PLANAR DIFFUSED UJT's

A. Conducting Hemisphere Embedded in Semi-infinite Solid

The spreading resistance from a conducting hemisphere embedded in a semi-infinite solid of homogeneous resistivity is given by [9] as

$$R_S = R_{BB} = \rho \int_a^\infty \frac{dr}{2\pi r^2} = \frac{\rho}{2\pi a} \qquad (13)$$

whee a is the radius of the conducting hemisphere and r is the radial distance from the spherical center point. This method also yields an expression for η as follows:

$$\eta = \frac{R_1}{R_{BB}} = \frac{\int_a^b \frac{dr}{2\pi r^2}}{\int_a^\infty \frac{dr}{2\pi r^2}} = \frac{b}{b + a} \qquad (14)$$

where b is the separation distance between base one and the emitter.

B. Point Charge Reflected Images Model

In order to examine the voltage distribution at the die surface more closely, which controls η, the point charge reflected images model is introduced to account to the fact that the metalized chip's underside lies in a plane at a finite distance from the B_1 enhancement layer. As illustrated in Fig 4. the semiconductor region is bounded above by a dielectric to dielectric plane boundary and below by a dielectric to conductor plane boundary, but unbounded in the radial direction.

The radial component of electric field E_r at the upper boundary at distance r is

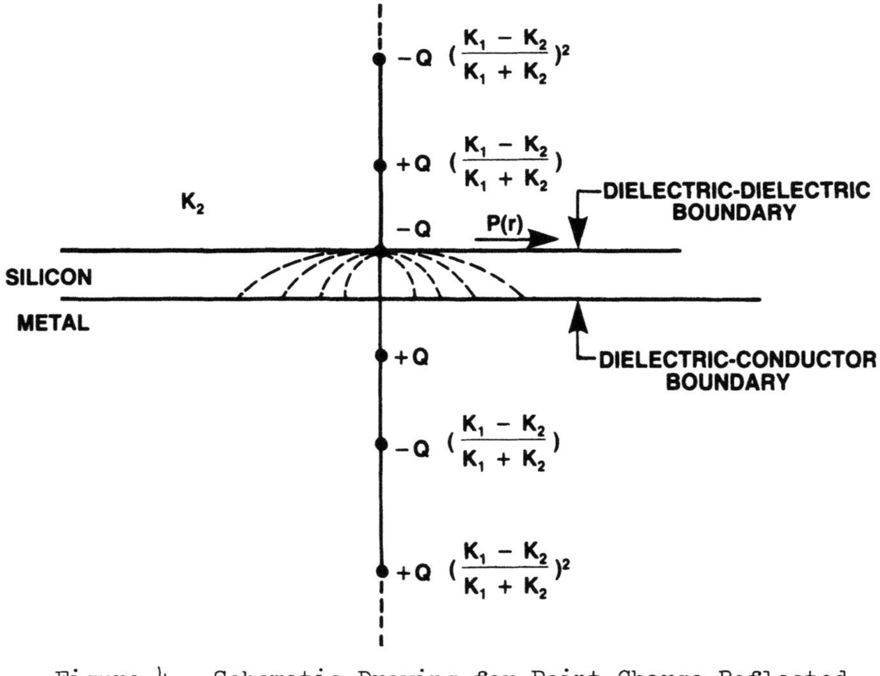

Figure 4. Schematic Drawing for Point Charge Reflected Images Model.

$$E_r = \left[\frac{Q}{4\pi\epsilon_{Si}} \frac{-1}{r} + \frac{r}{(4t^2 + r^2)^{3/2}} + \frac{\left(\frac{K_1 - K_2}{K_1 + K_2}\right)}{(4t + r^2)^{3/2}} - \frac{\left(\frac{K_1 - K_2}{K_1 + K_2}\right)}{(16t^2 + r^2)^{3/2}} - \frac{\left(\frac{K_1 - K_2}{K_1 + K_2}\right)^2}{(16t^2 + r^2)^{3/2}} + \cdots \right] \quad (15)$$

The potential U_r is found by integrating term by term

$$U_r = \int E_r \, dr \quad (16)$$

$$U_r = \frac{Q}{4\pi\epsilon_{Si}} \left(\sum_{n=o}^{\infty} (-1)^n \cdot \left(\frac{K_1 - K_2}{K_1 + K_2}\right)^n \cdot \left[(2nt)^2 + r^2\right]^{-\frac{1}{2}} + \sum_{n=o}^{\infty} (-1)^{n+1} \cdot \left(\frac{K_1 - K_2}{K_1 + K_2}\right)^n \cdot \left\{ \left[2(n+1)t\right]^2 + r^2 \right\}^{-\frac{1}{2}} \right) + C_1 \quad (17)$$

where

K_1, K_2 = dielectric constants of materials 1, 2
r = radial distance from axis
t = thickness of semiconductor region
Q = point charge
ϵ = permittivity of the semiconductor material
C_1 = constant of integration

C. Two Conducting Discs of Different Radii Surrounded by a Dielectric Medium

A solution to the problem of a single conducting disc surrounded by a dielectric medium bounded at infinity by a conducting sphere at zero potential is given by [8]. By the method of superposition of potentials a solution was derived for two conducting discs of different radii, axially entered, parallel to each other and separated by a distance x, as illustrated in Fig 5. The solution is given with numerical constants based on +10 volts potential applied to the lower disc and the upper disc held at zero potential as

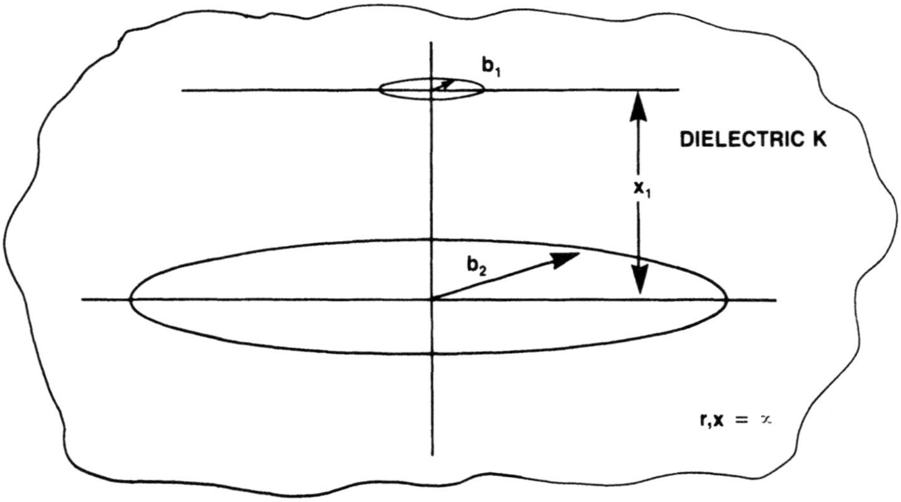

Figure 5. Schematic Drawing for Two Discs of Different RADII Surrounded by a Dielectric Medium.

$$U = (9.7569) + (0.42523)\tan^{-1} \frac{\sqrt{2} \cdot b_2}{\sqrt{r^2 - b_2^2 + \sqrt{(r^2 - b_2)^2 + 4x^2 b_2^2}}} - (6.4709)\tan^{-1} \frac{b}{\sqrt{r^2 - b_1^2}} \qquad (18)$$

where

U = surface potential at distance r in the plane of the upper disc
b_1 = radius of upper disc
b_2 = radius of lower disc
x = separation distance between the two discs

D. Three Dimensional Electrolytic Tank

An analogous solution to the three dimensional problem of the distribution of potential in a vertical planar UJT was obtained in another, more physical, manner: a sealed electrolytic solution replica of a UJT die, with the emitter region omitted, was probed for AC voltage at the liquid surface. The base 1 enhancement was simulated by a small copper disc appropriately scaled and partially immersed in the solution to approximate the base 1 diffusion depth and supported across the tank by a plastic beam. The tank was made of polypropylene material. (Fortunately a tank was available which was of the correct rectangular proportions, and did not have to be fabricated). A sheet of copper was cut to fill the bottom of the tank, to which a covered wire was fastened. A sine wave signal generator was connected to the apparatus so that the small copper disc was at ground potential, while the copper sheet was subjected to 10 volts AC output from the signal generator. The tank was filled to a level proportional to the die thickness. A calibrated AC VTVM was used to monitor the probe potential. NaCl was used as the electrolyte in an aqueous solution. The frequency was varied over a wide range (60 HZ to 100 KHZ) to examine capacitance effects or ion aggregation effects.

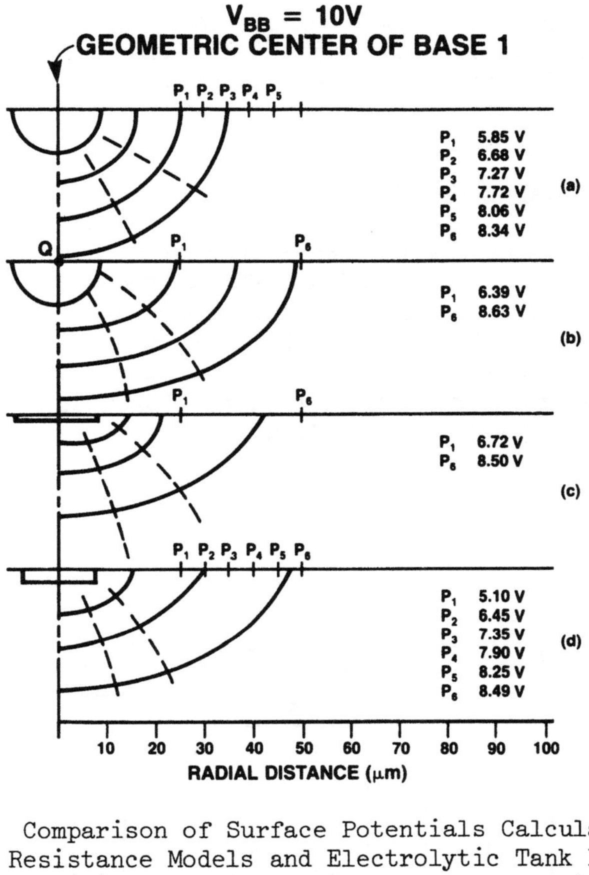

Figure 6. Comparison of Surface Potentials Calculated From Spreading Resistance Models and Electrolytic Tank Probe Measurements. (A) Conducting Hemisphere Embedded in a Dielectric. (B) Conducting Boundary Formed by Reflected Images of Q. (C) Two Conducting Discs of Different Sizes Embedded in a Dielectric. (D) Conducting Disc at Surface of Electrolyte Solution in Tank Held at Zero Potential While Large Conducting Metal Plate at Bottom of Tank Held at + 10 Volts (V_{BB}).

E. Comparison of Models

In Fig 6. results obtained from models in A, B, and C are compared, at V_{BB} = 10 volts, to measurements made with electrolytic tank model C in the vicinity of where the emitter is typically located in the vertical planar structure. These values (x 10^{-1}) are the expected values of η predicted by the respective models.

All the models predict η to within a useful first order approximation of measured values on real devices. However the distribution of η on real devices is generally centered slightly lower than predicted. This may be attributed to the distortion of the equipotential boundaries by the emitter diffusion and the metal-oxide overlay which have opposite effects. Calculations were made which indicate the distortion caused by the emitter diffusion predominates the vertical electric field components in that region as is illustrated in Fig 7.

SPREADING RESISTANCE PROBE MEASUREMENTS

Measurements made with a spreading resistance probe on the surfaces of 50 to 70 ohm·cm FZ and NTD wafers typical of those used in fabricating UJT's are presented in Fig 8 and Fig 9 respectively. Major striations of doping concentration are evident in Fig 8, whereas in the NTD wafer in Fig 9 the existing striations are suppressed by transmutation overdoping. (Greater suppression is achievable with greater overdoping.)

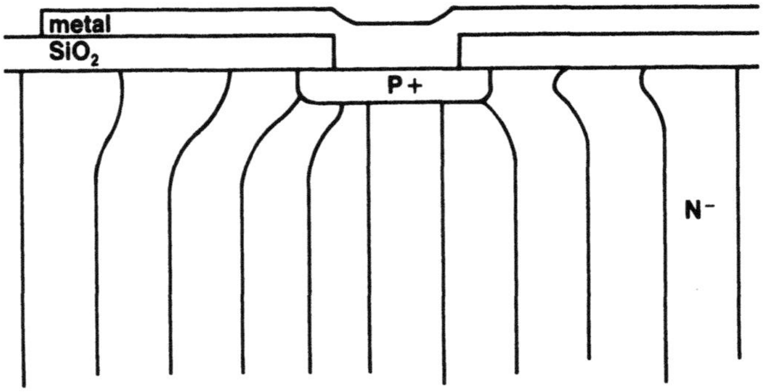

Figure 7. Representation of Distortion of Equipotential Boundaries near Emitter Junction with Metal Overlay.

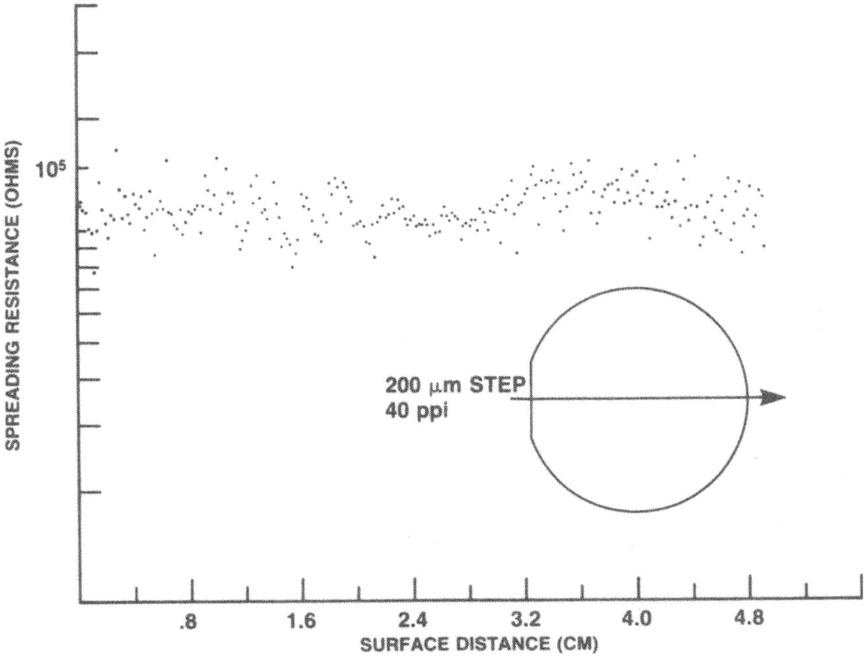

Fig. 8. Spreading Resistance Measurements Across FZ Silicon Substrate

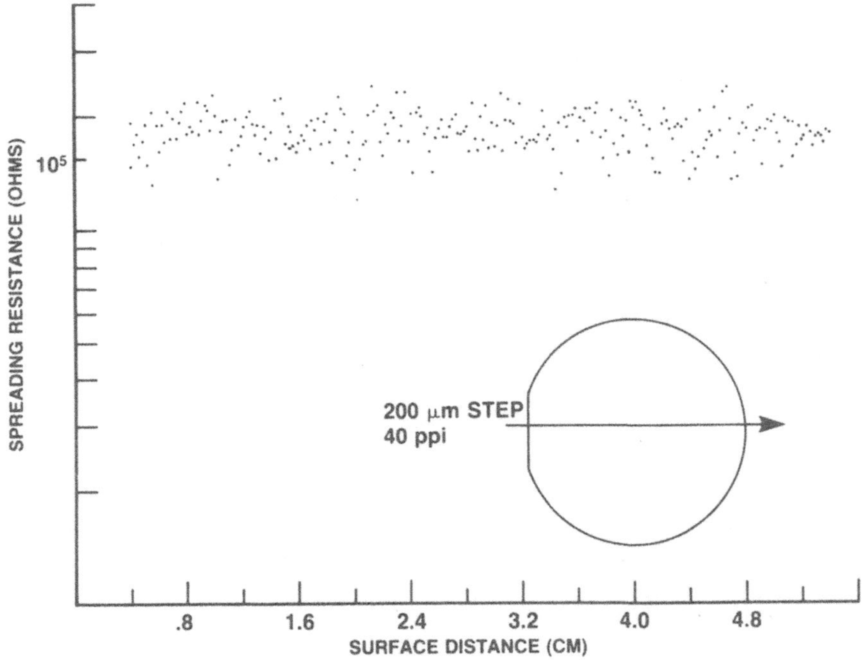

Fig. 9. Spreading Resistance Measurements Across NTD Silicon Substrate

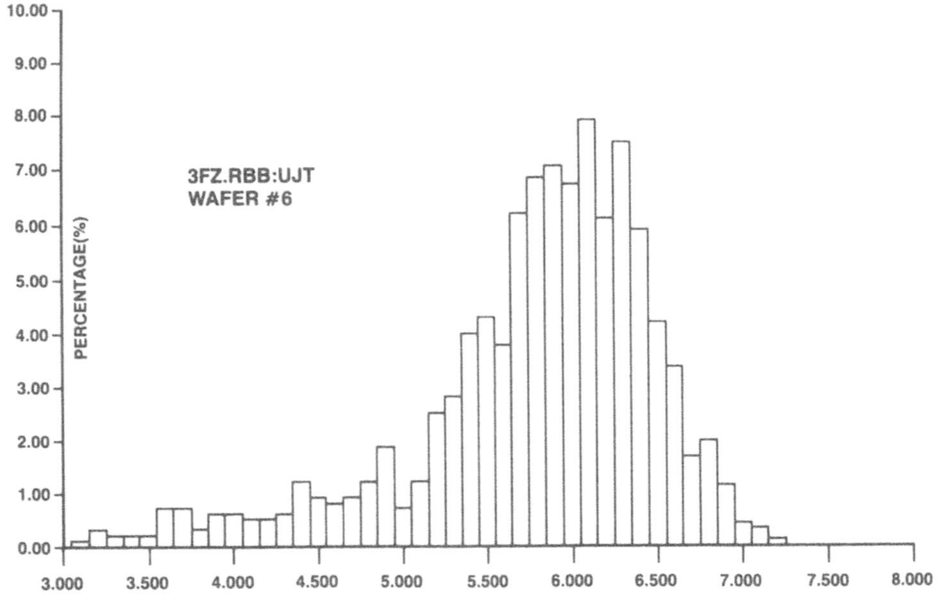

Fig. 10a. Frequency Distribution of Interbase Resistance (R_{BB}) of Devices Manufactured on FZ Silicon Substrate.

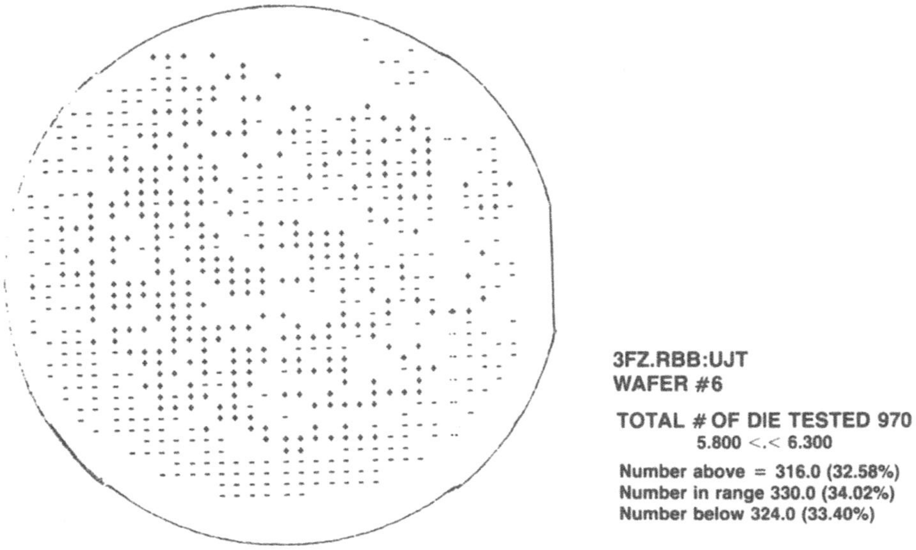

Figure 10b. Spatial Distribution of Interbase Resistance Across a FZ Silicon Wafer.

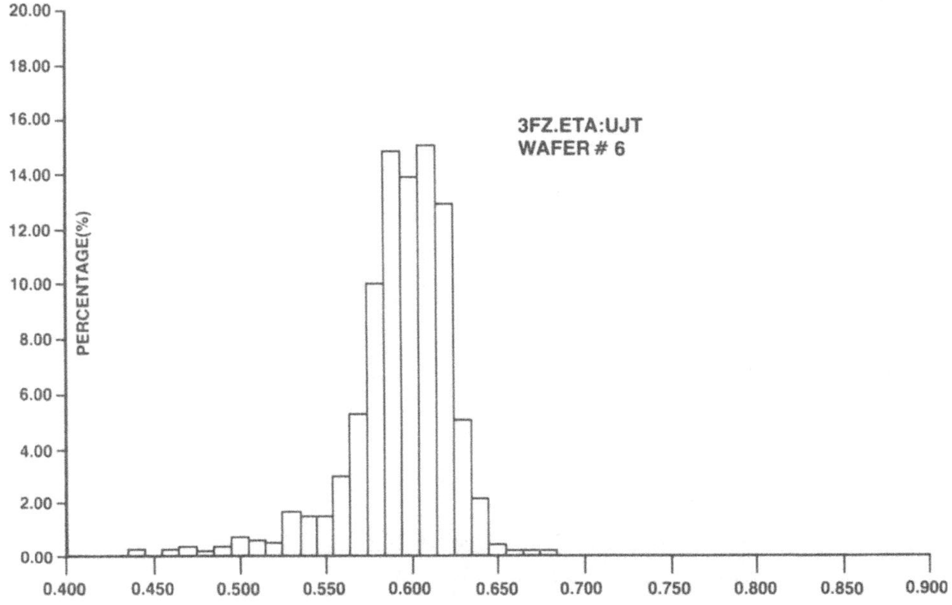

Figure 10c. Frequency Distribution of Intrinsic Stand-Off Ratio (η) of FZ-Silicon Devices.

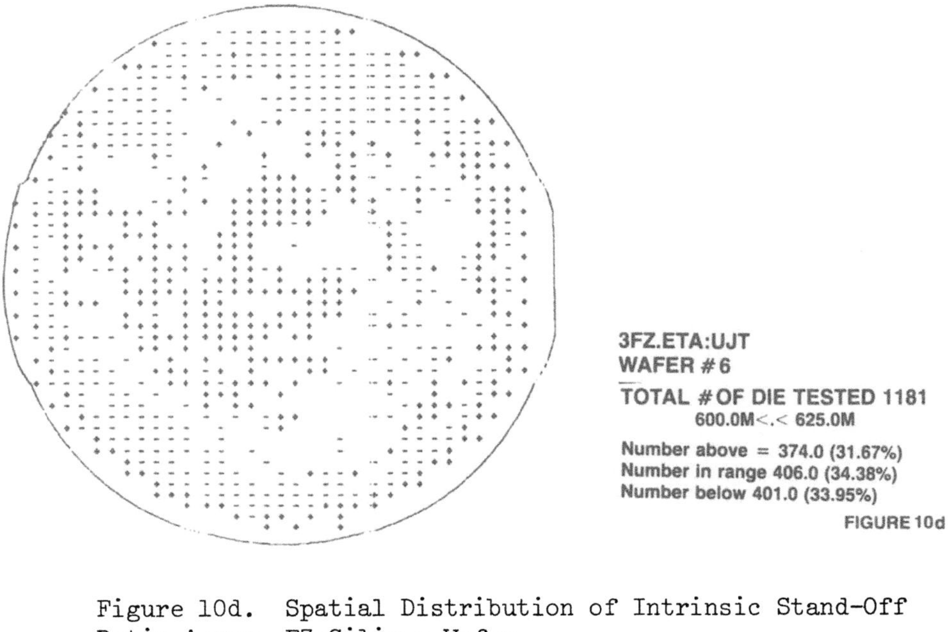

Figure 10d. Spatial Distribution of Intrinsic Stand-Off Ratio Across FZ-Silicon Wafer.

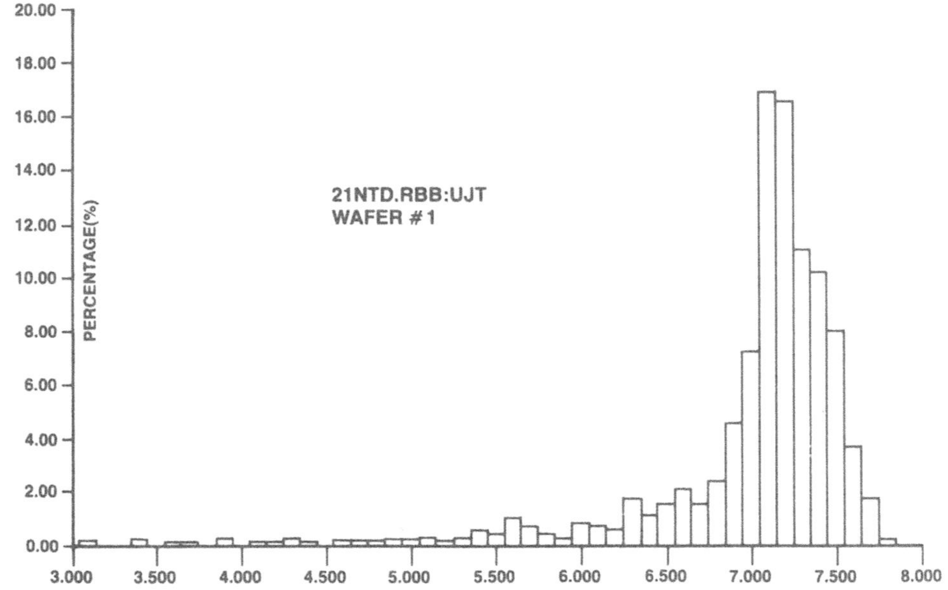

Figure 11a. Frequency Distribution of Interbase Resistance (R_{bb}) of Devices made from NTD-Slices.

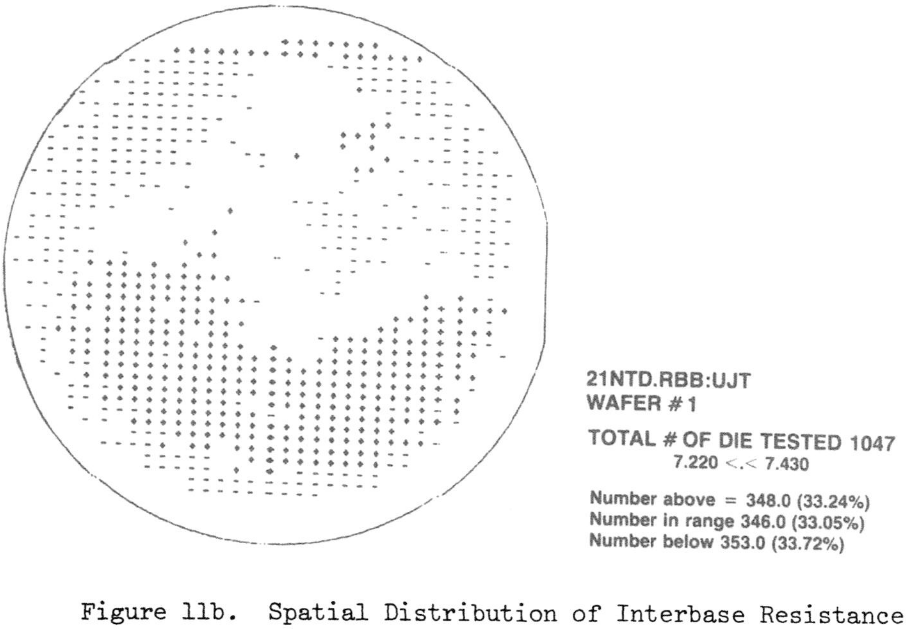

Figure 11b. Spatial Distribution of Interbase Resistance Across a NTD-Silicon Wafer.

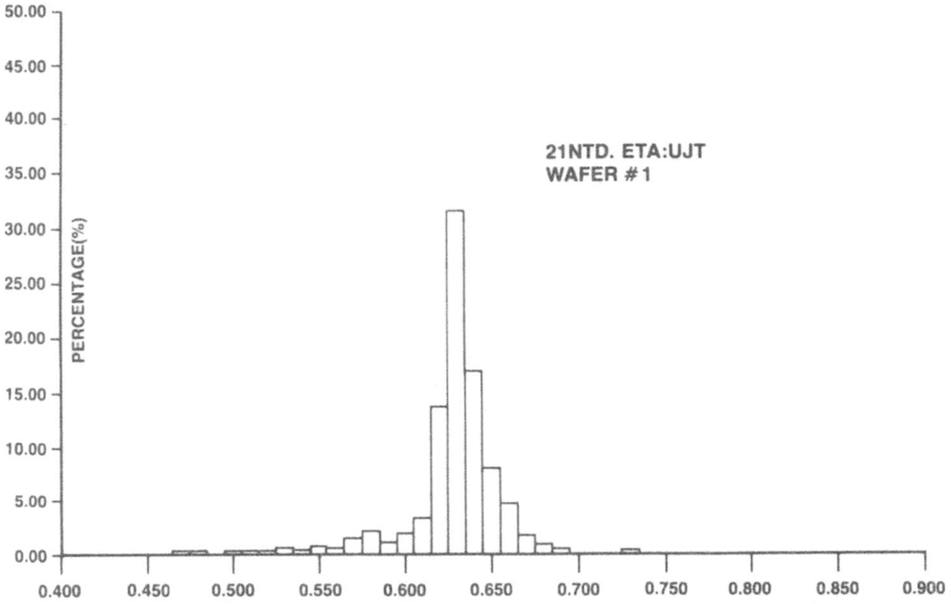

Figure 11c. Frequency Distribution of Intrinsic Stand-Off Ratio (η) of NTD-Silicon Devices.

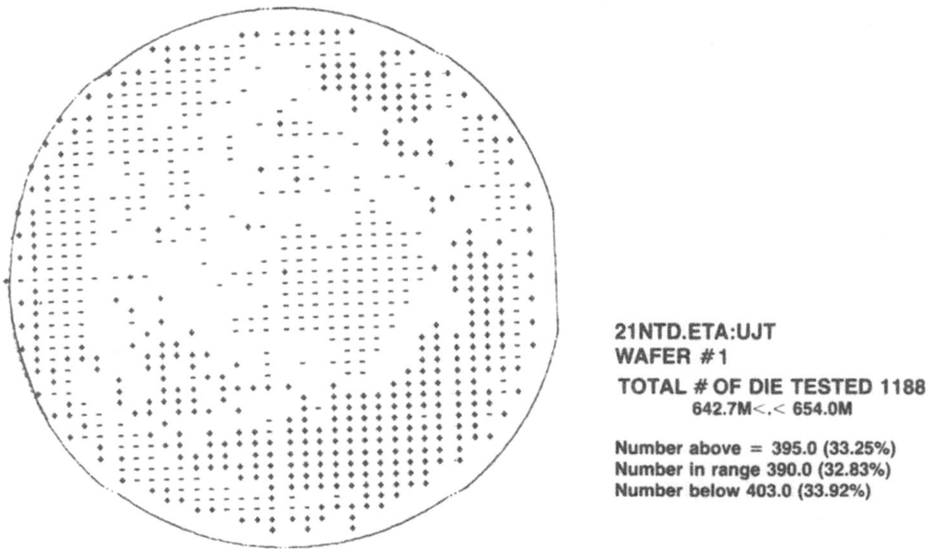

Figure 11d. Spatial Distribution of Intrinsic Stand-Off Ratio Across a NTD-Silicon Wafer.

It was observed in several spreading resistance probe measurements using 40μm steps that the 50 to 70 ohm·cm FZ wafers typically exhibit local resistivity gradients over short distances comparable to the base to emitter separation distance, (herein referred to as microgradients) of about 0.2%/μm, while NTD wafers in that resistivity range typically were found to exhibit microgradients of about 0.1%/μm. Larger scale variations in ρ are also evident in Fig 8 and Fig 9.

From the models given for distribution of potential, η varies only with microgradients and geometry; macrogradients have little effect on η. R_{BB} is affected by both microgradients and macrogradients. Consequently individual die on a wafer which give normal η readings ensure that R_{BB} measurements taken on the same die depend on only local average resistivity and the geometry of the base 1 diffusion as in (13). The N^+ enhancement radius a is measured with a microscope using a filar eyepiece. This technique was used frequently to examine variation in ρ both across wafers and between wafers.

EXPERIMENTAL

A wafer processing experiment was designed and carried out to directly compare NTD silicon to FZ silicon in the range of 50 to 70 ohm·cm in fabricating UJT's. Twenty wafers each of NTD and FZ were processed in the same lot through all steps. The wafer groups were deliberately interspersed with each other at each batch operation to avoid any processing bias. Extreme care was taken to control the surface geometries to $< \pm$ 2%. Wafers whose geometry variation exceeded 2% were excluded from the characterization sample.

The finished wafers were electrically characterized on a Tektronix S3260 computerized wafer prober, which had available hardware and software capability for plotting wafer maps and histograms. Fig 10 a - d histograms and wafer maps of R_{BB} and η are presented for a typical FZ UJT wafer; similar information is presented in Fig 11 a - d for a typical NTD wafer. The organization of Fig 10 and Fig 11 is as follows:

- a distribution of R_{BB} (KΩ) on wafer.
- b wafer maps of R_{BB} where distribution is approximately divided into thirds by +, ·, - symbols.
- c distribution of η on wafer.
- d wafer map of η where distribution is approximately divided into thirds by +, ·, - symbols.

A preliminary investigation into the possible use of NTD wafers for fabricating UJT wafers was conducted by the author approximately five years before this writing. A conversion to NTD wafers for UJT's was made in manufacturing based on the results obtained. An immediate and substantial yield improvement was realized. The more recent experimental results given here clearly show that despite other process improvements made in the interim, the advantages derived from improved doping unifomity remain very significant.

CONCLUSIONS

A refined equivalent circuit model is given which together with data from real devices shows the dependence of the negative resistance region upper boundary parameter I_V on η. The model can also be used to predict the negative resistance lower boundary parameter I_P.

Various mathematical models have been presented to describe the distribution of potentials in the vertical planar UJT structure. These are compared numerically to an analogous electrolytic tank model which simulates the die geometry and is probed at the surface for distribution of potentials. First order agreement is obtained between all the models, so that a choice of one or another may be based on assumptions preferred.

It has been shown that improved homogeneity of resistivity in wafers used in fabrication of UJT's results in better control of important electrical parameters. It has also been shown that NTD silicon wafers fulfill the homogeneity requirements toward substantial yield improvements on UJT wafers.

Histograms and maps plotted of R_{BB} and η parameters within wafers show clearly the resistivity striations in FZ and NTD wafers confirming the proposal that UJT wafers carefully processed are themselves an excellent vehicle for evaluating resistivity uniformity.

REFERENCES

1. W. Shockley, G. L. Pearson, and J. R. Haynes, "Hole Injection in Germanium - Quantitative Studies and Filamentary Transistors," Bell Syst. Tech. J., 28: 344 (1949)
2. "Theory and Characterization of The Unijunction Transistors," Application Note AN-293, Motorola Semiconductor Products, Inc., (1972)
3. L. E. Clark, "Now, New Unijunction Geometries," Electronics, 38, pp. 93 - 97 (1965)
4. S. M. Sze, "Physics of Semiconductor Devices," Wiley & Sons, New York (1964)

5. Alvin B. Phillips, "Transistor Engineering," McGraw-Hill, Inc. (1962) pp. 103 - 108
6. A. S. Grove, "Physics and Technology of Semiconductor Devices," John Wiley & Sons, New York (1967) pp. 180 - 191
7. A. S. Grove,"Physics and Technology of Semiconductor Devices," John Wiley & Sons, New York (1967) pp. 214 - 228
8. Alvin B. Phillips, "Transistor Engineering," McGraw-Hill, Inc. (1962) pp. 234 - 237
9. R. Holm, "Electrical Contacts Handbook", 3rd Ed., Sringer, Berlin (1958)

NTD SILICON BEHAVIOUR DURING DIFFUSION HEAT TREATMENT AND HIGH POWER DEVICES OPTIMISATION

A. Senes

Thomson C.S.F. Division Semiconducteur
37 100 Tours - France

INTRODUCTION

NTD silicon is now extensively used in the high power device manufacturing because of its high degree of resistivity homogeneïty on a wafer and of the narrow resistivity distribution of the wafers issued from the same ingot. This silicon has led to large improvements in the device parameter design trade-offs and was one of the main technological progress in that field for the five past years. Very performing large area devices have only been possible to obtain on an industrial scale owing to its use.

The NTD silicon commercially available has good resistivity homogeneïty and minority carrier lifetime. But are these parameters going to remain stable during the long time high temperature diffusion treatment of the high power devices manufacturing ?

And is there any influence of the cristal elaboration conditions such as irradiation fluence, cadmium ratio, pulling conditions residual impurities, annealing treatment etc ... upon these parameters behaviour ?

Those were some of the questions we were wondering about when we began to use NTD silicon.

RESISTIVITY EVOLUTION DURING HEAT TREATMENT

Three NTD silicon cristals from different suppliers have been used for that experiment and one float zone conventionally doped cristal was added as a comparaison. 25 wafers from each cristal were cut lapped and chemically polished to 540 μm ± 10 μm. Resistivity of each wafer was measured in five points by four points probe. One

Table 1 : Charactics of 33 mm diameter silicon wafers

CRISTAL SUPPLIER	A NTD	B NTD	A + IRRADIATION NTD	C CONVENTIONAL FZ
4 POINTS RESISTITIVTY mean value Ω CM	118	116	145	132
STANDART DEVIATION ON 25 WAFERS ± %	1.7	1.7	2	15
SPREADING RESISTANCE RESISTIVITY Ω CM	120	115	140	137
$\frac{\Delta R_S}{R_S}$ ± %	5	6	3	25

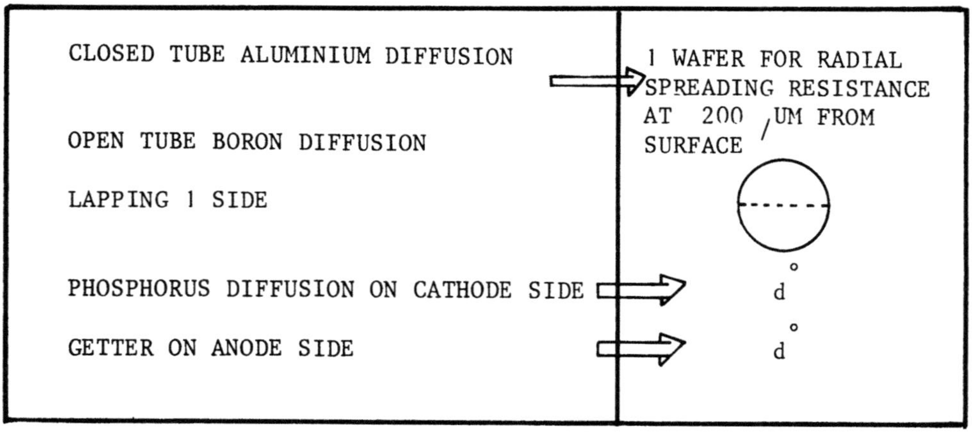

Fig. 1 : Fabrication proeedure for high voltage 33 mm rectifiers

wafer of each cristal was mecanically polished and a spreading resistance measurement was recorded on a diameter. The values we got are summed up on TABLE I.
No information upon the cristals history was available from the suppliers except for cristal number 3. It was a high resistivity float zone cristal (2000 Ω cm) from supplier A and was irradiated with a cd ratio of 3000 and annealed at 700° C during 2 hours.

All the wafers were processed at the same time during the fabrication process presented on fig. 1, in order to obtain 33 mm diameter high voltage rectifiers. One wafer from each cristal is removed after the different high temperature diffusion treatments ; 200 µm are mecanically polished on one face in order to eliminate the diffused layers and to reach the N type substrate. A spreading resistance measurement on a diameter is then performed. The results presented in table II show that resistivity decreases of 10 to 20 % after the closed tube aluminium diffusion, but this phenomenon is not specific of NTD silicon, it is also observed on classical float zone material. The resistivity comes back to its initial value after phosphorus diffusion and remains stable during gettering process. The most important resistivity variation is observed on FZ material, but as a different wafer is used for each measurement, the variation may come from the initial rather large resistivity distribution on the wafers. The spreading resistance variation ($\Delta R_S/R_S$) increases by a factor of 2 during P and N diffusion but comes back to its initial value after gettering. No striation has been observed on NTD silicon which is not true for FZ cristal ; in that case, we recorded classical resistivity striations on the wafer periphery, but it remains stable in amplitude during the diffusion treatment. Twenty wafers from each cristal were processed to obtain 30 mm diameter rectifiers and were electrically tested. The results presented in table III show the highest breakdown voltage obtained on NDT silicon rectifiers and the small distribution of the value (± 2 %) compared with the ones obtained by classical float zone cristal ($>$ ± 6 %).

High level carrier lifetime measured on devices by the OCVD method indicates higher value on NTD material than on classical FZ which is not correlated with the original lifetime measured on the cristals by the photo decay method.

INFLUENCE OF IRRADIATION CADMIUM RATIO

The irradiation cadmium ratio indicates the proportion of thermal neutrons to fast neutrons. It is of main importance to know whether the fast neutrons quantity received by the cristal may influence some silicon parameters or some electrical characteristics of the high power devices. It has been shown elsewhere (1) that cd ratio has no evident influence on resistivity and lifetime when annealing temperature is above 750° C.

Table 2 : Radial spreading resistance resistivity in the N type substrate 200 µm below the surface

$$\frac{\Delta R_S}{R_S} = \frac{2\ (R_{S\ max} - R_{S\ min})}{R_{S\ max}\quad R_{S\ min}} \times 100$$

		NTD cristal 1	NTD cristal 2	NTD cristal 3	FZ cristal 4
SUBSTRATE	$\bar{\rho}\ \Omega cm$	120	115	140	137
	$\frac{\Delta R_S}{RS}$	5 %	6 %	3 %	25%
AFTER ALUMINIUM DIFFUSION	$\bar{\rho}\ \Omega cm$	100	107	110	122
	$\frac{\Delta R_S}{R_S}$	10 %	15 %	12 %	30 %
AFTER PHOSPHORUS DIFFUSION	$\bar{\rho}\ \Omega cm$	120	117	142	150
	$\frac{\Delta R_S}{R_S}$	8 %	10%	9 %	28%
AFTER GETTER	$\bar{\rho}\ \Omega cm$	122	127	143	145
	$\frac{\Delta R_S}{R_S}$	3,2%	4,4%	3,8%	25%

Table 3 : Electrical characteristics of 30 mm diameter rectifiers

	NTD Cristal 1	NTD Cristal 2	NTD Cristal 3	FZ Cristal 4
BREAKDOWN VOLTAGE < 10 mA 25°C	3620 V ± 40 V	3900 V ± 50 V	4050 V ± 70 V	3250 V ±210 V
HIGH LEVEL CARRIER LIFETIME (O.C.V.D.) μs	50 - 60	45 - 60	50 - 65	20 - 30
CARRIER LIFETIME ON CRISTAL PHOTO DECAY μs	80 - 120	300 - 600	400 - 600	800 - 1200

In order to evaluate the cadmium ratio influence on device manufacturing we used three float zone high resistivity, 33 mm diameter 20 cm long cristal ingots. Resistivity was above 3000Ω.cm N type and minority carrier lifetime above 2 ms. The cristals were dislocation and swirl free, the carbon concentration was lower than 2.10^{16} at/cm^3.

The ingots were cut into four pieces, 50 mm long, numbered from 1 to 12. One piece of each ingot was irradiated with a different cd ratio to a target fluence of 2 10^{17} n/cm^2. R value of 3000 was obtained in a D_2 O_2 graphite reactor at Saclay and value of 3; 80 and 500 were obtained in different positions at the swimming pool reactor Melusine at Grenoble. The thermal and fast neutron fluence was measured for each silicon piece. The ingot pieces are cleaned and then annealed at 700° C in a silicon tube under Nitrogen for 2 h and cooled at 1°C/mn to 300° C and then rapid cooling to room temperature;4 points probe resistivity is measured on the ingot piece ends, as well as the minority carrier lifetime. Each ingot piece is cut and wafers are lapped on both sides ; 4 points probe resistivity is measured on 5 points on 10 wafers of each piece.

Table 4 : Irradiation conditions and measurements after annealing at 700° C - 2 Hours

CRISTAL N°	A				B				C			
PIECE N°	1	2	3	4	5	6	7	8	9	10	11	12
Irradiation place	EL3	Melusine			EL3	Melusine			EL3	Melusine		
R : th/fast neutron	3000	2,8	80	520	3000	3	82	510	3000	3,4	86	550
Thermal fluence x 10^{17} $n \cdot cm^{-2}$	2,02	1,94	227	2,63	2,02	1,98	227	2,64	2,02	2,10	2,20	2,61
ρ calculated $\Omega \cdot cm$	123,5	129	110	95	123,5	126	110	94,6	123,5	119	114	95,7
Resistivity on piece $\Omega \cdot cm$	95 100	130 150	90 95	82 81	95 100	105	90	80 86	100	135	92	81
Lifetime μs	500 700	600	900 500	600 800	220 420	380 680	250 400	150 200	2100 2400	280 400	800 1000	1600 2200
Resistivity on wafer $\Omega \cdot cm$	126,6 ±1,1	134,5 ± 5,6	111,1 ± 1,3	95,9 ±0,9	126,1 ± 0,9	114,7 ± 4,6	108,6 ± 0,8	96,1 ±0,9	137,4 ± 2	136 ±8,7	109,6 ± 1,7	100,5 ± 2,8

The results are summed up in table 4. It can be noticed a good correlation between the thermal neutron fluence (Ø) and the resistivity measured on wafers, we obtain the relation :

$$\rho\, \emptyset = (2,55 \pm 0,10)\ 10^{19}$$

where ρ is the final resistivity in ohm.cm

and $\emptyset$ is the thermal neutrons fluence n_{th} cm^{-2}.

For pieces irradiated with a cd ratio of three we notice a higher resistivity value than the one calculated with the preceding relation and also a higher standard deviation ± 4 % instead of ± 0,8 % This point can be explained by a too low annealing temperature. The minority carrier lifetime doesn't seem to be correlated with the cd ratio, except for cristal C for which carrier lifetime increases with R.Double DLTS analyses were performed on 4 wafers with different cd ratio. The wafers were chemically polished then 5 mm diameter platinuim schottky diodes were evaporated. No trap center has been observed. The detection limit of the method is 1.10^{10} cm^{-3}.

Forty wafers of each cristal pieces were processed as shown in fig. 1 to obtain 30 mm diameter rectifiers. Resistivity in the N central region was measured by spreading resistance on a bevel after gettering process. A comparaison of the values with the ones obtained on wafer before diffusion process is presented in figure 2.

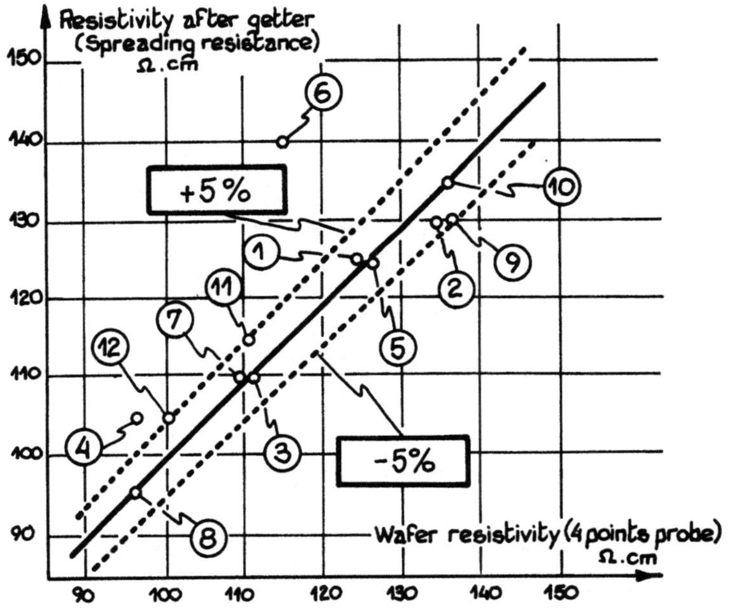

Figure 2 : Resistivity before and after diode process for the cristal pieces.

Except for piece number 6, all resistivities after diffusion are to ± 5 % equal to the starting wafer resistivity. Taking into account that spreading resistance accuracy is approximatively ± 3 %, it can be concluded that resistivity remains stable during the long time high temperature diffusion cycles of high power rectifiers.

The rectifiers were assembled and electrically measured. Breakdown voltage versus wafer resistivity is plotted in figure 3. The mark length gives the 90 % distribution of the devices. The full line represents the relation obtained by M. HILL and all (2) and the dotted line is the calculated curve obtained by KOKOSA and DAVIES (3) with the ionisation coefficients published by LEE. For piece number 6 again the breakdown voltage is higher than that is expected from the wafer resistivity (115Ω. cm) and correspondonds better to the resistivity measured after the final diffusion treatment (140Ω. cm). High level carrier lifetime measured on the diodes are in the range of 40 to 60 microseconds on all the cristal pieces. These values are not correlated with the lifetime measured after annealing and are not dependent on cd ratio.

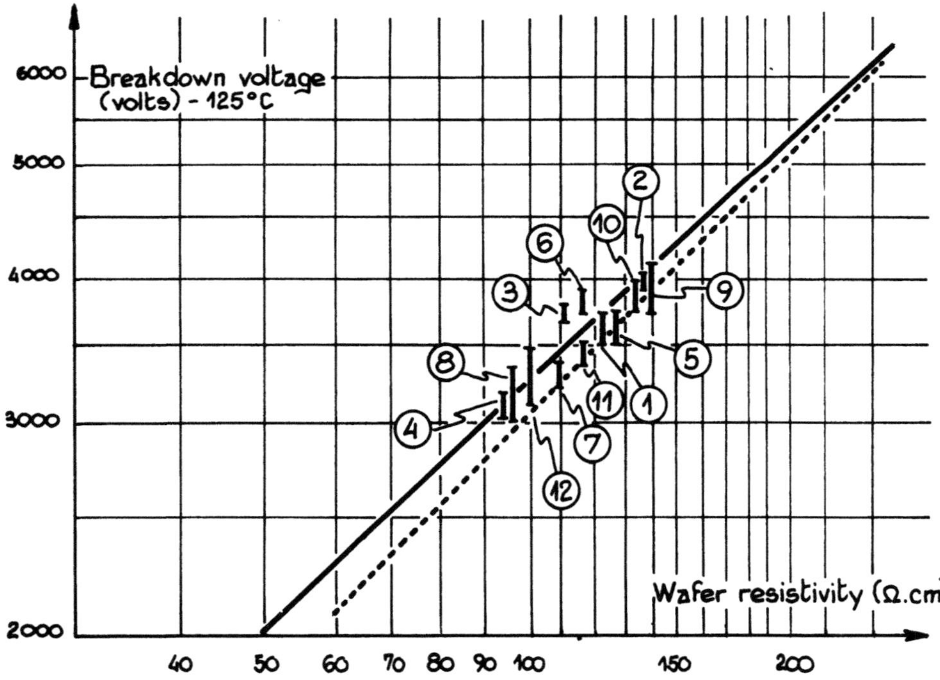

Figure 3 : Breakdown voltage versus wafer resistivity.

ABNORMAL LIFETIME ON HIGH POWER THYRISTOR

On 40 mm and 50 mm diameter high voltage thyristors manufactured with NTD silicon we observed on some batches, devices with too high forward voltage drop. The caracteristics analyse showed that very low carrier lifetime value in N type central base was responsible for it. Carrier lifetime was between 2 to 4 µs instead of 15 to 30 µs in general case. Trap center analyses by DLTS measurement were performed on diodes obtained from the PNP structures by polishing one P region and alloying a gold antimony contact on the N region at 600°C under hydrogen ambiant. One trap center appeared in very large quantity and was found out to be an electron trap with the following characteristics :

Energy level : $E_c - E_T = 0{,}25 \pm 0{,}01$ ev
Electron emission rate : $e_n = 7{,}94\ 10^{10}\ (T/300)^2\ \exp - 0{,}25/kT.s^{-1}$
Trap concentration : $N_T = 1$ to $3\ 10^{13}\ cm^{-3}$

This trap center is found on all the wafers issued of the same NTD cristal ingots after the P type gallium diffusion. This diffusion is performed in a scelled quartz tube under argon partial pressure at 1240° C for 60 hours with 1°C/min cooling rate. The trap center is found each time the wafers of the involved ingots are processed but it is not found on wafers issued from other ingots (NTD or classical FZ) processed together. It appears on cristal resistivity above 95Ω cm and has been found on 7/15 cristals from one supplier 5/6 cristals from another supplier and 0/7 and 0/2 from two other ones. The carrier lifetime of the cristal before processing was always above 200 µs and even on some of them above 1 ms. We tried many gettering process to remove the trap center but the only efficient way was to cool down the wafers from diffusion temperature at a rate slower than 15°C/hour.

From the trap center energy level, and emission rate and also from the way of annealing it, we believe we are concerned with the so called quench-in center describe by C.T. Sah (4). It seems that some NTD cristals are much more sensitive than some others to cooling rate after long time high temperature diffusion. Presently it is difficult to say whether irradiation and annealing conditions are involved or whether the cristal characteristics such as carbon or oxygen content may favour, after neutron irradiation, the creation of these quench in centers.

HIGH POWER DEVICE OPTIMISATION

1/ Rectifiers

For high power rectifier, silicon wafer above 33 mm are used and only one device is obtained per wafer. With classical doped FZ silicon the maximum resistivity striation amplitude takes place on the wafer periphery.

As the whole wafer is used for the device, the lower resistivity region of the wafer will give the reverse breakdown voltage. Figure 4 shows breakdown voltage versus silicon resistivity obtained by KOKOSA (3) on experimental small diodes with FZ silicon and the values we obtained on experimental 30 mm diameter diodes with that material. Also we show the experimental results on diodes using NTD silicon, manufactured with the same process. An improvement of about 40 % on breakdown voltage is observed between the two silicon qualities.

This point allows the use of lower resistivity for a given breakdown voltage and also to decrease the central N region thickness. An improvement in forward voltage drop, recovery time trade-off is obtained. Table 5 gives a comparaison for a 2500 V, 25 mm cathode diameter diodes with the two silicon types, after electron irradiation in order to obtain recovery charges lower than 200 μcb.

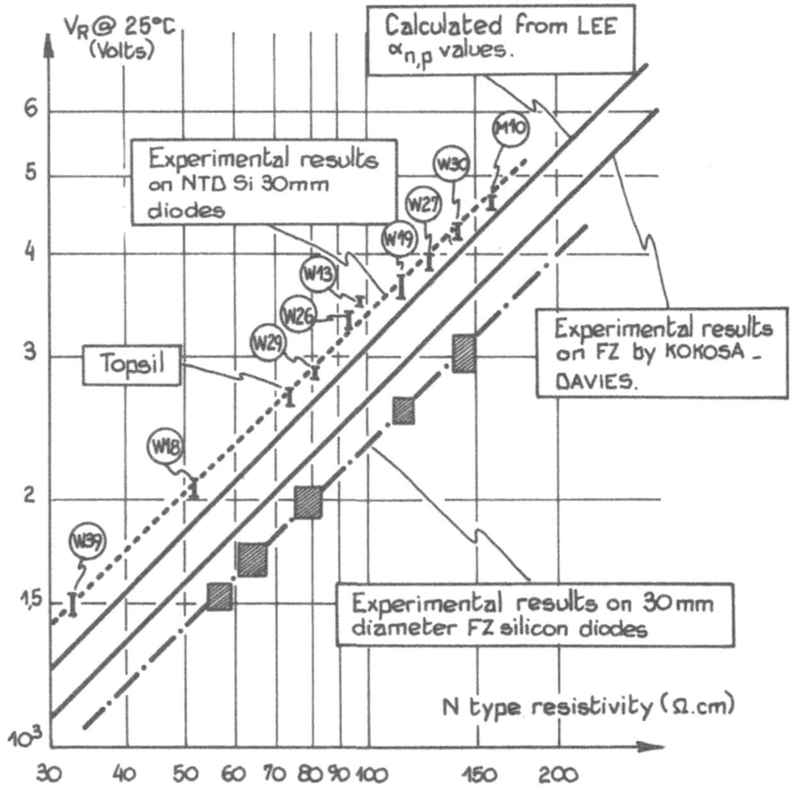

Figure 4 : Reverse breakdown voltage versus N type silicon resistivity.

Table 5 : Comparaison between classical FZ and NTD silicon 30 mm, Rectifiers

	Classical FZ		NTD	
N Resistivity Ω.cm	120 - 130		80	
N region thickness μm	310		240	
Wafer thickness μm	420		350	
Blocking voltage 5 mA - 25° C - V	2500	2800	2800	2950
Forward voltage drop 1500 A V	1.8	2.1	1.5	1.6
Recovery charges $\frac{dI_R}{dt}$ = 25 A/μs - μcb	150	200	150	200
Leakage current 2500 V ; 150° C ; mA	20 ; 30		12 ; 17	

2/ High power thyristors

The device designer has to deal with three main electrical characteristics : blocking voltage, forward voltage drop and turn off time. The need of performing devices for ralways traction for instance, contrains to imporve each of these characteristics, that means : to decrease forward voltage drop (V_{TM}) and turn-off time (tq) and to increase blocking voltage. The technological parameters we can act upon are chiefly :

- starting silicon resistivity
- wafer thickness
- impurity concentration profile
- carrier lifetime in N base
- gate - cathode geometry.

Interaction between these technological parameters and electrical characteristics already has been discussed widely. Decrease of V_{TM} and tq is obtained by a decrease of the N base region thickness and the choice of the carrier lifetime killer type and quantity. Limitations come from the blocking voltage which is generally specified as a maximum leakage current for a given applied voltage and temperature. Thyristors during blocking voltage, can be assimilated to a PNP transistor with two symetrical P zones and a N type homogenous base. When the depleted region is much smaller than the N base thickness, then blocking voltage is limited by the PN junction breakdown voltage. But when the depleted region approaches the base thickness, then the PNP gain increases, amplifying the PNP junction leakage current and decreases the blocking voltage for a given leakage current (fig. 5).

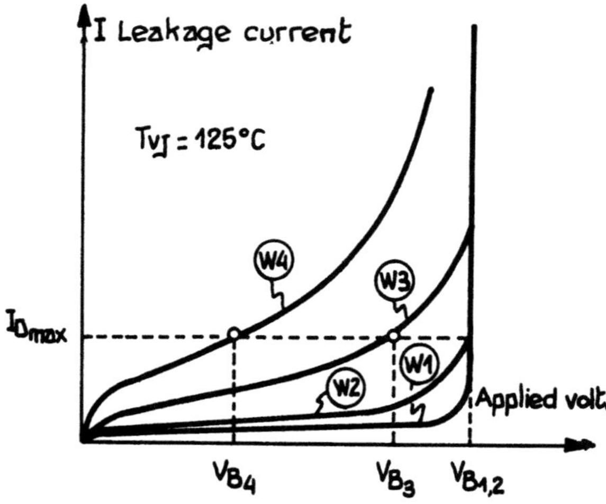

Figure 5 : Leakage current versus applied voltage for thyristors for a N base resistivity and decreasing base thickness W $W_1 \geq W_2 \geq W_3 \geq W_4$

Leakage current will depend on : the applied voltage ; the not depleted N base thickness ; minority carrier diffusion length in this base and generation current in the depleted N region. A simple analyse shows that blocking voltage (V_B) for a given leakage current follows the relation : $V_B = X_i^2/\rho_N$ [1] .

Where ρ_N is the N type base resistivity and X_i a constant depending on technological and design parameters. For a given structure if, starting from low values, N resistivity is increased, then blocking voltage increases like the PN junction breakdown voltage until point A on figure 6. Then it decreases with relation [1]. Point A position gives the highest blocking voltage for a resistivity value :ρ_{NA} when all the other parameter are fixed. Blocking voltage in point A increases when x_i value increases that it to say when :

x_i increases when :
- W_N ↗ : N base thickness
- τ_{sc} ↗ : space charge generation lifetime
- L_P ↘ : minority carrier diffusion length
- J_D ↗ : leakage current density
- T ↘ : junction temperature.

Optimisation of N base resistivity and thickness for a aimed blocking voltage will depend on trap center type used to lower the carrier lifetime. If gold diffusion is used then L_p will be low and τ_{SC} low too. Each variation has a contracditory effect on x_i ; but generally for junction temperature above 125° C, τ_{SC} decrease will have the main effect and x_i will decrease. An increase of W_N and ρ_N will be then necessary to maintain blocking voltage.

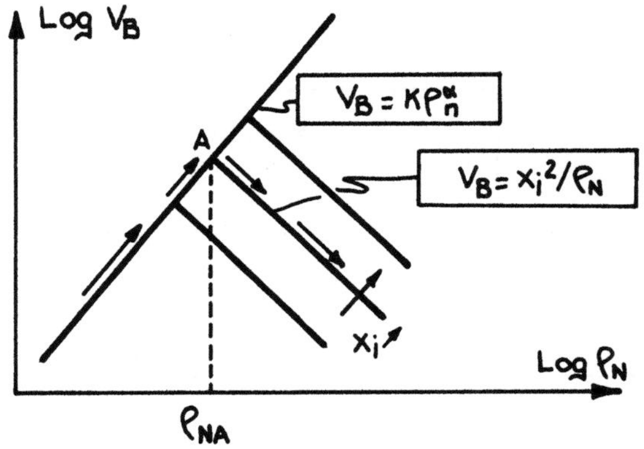

Figure 6 : Blocking voltage versus N base resistivity.

If electron irradiation is used, then Lp is decreased but τ_{sc} is generally inchanged, x_i is then increased and it possible to use lower ρ_N and W_N values.
Such an optimisation was not possible with classically doped float zone silicon because of its high resistivity variation. Mean wafer resistivity and wafer thickness had to be choosen too high to take into account the lowest and the highest resistivity value. NTD silicon wafers with tight resistivity allows to optimise the wafer thickness and resistivity to reach point A position (figure 6).

We obtained on an industrial scale, 50 mm diameter thyristors, for instance, with electrical characteristics as presented in table VI which could not be obtained with classical FZ silicon.

Table 6 : Electrical characteristics for 50 mm diameter NTD silicon thyristors.

Characteristics	Conditions	Limits
Forward blocking voltage	25° C $I_D < 5$ mA	$>$ 1500 V
Forward leakage current	125° C $V_D = 1400$ V	$<$ 15 mA
Reverse blocking voltage	25° C $I_R < 5$ mA	$>$ 1500 V
Reverse leakage current	125° C $V_R = 1400$ V	$<$ 15 mA
Forward voltage drop	$I_{TM} = 1500$ A 25° C	$<$ 1,8 V
Reverse recovery charges	$dI_R/dt = 25A/\mu s$	$<$ 200 μcb
Turn off time	$dV/dt = 20V/\mu s$ 125° C	$<$ 25 μs

SUMMARY

The long time, high temperature diffusion heat treatments of high power devices manufacturing, keeps the resistivity homogeneity of NTD silicon as it is observed after irradiation annealing. With the accuracy of spreading resistance measurement no significant resistivity variation has been observed whatever irradiation conditions, if the annealing temperature is above 750° C. Minority carrier lifetime measured on final devices was generally equal or higher with NTD silicon than with conventional FZ silicon and seems more dependent on the diffusion process than on the irradiation and annealing condition.

Nevertheless we observed that some high resistivity cristal were very sensitive to cooling rate at the end of the diffusion process. The quenched-in centers created could be only removed after a long time at diffusion temperature and a very slow cooling rate. Improvement of 40 % on diode breakdown voltage is observed on large diameter devices. Very performing thyristors are now fabricated with good yield owing to the use of NTD silicon and to the optimisation of the structural parameters it allows.

ACKNOWLEDGEMENTS

I would like to thank Mr. J.L. TEZNER for the constant interest on this work.we are indebted to O. MALMROS from Topsil for providing high resistivity cristals. This work has been supported by the french D.G.R.S.T.

REFERENCES

1/ A. SENES, G. SIFRE, M. BREANT - Semiconductor Silicon 1977 Electrochem. Soc. Princeton. N.J. p. 135 (1977)

2/ M.J. HILL, P.M. VAN ISEGHEM, W. ZIMMERMANN, I.E.E. trans. on ED 23 809 (1976)

3/ KOKOSA - DAVIES - I.E.E.E. Trans. on ED 13 874 (1966)

4/ C.T. SAH, C.T. WANG - Journal of Applied Physics 46 1767 (1975)

AN OPTIMIZATION OF BLOCKING CHARACTERISTICS OF HIGH VOLTAGE THYRISTORS USING NTD CRYSTAL

A. Watanabe, M. Naito, T. Yatsuo, M. Okamura, and K. Morita*

Hitachi Research Laboratory, Hitachi, Ltd., 4026, Kuji, Hitachi, Japan

* Hitachi Works, Hitachi, Ltd., 3-1-1, Saiwai-cho, Hitachi, Japan

ABSTRACT

A new approach for improving the blocking characteristics of a high power thyristor was investigated using the NTD silicon crystals. It is found that the blocking capability can be improved by the full ruduction of the doping concentration in the p-type diffused layers. This is due to the reduction of the injection efficiency and the avalanche multiplication.

INTRODUCTION

In HVDC transmission systems, raising up the blocking capability of thyristors is highly required to improve the cost performance and reliability of the equipments by reducing the number of thyristors used in series. The improvement of the blocking capability of thyristors has so far been carried out by increasing the resistivity and the width of the n-base. However, the blocking voltage of around 5 kV is considered to be the upper limit by this approach,[1] because the forward conduction and the dynamic characteristics become very poor for wider n-bases. Therefore, some new approaches are necessary to develop higher voltage thyristors.

On the other hand, the blocking operation of high power thyristors was recently analyzed theoretically, and it was found that the blocking characteristics depend considerably upon the diffusion profile of the p-emitter.[2]

This paper studies the possibility of improving the blocking characteristics by a suitable choice of the doping profile of the p-type diffused layers. First, a theoretical investigation is done to determine the outline of the experiment, and the theorectical prediction is examined by the experiment. The NTD crystals, which have good uniformity in resistivity, are used in the experiment to get a clear result. The experimental results are discussed finally.

THEORETICAL PRELIMINARIES

Blocking operation of a model 4 kV thyristor is investigated with the one-dimensional model of a thyristor described in ref.3, to know what an influence the impurity profile of the p-type diffused layers has on the blocking characteristics.

The model thyristor has a substrate doping of 2.5×10^{13} cm^{-3} and Gaussian type diffused layers; an n-emitter diffusion with the surface concentration of 5×10^{20} cm^{-3} and the depth of 30μm, a p-base diffusion with 1×10^{17} cm^{-3} surface concentration and 100μm depth, and a 100μm deep p-emitter diffusion with the surface concentration varying in the range from 1×10^{15} to 1×10^{20} cm^{-3}. The n-base width is 750μm, and the device area is 1 cm^2.

In general, a carrier lifetime depends on the doping concentration, and is short when the concentration is high. However, the quantitative relationship between them depends upon the individual diffusion process and is not known yet. Therefore, the lifetime in the p-emitter having the surface concentration Ns are tentatively assumed to be given by a formula

$$\tau = \tau_0\{\bar{N}(10^{17})/\bar{N}(Ns)\}^{b} \qquad (1)$$

where τ_0 is the lifetime for $Ns=1\times10^{17}$ cm^{-3}, $\bar{N}(Ns)$ the average doping concentration within 10μm distance from the junction, for the p-emitter with the surface concentration of Ns, in which the carrier injection is significant, and b a parameter. The lifetimes are incorporated in the SRH formula to give the carrier recombination rate. The values of lifetimes were obtained by fitting the calculated leakage current of a thyristor to the measured one.[2] They were τ_{po}=1ms, τ_{no}=2ms in the n-base, and τ_{po}=5μs, τ_{no}=10μs in the p-emitter and in the p-base. These values for the p-layers were used for τ_0 in eq.(1). As for the ionization coefficients of carriers, values obtained by Ogawa[4] were employed.

Using the above mentioned model, leakage current in a forward blocking state at a high temperature was calculated as a measure of the blocking capability, since the rated blocking voltage is usually defined by the voltage at which the leakage current becomes a

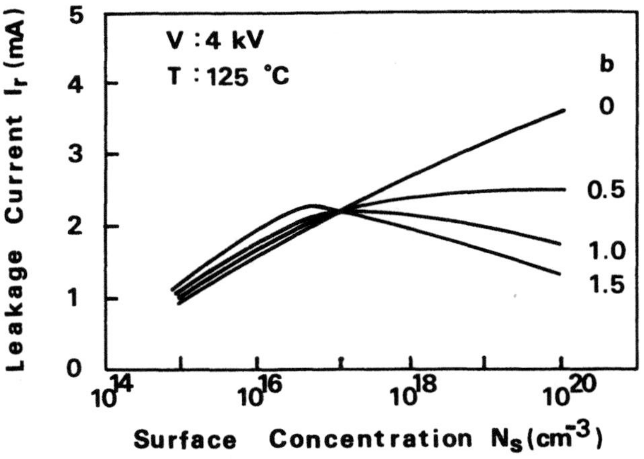

Fig. 1. Calculated leakage current as a function of the surface concectration Ns of the p-emitter. Applied voltage is 4 kV and the temperature is 125°C.

specified value at a high temperature. Reverse blocking characteristics are much the same as forward ones because of the shorted n-emitter.

Fig. 1 shows the calculated result. As shown in the figure, the leakage current depends considerably upon the doping concentration of the p-emitter. For example, the leakage current Ir at Ns=1×10^{15} cm^{-3} becomes as small as 27% of that at Ns=1×10^{20} cm^{-3} when b=0, and Ir has a broad maximum around Ns=1×10^{17} cm^{-3} and reduces to about half of the maximum value, at Ns=1×10^{15} or 1×10^{20} cm^{-3} when b=1.5. The notable point here is that the leakage current can be reduced by lowering the surface concentration of the p-emitter irrespective of the choice of b, for Ns less than about 3×10^{16} cm^{-3}. This result is very interesting, since the p-emitter profile has so far been considered to have little effect on the blocking characteristics.

The reason why the leakage current depends on the p-emitter profile was investigated by calculating the common-base current gains of the transistors composing a thyristor. The current gain of the n-p-n transistor section was negligibly small as the order of 10^{-3} because of the shorted n-emitter, and thus, this transistor section did not contribute to the operation of the thyristor. For the p-n-p transistor section, the current gain α was decomposed into its components β and γ, where β is the n-base transport factor and γ the injection efficiency of the p-emitter. The result

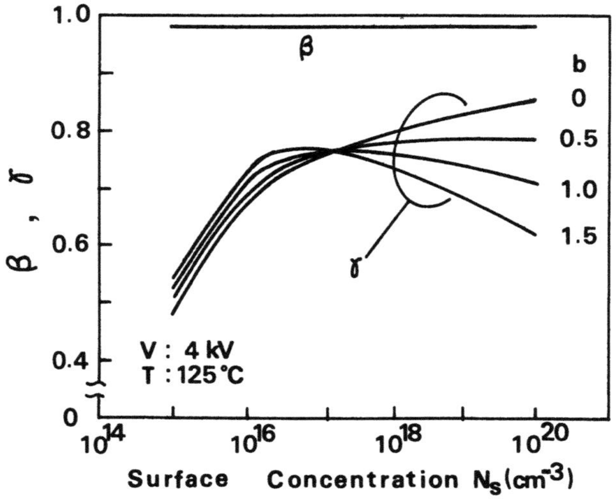

Fig. 2. Transport factor β and injection efficiency γ as a function of Ns. Applied voltage is 4 kV and temperature is 125°C.

is shown in Fig. 2. As seen in the figure, the injection efficiency γ is much smaller than the transport factor β, and hence the current gain $\alpha = \beta \cdot \gamma$ is determined by γ rather than by β. This is just contrary to the conventional assumption[1,5] for power thyristors that $\gamma = 1$ and α is determined by β. In this way, the Ns dependence of Ir shown in Fig. 1 directly results from the Ns dependence of γ.[2]

The reduction of doping level of p-layers is also acceptable from the viewpoint that it would result in lower avalanche multiplication.

EXPERIMENTAL PROCEDURE

In the following, the devices with differing doping levels in the p-type diffused layers were fabricated, and the theoretical prediction obtained above was examined.

Although a thyristor is a p-n-p-n four layer devices, its operation in the blocking state is essentially the same as that of a p-n-p transistor because of the shorted n-emitter. Therefore, p-n-p devices were fabricated in this work to simulate thyristors.

As for the starting wafer, the NTD crystals, which give little variation in resistivity, were used in order to obtain a clear result. The FZ crystal has the resistivity variation of more than ±20% in a wafer,[6] which corresponds to the uncertainty of the avalanche breakdown voltage of more than 1 kV for devices with the blocking voltage of several kilo-volts. Hence, the FZ crystals are entirely inadequate for the detailed experiment.

The width and the resistivity of the n-base of the samples were 1000μm and 290 Ω cm, respectively. Aluminum was diffused to form p-layers. Then, the p-emitter and the p-base were formed simultaneously. In order to examine the effect of doping level, samples of three different doping profiles were prepared. The doping profile, measured with the successive etching and four-point probe method, are shown in Fig. 3. They are well approximated by the exponential function exp(-x/L) near the p-n junction, with the values of L shown in the figure. The quantity L can be used as a measure of the doping concentration. Concentration is low when L is large.

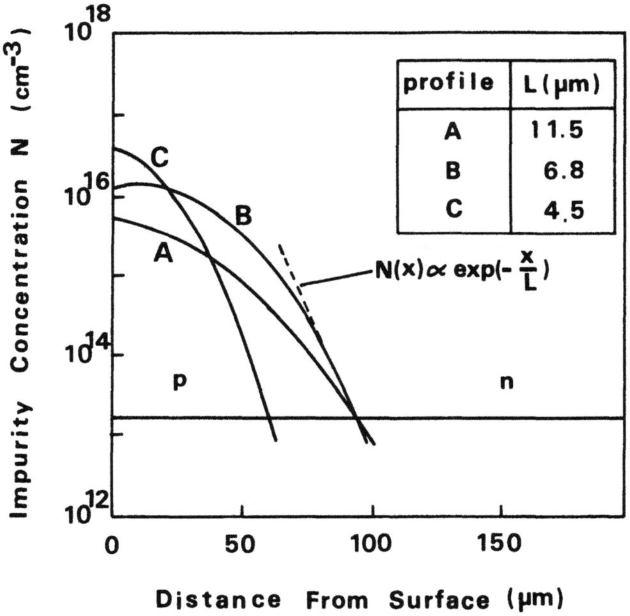

Fig. 3. Three types of doping profiles for p-emitter and p-base used in the experiment.

After diffusing the p-layers, pellets with 24mm diameter were cut from the wafers, and aluminum was evaporated and sintered on both sides of the pellets to form ohmic contacts. Then, the pellet edge was beveled to the angle of 45° and coated with silicone rubber to protect against surface breakdown.

RESULTS AND DISCUSSION

Current-voltage curves of the samples are shown in Fig. 4 and in Fig. 5. These are the typical results for the samples having the profiles A, B, or C. Fig. 4 and the data for T=90°C in Fig. 5 clearly show that the breakdown voltage certainly increases as the length L increases, i.e., as the doping concentration decreases. As for the leakage current at the voltages below the breakdown voltage, those of B and C are inverted, and the sample having the profile B shows poorer blocking characteristic than the sample having C. However, the leakage current of A is small showing good blocking characteristic as well.

The blocking voltages, which are defined here as the voltages for the leakage current density of 2.5mA/cm², are shown in Fig. 6. The blocking voltage falls off remarkably for temperatures higher than about 125°C. At temperatures below 125°C, the blocking voltage is high in order of A, B, C, while, at higher temperatures, the

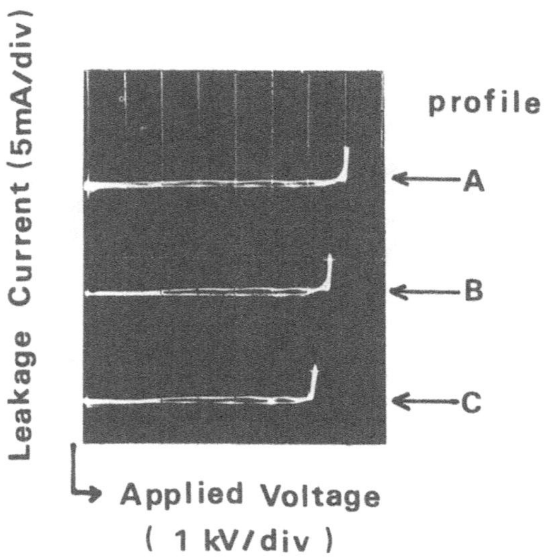

Fig. 4. Current-voltage curves of the samples at 20°C. Typical results for the profiles A, B, and C are shown.

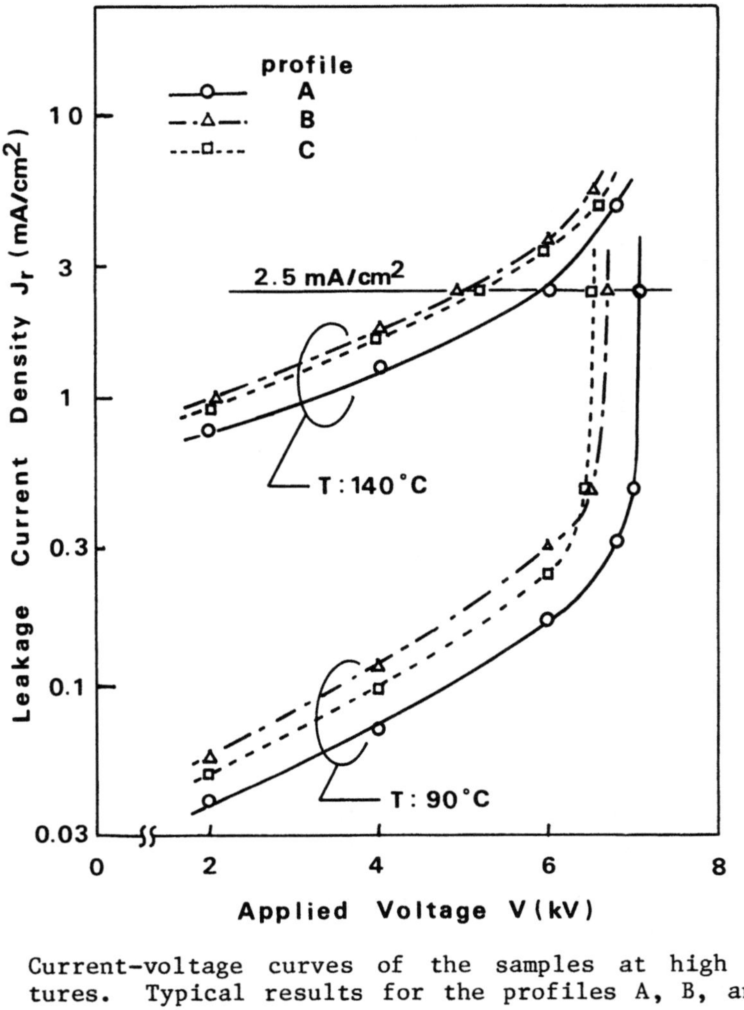

Fig. 5. Current-voltage curves of the samples at high temperatures. Typical results for the profiles A, B, and C are shown.

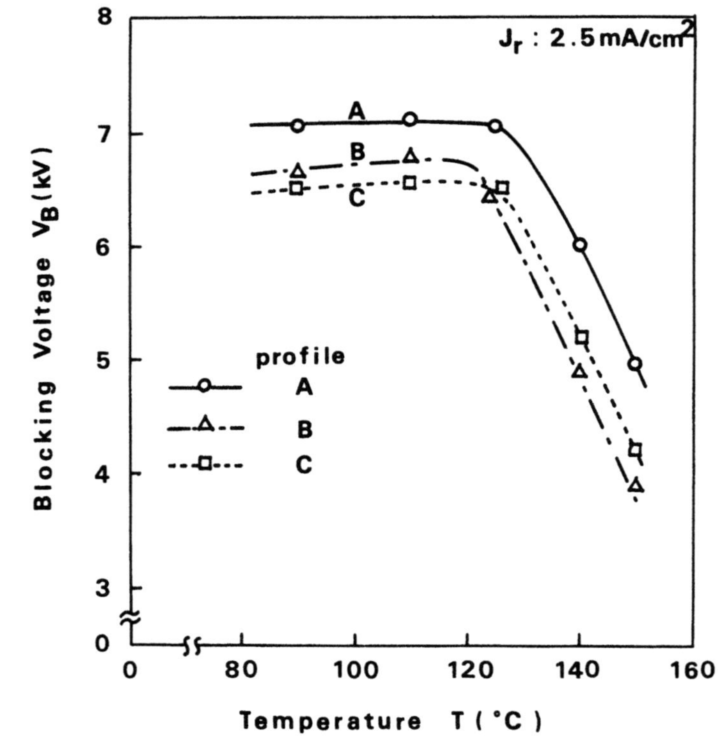

Fig. 6. Temperature dependence of the blocking voltage.

order of V_B for the samples having the profiles B and C is inverted and V_B for B comes the lowest. This is because the blocking voltage is determined essentially by the leakage current at high temperatures, while at low temperatures, it is near the breakdown voltages, as shown in Fig. 5.

In the below, the reason why the magnitude of the leakage current is in order of A, C, B will be discussed. The leakage current can be decomposed into the generation current in the depletion region and the current amplification factor due to the transistor action. The current continuity equation gives the formula

$$Jr = P \cdot Jg \quad (2)$$

where Jr is the leakage current, Jg the generation current, and P

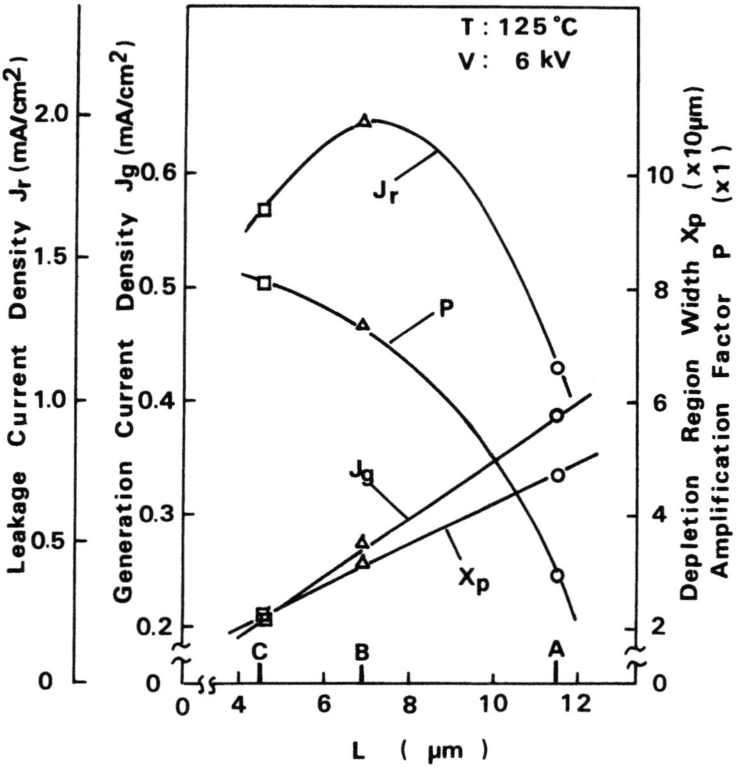

Fig. 7. Comparison of Jg and P for three types of profiles, where Jg is the generation current in the depletion region and P is the current amplification factor due the the transistor action, which are multiplied to give the leakage current Jr. Quantity x_p is the depletion region width in the p-base. Applied voltage is 6 kV.

is the current amplification factor;

$$P = \frac{M}{1-M\cdot\alpha} \tag{3}$$

with M the avalanche multiplication factor. These factors P and Jg are calculated in the following way. For a given applied voltage, the depletion region width is calculated in turn giving Jg. Then, the factor P is obtained from eq.(2) using the measured leakage current. In calculating Jg, carrier lifetimes were assumed to be the same for all three profiles A, B and C for simplicity, and they were 15μs in the p-base and 3ms in the n-base, corresponding to the lifetime $\tau_{p0}+\tau_{n0}$ used in the theoretical consideration.

The result is shown in Fig. 7 for T=125°C as an example, where x_p is the width of the depletion region in the p-base. As shown in the figure, Jg increases with increasing L as a result of the increase in x_p. The magnitude of Jg is almost determined by x_p irrespective of the width of the depletion region in the n-base, because the lifetime in the p-base is much shorter than that in the n-base. The factor P decreases with increasing L. This is consistent with the experimental result that the breakdown voltage is higher for larger L, and is a natural result because M and γ decreases as the value of L increases. The dependence of Jr upon L can be understood from the variation of Jg and P with L. Namely, the current Jr and the factor P have different quantitative dependencies on L, which results in the non-monotonic behavior of Jr. For the sample having the profile B, the value of P is not so small while Jg becomes large, resulting in large Jr. When the value of L becomes as large as that of A, the factor P decreases remarkably, and the smallness of P prevails over the large Jg resulting in small Jr.

CONCLUSION

A possibility of improving the blocking characteristics of a high power thyristor by lowering the doping concentration in p-layers was pointed out by a theoretical consideration. Based on this, samples with different doping levels were fabricated and their blocking characteristics were measured.

It is found that the bloking voltage at low temperatures certainly increases with decreasing doping concentration. On the other hand, the blocking voltage at high temperatures does not always increase for the samples with lower level p-layers, because of larger generation current resulting from wider depletion region. However, the full reduction of the doping concentration is found to improve the blocking capability at high temperatures as well as at low temperatures, by the reduction of the injection efficiency and avalanche multiplication.

ACKNOWLEDGEMENTS

The authors would like to thank Drs. T. Takasuna, Y. Takeda and Mr. S. Kawai for their encouragement and guidance throughout this work. They also would like to thank Dr. N. Monma for his contribution to the diffusion process and Mr. T. Yokota for preparing the samples.

REFERENCES

1. T.Ogawa, et al., Electrical Characteristics of Ultrahigh-Voltage Thyristors and Related Problems, IEEE Trans. Industry Applications, IA-10:112 (1974).
2. M.Naito, et al., NUMERICAL ANALYSIS OF BLOCKING CHARACTERISTICS FOR THYRISTORS, Tech. Digest, 1978 IEEE IEDM: 560.
3. M.Naito, et al., One-dimensional Analysis of Turnoff Phenomena for a Gate Turnoff Thyristor, IEEE Trans. Electron Devices, ED-26:226 (1979).
4. T.Ogawa, Avalanche Breakdown and Multiplication in Silicon p-i-n Junctions, Japanese J. appl. Phys., 4: 473 (1965).
5. Y.C.Kao and D.J.Page, THEORETICAL LIMITS FOR HIGH VOLTAGE RECTIFIERS AND THYRISTORS, Tech. Digest, 1977 IEEE IEDM: 313.
6. M.J.Hills, et al., Preparation and Application of Neutron Transmutation Doped Silicon for Power Device Research, IEEE Trans. Electron Devices, ED-23: 809 (1976).

THE FABRICATION AND CHARACTERIZATION OF SPREADING RESISTANCE TEMPERATURE SENSORS USING NTD SILICON

Carl Derrington

Motorola Semiconductor Group Inc.
Semiconductor Research and Development Laboratories
Phoenix, AZ 85008

ABSTRACT

The design and fabrication of a silicon thermistor exhibiting a positive temperature coefficient are presented. The device performance utilizes the temperature dependence of the electrical resistivity of silicon as a fundamental property and the spreading resistance technique for monitoring the change in resistivity. It was found that the resistance of these parts is given by $R = \rho/2\pi a \cdot K$ where ρ is the resistivity of the starting material, 2a is the diameter of the contact and K is an empirical constant close to unity. The nominal resistance value can be controlled by choosing the correct substrate resistivity, contact geometry and enhancement diffusion depth. By using NTD silicon wafers as starting material it was found that over 80% of the die are within ±1% of the nominal resistance value. The temperature vs. resistance function was found to deviate slightly from linearity and when plotted logarithmically gave a TCR of 0.65%/^{o}C over the temperature range of -40°C to 150°C. An interesting relationship for designers to take note of is that regardless of nominal resistance at a given temperature, if one takes resistance ratios (i.e., R_{125}/R_{25} or R_{-40}/R_{25}, etc.) one gets remarkably consistent results. This implies that the unit to unit tracking of the temperature vs. resistance slope is excellent and thus allows for the design of systems that require only a single point calibration.

INTRODUCTION

The spreading resistance technique has long been used for determining local resistivity and resistivity variations in silicon (1). Many discussions have appeared in the literature discussing the technique and the problems associated with it (2). In this paper the fabrication of temperature sensors using spreading resistors with precisely controlled geometries in NTD silicon is presented.

It has been shown (3) that for a perfectly conducting hemisphere embedded in a homogeneous semi-infinite solid of resistivity ρ, the spreading resistance in the semi-infinite solid is given by:

$$R_s = \rho/2\pi a \qquad [1]$$

where a is the radius of the embedded hemisphere. In the conventional spreading resistance method a probe is pressed directly onto the silicon surface. In this case a more appropriate relationship describing the spreading resistance is:

$$R_s = \frac{\rho}{4a} \qquad [2]$$

and the actual measured resistance R_c is given by (4)

$$R_c = K\, R_s \qquad [3]$$

where K is an empirical factor. Kramer and Van Ruyven (5) have published a compilation of K factors from different laboratories and on material of varying resistivities and have shown that K varies by almost an order of magnitude.

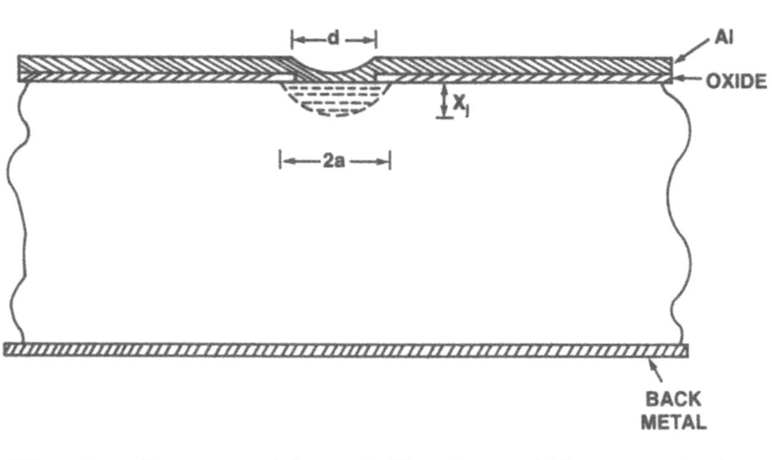

Fig. 1. Cross-section of Aluminum-silicon contact

In designing and fabricating a positive temperature coefficient thermistor we have used the temperature dependence of the electrical resistivity of silicon as a fundamental property and the spreading resistance technique for monitoring the change in resistivity.

EXPERIMENTAL

In contrast to the conventional spreading resistance method, in this study contact was made to n-type silicon with an n^+ enhancement-aluminum contact. A cross section of the aluminum-silicon contact is shown in Figure 1. It has been shown (1) that with this technique the radius of the contact, a, is given by:

$$a = d/2 + 2/3\ X_j \qquad [4]$$

where d is the diameter of the via in the oxide layer and X_j is the depth of penetration of the n^+ region. Two factors that have a large influence on the resulting spreading resistance, R_s, of a die are the resistivity of the bulk of the material and the size of the contact. Using standard photolithographic techniques and choosing diffusion temperatures such that diffusion times were in excess of 60 minutes it was found that the contact geometry could be well controlled. In order to control the resistivity of the starting material NTD-silicon was chosen. Earlier work (6) has demonstrated that both the radial resistivity fluctuations and the wafer-to-wafer resistivity fluctuations of NTD silicon is as low as 2%, compared with >10% radial resistivity and 20% wafer-to-wafer resistivity fluctuations for FZ material. In this study material of 5.0 Ωcm to 10.0 Ωcm was used. The via size, d, was varied between 12.7 μ to 50.8 μ in 2.54 μ intervals and the n^+ penetration depth, X_j varied between 2.0 μ to 3.5 μ. The die size was 381 μ x 381 μ x 305 μ.

The temperature dependence of the spreading resistance was monitored by placing plastic encapsulated (TO-92) parts in constant temperature oil bath (Lauda K4R ±0.2°C), forcing 1.0 mA constant current and measuring the voltage drop.

The fabrication of silicon thermistors utilizing the spreading resistance technique requires the ability to make devices with a predetermined nominal resistance at a given temperature. Devices fabricated using the method described above were found to follow the relationship:

$$R = \frac{\rho}{2\pi a} \cdot K \qquad [5]$$

where K is an empirical factor between 0.95 and 1.05. Figure 2 shows typical results of the resistance of devices at room temperature as a function of 1/2a for a constant X_j = 3.5 μ and a ρ = 6.0 Ωcm. It is interesting to note that the factor K shows a

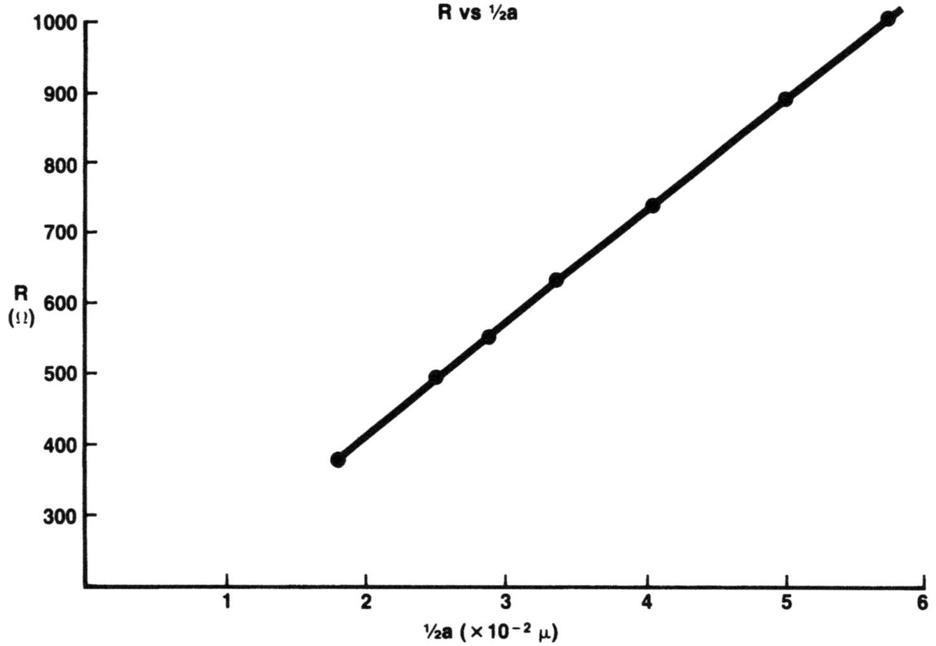

Fig. 2. Experimental spreading resistance values plotted against a/2

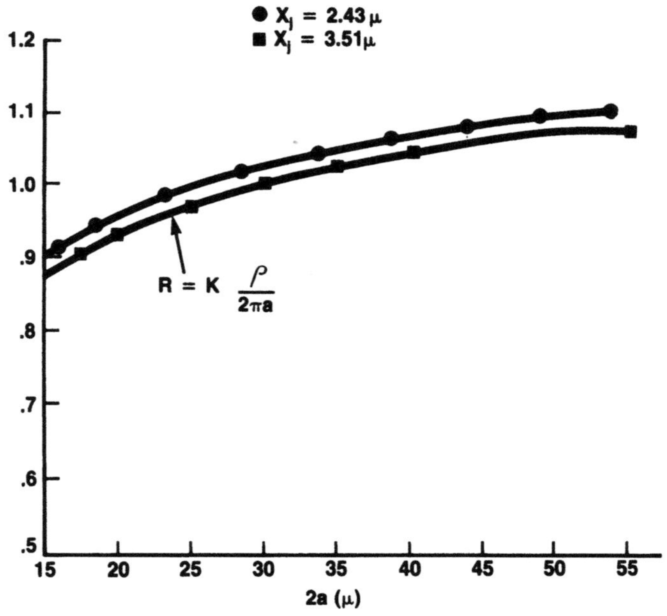

Fig. 3. Plot of the K factor as a function of (2a)

slight dependency on 2a (Figure 3). It is not clear at this time the nature of this dependency. However, it is clear that by simply knowing the resistivity of the starting material and carefully controlling the contact via size and n^+ penetration depth thermistors with a predetermined nominal resistance at a given temperature can be fabricated routinely. The importance of choosing NTD silicon for starting material can be seen in Figures 4, 5 and 6. These charts show the frequency distribution of resistance values of devices fabricated in a single lot of wafers. By carefully controlling the processing parameters it is not only possible to fabricate devices with a predetermined nominal resistance but it is also possible to obtain over 80% die yield that are ±1% of nominal. This is indeed an important consideration when fabricating temperature sensors since it is highly desirable to have part-to-part interchangeable with a minimum of calibration and still maintain the accuracy of the system.

If one assumes that the temperature dependence of a spreading resistor comes from the temperature dependence of the mobility of charge carriers (in this case the majority carriers are electrons) then the following is true:

$$R = \frac{1}{ne\mu 2\pi a} \cdot \qquad [6]$$

According to Sze (7) the temperature dependence of the mobility of electrons is given by:

$$\mu = A\, T^{-5/2} \qquad [7]$$

where A is a constant if m, the effective mass, is not a strong function of T. Substituting [7] into [6] one obtains:

$$R = \frac{T^{5/2}}{Ane2\pi a} \qquad [8]$$

or R is proportional to $T^{5/2}$. A plot of ℓn R vs. ℓn T is shown in Figure 7. The slope of this line shows the temperature dependence of R and is equal to 2.83 which agrees well with the $T^{5/2}$ dependence. Figure 8 is a plot of R vs. T of a typical spreading resistance temperature sensor. The solid line in this figure is a least squares straight line fit of the type $y = Bo + B_1X$ and the dashed line a polynominal linear regression fit of the type $y = Bo + B_1X + B_2X^2$. As can be seen from the value of $B_2 = 0.0145$ the nonlinear contribution is small. This nonlinearity however is large enough to introduce large temperature errors if one assumes the device is linear. This nonlinearity can be easily compensated for by placing a shunt resistor in parallel to the spreading resistance temperature sensor in a signal conditioning circuit. A more critical parameter to investigate is the part-to-part reproducibility, i.e., how well one part tracks with temperature compared with another. A straightforward method for determining this

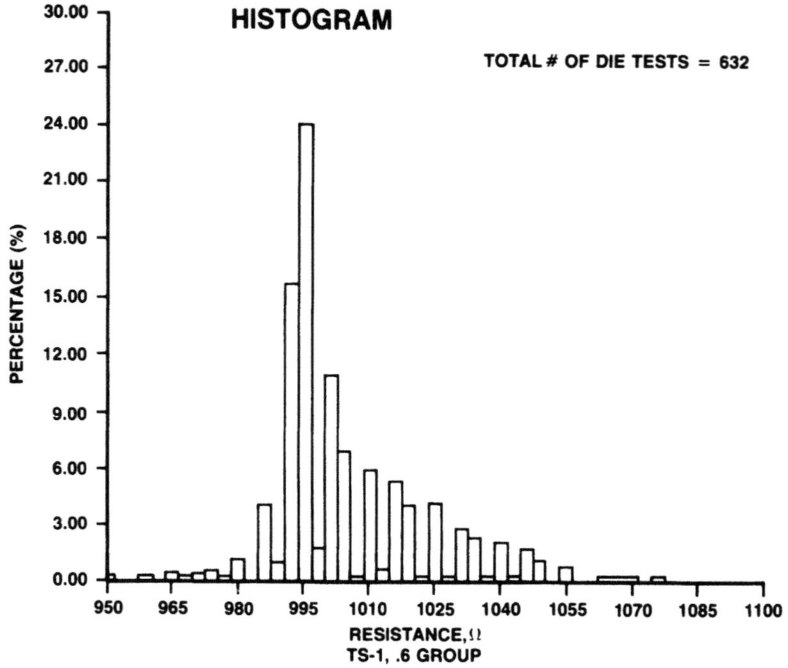

Fig. 4. Frequency distribution of spreading resistance values of devices from one wafer (TS-1, .6 group).

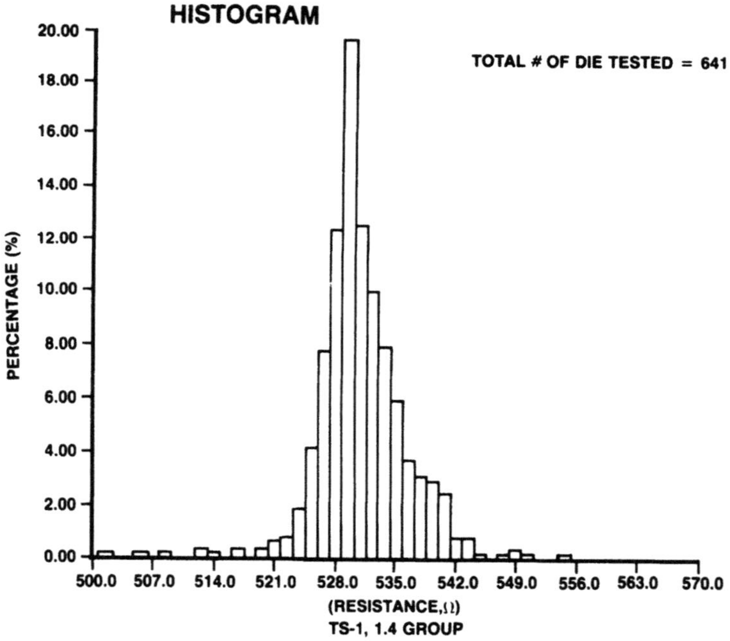

Fig. 5. Distribution of device resistivities (TS-1, 1.4 group)

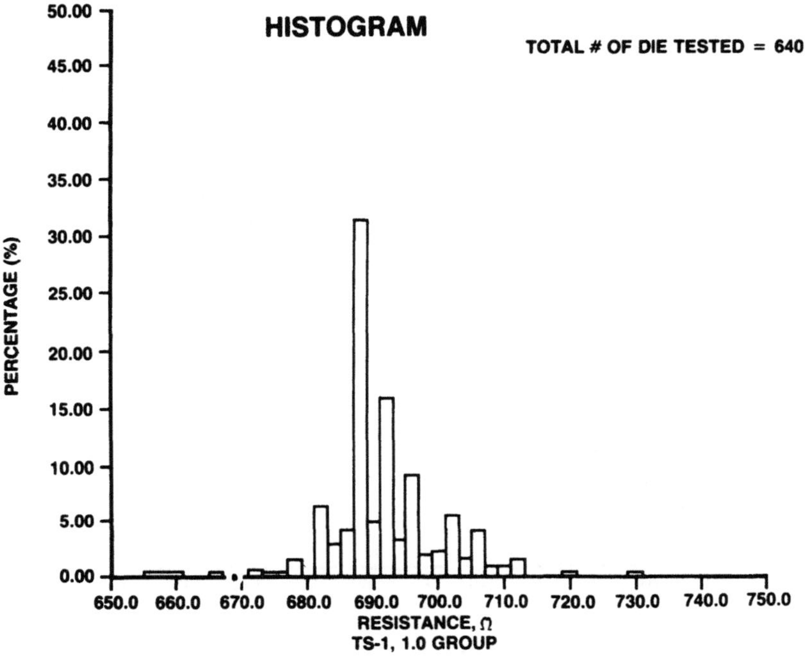

Fig. 6. Distribution of device resistivities (TS-1, 1.0 group)

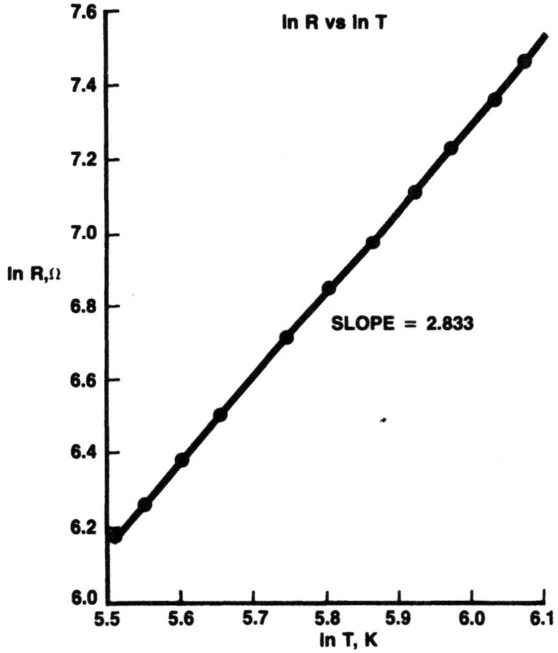

Fig. 7. Relationship between ln R (Ω) and ln T (K)

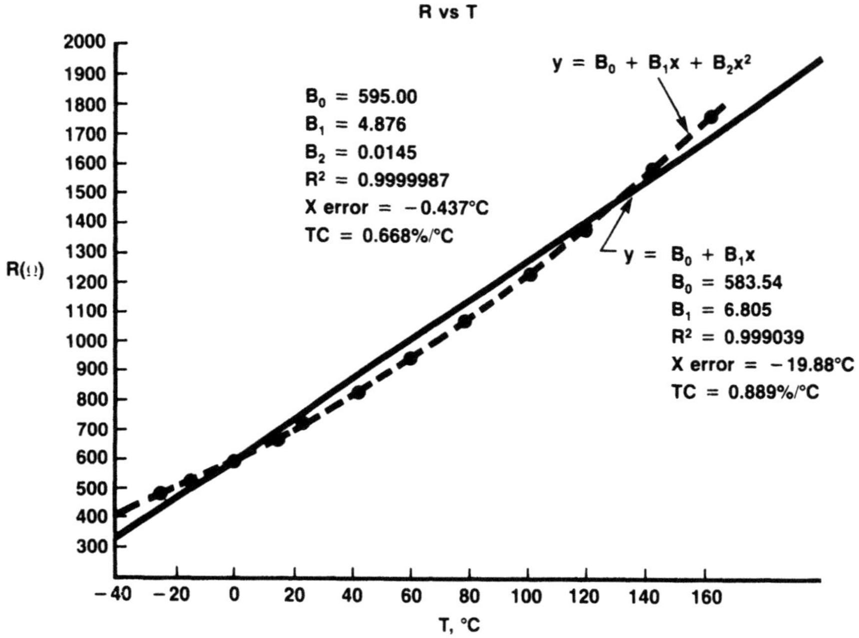

Fig. 8. Device resistivity as a function of device temperature. Also shown are polynominal approximations to experimental curve.

is to take resistance ratios of different devices (8). This ratio is simply the ratios of two resistance values taken at two different temperatures. Table I gives these resistance ratios of spreading resistance temperature sensors. It can be seen from the small σ that excellent part-to-part resistance vs. temperature tracking is achieved. This ratioing technique allows for part-to-part interchangeability, regardless of the nominal resistance, with only a single point calibration. By using this technique temperature errors as small as $\pm 2^{0}$C can be achieved.

CONCLUSIONS

Silicon thermistors exhibiting a positive temperature coefficient have been fabricated. It was found that by using an n^+ enhancement diffusion-aluminum contact and carefully determining the contact geometry devices (i.e., via size & X_j) followed the relationship:

$$R = \frac{\rho}{2\pi a} \cdot K$$

where K is an empirical factor close to unity and 2a is the contact radius. The choice of NTD silicon as starting material makes it possible to fabricate devices with a die yield of over 80% of ±1%

TABLE I. Ratios of device resistance values at various temperatures and spread (σ) of experimental values.

$R_{(-25)}/R_{25}$	σ	R_{125}/R_{25}	σ	R_{160}/R_{25}	σ
0.675	.001	1.937	.007	2.461	.009

of nominal. The resistance vs. temperature dependence follows the $T^{5/2}$ dependence that electron mobility exhibits as would be expected. While spreading resistance temperature sensors are far more linear than other types of thermistors, the nonlinearity is still sufficient to introduce large temperature errors unless linearity is established by using a shunt resistor in parallel in an external conditioning circuit. By using a resistance ratio technique it is possible to have part-to-part interchangeability giving temperature errors of ±2.0°C in the temperature range of -40°C to 150°C with only a single point calibration.

REFERENCES

1. J. Krausse, Natl. Bur. Std. Special Publication 400-10, June 13, 1974.
2. "Methods to Experimental Physics," Vol. 6, K, Lark-Horovitz and V. A. Johnson, Ed. (1959).
3. R. Holm, "Electric Contacts Handbook" 3rd Ed., Sringer, Berlin (1958).
4. R. G. Mazur and D. K. Dickey, J. Elect. Chem. Soc., 113 (3), 255 (1966).
5. P. Kramer and L. J. Van Ruyven, Sol. St. Elec. 15, 775 (1972).
6. C. J. Varker and M. Liaw, private communication.
7. S. M. Sze, "Physics of Semiconductor Devices," Wiley & Sons, New York (1964).
8. K. C. Wiemer, W. T. Matzen, S. J. Tribble, J. G. McIntire, and T. Bialek, SAE International Automotive Engineering Symposium, Paper No. 770395 March (1977).

CHARACTERIZATION OF NTD SILICON CRYSTALS BY THE PHOTOLUMINESCENCE TECHNIQUE

Michio Tajima

Electrotechnical Laboratory, Ibaraki 305, Japan

Atsushi Yusa

Komatsu Electronic Metals Co., Ltd., Kanagawa 254, Japan

ABSTRACT

The neutron-transmutation-doped (NTD) silicon crystals with a controlled neutron dose have been used as standard samples for the calibration of the photoluminescence (PL) method for the quantitative impurity analysis. The revised calibration curve for determining the P concentration in the range between 10^{11} and $10^{15}cm^{-3}$ is reported. The calibration curve for B is also revised using the results obtained by the accurate PL and resistivity measurements. It is demonstrated that the concentrations of the transmuted P^{31} and background shallow impurities in commercial NTD crystals are determined precisely by using these calibration curves.

In order to investigate the properties of radiation-induced defects, neutron irradiation has been performed by four different reactors with the thermal to fast neutron flux density ratio varying from 5 to 1000, followed by heat-treatment at 380-1200°C for 10min-30h. Depending on the stage of annealing, many characteristic PL lines associated with the defects can be observed. The appearance of these lines is correlated with the fast neutron dose. It should be noted that the PL spectral change due to the irradiation and subsequent annealing is quite drastic in comparison with the resistivity change.

1. INTRODUCTION

Neutron transmutation doping (NTD) has been established as a key technique for improving the performance and yield of high power devices and detectors because of the advantages of doping homogeneity and controllability.[1,2] However, some problems are pointed out as to the radiation-induced defects and residual impurities.[1-3] Photoluminescence (PL) has been regarded as a promising technique for characterizing small amount of shallow impurities and various types of defects.[4-6] However, the validity of this method for the quantitative analysis of trace impurities was not confirmed, because there are no other analytical methods to determine the absolute concentration less than $10^{14}cm^{-3}$.

In this study we take advantages of the two techniques complementarily for solving the respective problems. First we describe the PL measurement techniques in Sec. 2. Then we report the calibration of the PL method for the quantitative impurity analysis by using the NTD technique. Conversely, Sec. 4 deals with the characterization of heat-treatment behavior of radiation-induced defects in NTD crystals by the PL technique.

2. PL MEASUREMENT TECHNIQUE

The samples for the PL analysis were cut into small rectangular pieces, $\sim 12x6x0.4mm^3$ in size. The samples were etched with a solution of 4 parts HNO_3 and 1 part HF for 1-2min before the measurement. They were immersed in liquid helium in a glass Dewar.

The 514.5 nm line from an Ar ion laser was used as an excitation source. The laser beam was unfocused on the sample with a beam diameter of about 2.5 mm at $1/e^2$-intensity-point, and the incident power was 300 mW. The PL from the sample was analyzed with a Spex 1701 grating monochromator (f/7; 0.75 m), and detected by an HTV 7102 photomultiplier (S1) cooled with dryice. The PL was measured in the photon-energy-region ranging from 0.98 to 1.16 eV. The photoresponse of the measurement system was not calibrated. The spectral resolution of the monochromator was set to 0.25 meV for the purpose of the quantitative impurity analysis, 0.69 meV for the defect analysis, and 0.125 meV for investigating a fine structure. Special precautions were taken to adjust the sample position precisely in the optical system with high reproducibility. The electrical signal was amplified with a lock-in technique, and resulting signal was transferred to an HP 9825A computer system in order to perform several data processings.

3. QUANTITATIVE IMPURITY ANALYSIS

3.1. Calibration of the PL Method

We reported that the impurity concentration can be determined by the PL analysis with using the calibration curves.[4] However, the reported calibration curves were obtained by a comparison of the PL intensity ratio with the resistivity data. Therefore, the errors are expected to become large in the low concentration range because of the difficulties in the reliable resistivity measurements. Moreover, it is quite difficult to obtain samples with various doping levels below 10^{13} cm^{-3}, because the controlled P doping is impossible in that concentration range.

The NTD technique makes it possible to perform a controlled and accurate P doping in the extremely low concentration range. The transmuted P concentration, $[P^{31}]$, can be obtained from the thermal neutron flux density, Φ_{th}, as follows[1];

$$[P^{31}] = 1.7 \times 10^{-4} \Phi_{th} \cdot t \quad (cm^{-3}) \qquad (1),$$

where t is irradiation time in second, and Φ_{th} is in cm^{-2} s^{-1}. Therefore, we can use the specially grown NTD crystals as "standard samples" for the calibration of the PL method for determining the P concentration.

3.2. Preparation of Standard Samples

In order to eliminate the effect of background impurities, the ultra-high-pure single crystals were used for a starting material for NTD. For this purpose a highly pure poly crystal was specially prepared and then float-zoned in vacuum for four times. The resistivity of the ingot was 106-110 kΩ cm. Judging from the PL analysis, the residual B and P concentrations of this crystal are the lowest of all the crystals which we have ever measured ($N_B \sim 1.3 \times 10^{11}$ cm^{-3}; $N_P \sim 6 \times 10^{10}$ cm^{-3}).

Wafers were sliced from the adjacent part of the ingot, and were subjected to neutron irradiation with a thermal neutron flux density, Φ_{th}, of 5×10^{11} cm^{-2} s^{-1}. The irradiation time was varied from 0.5 to 6 h (5 levels) to produce P^{31} with a concentration ranging from 1.5×10^{11} to 1.8×10^{12} cm^{-3}. After irradiation the wafers were annealed at 800°C for 23 h in a dry O_2 atmosphere.

The PL spectra of a starting material, a crystal after 0.5-h irradiation and that after 1-h irradiation are shown in Fig. 1 (a), (b) and (c), respectively. The symbol I in the spectrum denotes the intrinsic luminescence, and B and P are the extrinsic luminescence associated with B and P, respectively. The subscript indicates the phonon emitted, and the notations in parentheses denote the type of transition; FE, BE, and b_n denote free exciton, bound exciton, and

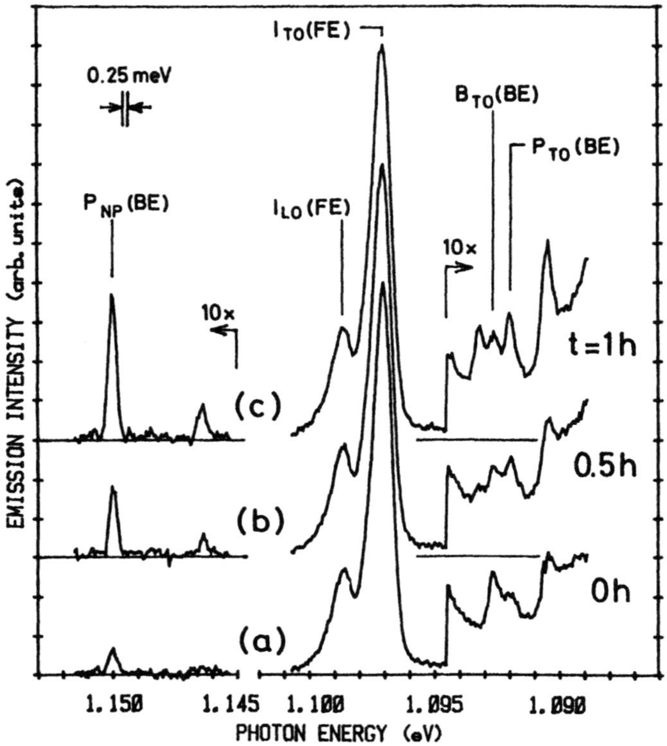

Fig. 1. PL spectra from highly pure Si crystals as a function of a neutron dose. (a) a starting material; (b) after 0.5-h irradiation; (c) after 1-h irradiation.

bound multiple exciton complex, respectively. All the spectra were normalized by the peak intensity of the respective I_{TO}(FE) component.

The intensity of P components increases with increasing irradiation time, while that of B components keeps constant. As reported in previous papers,[4,5] the PL intensity ratio between extrinsic and intrinsic components can be used as a measure of the impurity concentration. Therefore, the increase in the intensity ratio for P components indicates the increase in the P^{31} concentration by NTD, whereas the constant ratio for B components means the fact that the B concentration is not affected by NTD.

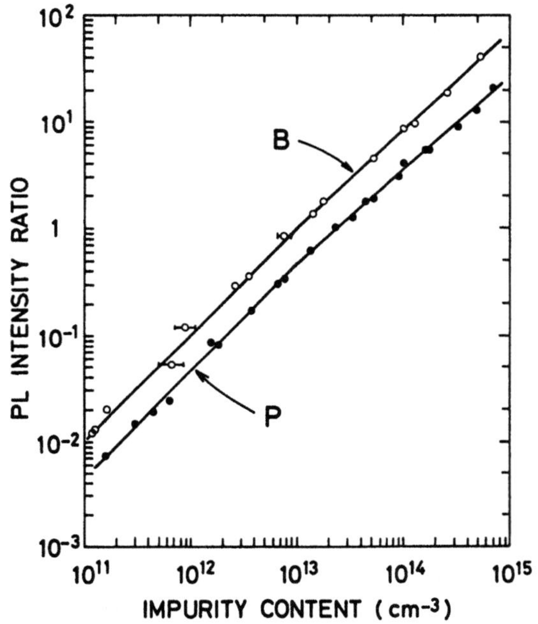

Fig. 2. Calibration curves for determining the P and B concentrations in Si from the PL analysis. The PL intensity ratios of P_{TO}(BE)/I_{TO}(FE) and B_{TO}(BE)/I_{TO}(FE) are plotted against the P and B concentrations, respectively. (submitted to Appl. Phys. Lett.)

3.3. Calibration Curves

The calibration curve for determining the P concentration from the PL analysis is shown in Fig. 2. The PL data on the standard samples described in 3.2 were used to make the calibration curve in the P concentration range below $2x10^{12}$ cm^{-3}. As for the concentration range higher than $2x10^{12}$ cm^{-3} , we used the PL data on the commercial NTD crystals which were so selected by the PL analysis that they contained quite low levels of background impurities. The P concentration, N_P, was calculated from the thermal neutron dose with using eq. (1), or from the resistivity, ρ, with using the relationship reported by Irvin[7];

$$N_P = 4.4 \times 10^{15} / \rho \qquad (2).$$

The calibration curve for B is also shown in Fig. 2. The samples used for making the calibration curve were checked by the PL analysis so that their background-impurity-levels were negligible. The B concentration, N_B, was obtained from the Irvin's relation[7];

$$N_B = 1.3 \times 10^{16} / \rho \qquad (3).$$

3.4. Characterization of Residual Impurities in NTD Crystals

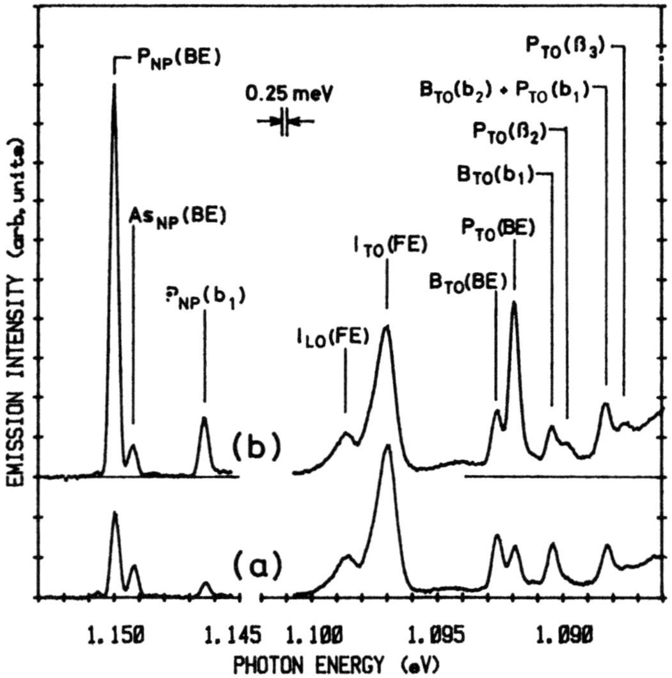

Fig. 3. A change in PL spectra due to NTD. (a) before irradiation; (b) after irradiation followed by annealing.

A typical example of PL spectrum of a commercial NTD crystal is shown in Fig. 3, where the PL of the starting material is also given for a comparison purpose. The spectra are so multiplied as to be equal magnitude at the I_{TO}(FE) peak.

In a starting material the B, P, and As components are clearly detectable. It is not uncommon in the PL analysis that commercially grown FZ crystals contain As and/or Al besides B and P.[16] The B and As components could also be detected after NTD. The PL intensity ratio between impurity and intrinsic components for P increases from 0.23 to 1.11 by NTD, while the ratios for B and As are kept constant at 0.36 and 0.22, respectively. That is, the intensity ratios for B and As are not affected by the drastic concentration change in coexisting P. This result indicates that the PL method for the quantitative impurity analysis can be applied not only to dominant impurity but to background impurities. Therefore, we can determine the P and B concentrations simultaneously by using the calibration curves in Fig. 2. As for Fig. 3, the B and P concentrations in the starting material are determined to be 3.6×10^{12} and 5.0×10^{12} cm^{-3}, respectively, and the P concentration after NTD is 2.7×10^{13} cm^{-3}.

At present, we have no other method that can make a quantitative and simultaneous analysis of donor and acceptor impurities in the ppb to ppt levels than the PL method. Therefore, the precise impurity analysis by the PL method is quite helpful for performing an accurate impurity control by the NTD technique.

3.5. Discussion

The calibration curves should be used for the PL data taken under the following experimental conditions: First, the excitation intensity of the laser beam is 300 mW with a beam diameter of about 2.5 mm. Second, the spectral resolution of the monochromator is 0.25 meV. Third, the S1 type photomultiplier is used as a detector. When applying the calibration curves to the quantitative impurity analysis, one should compare the PL results with the resistivity results for a few samples which contain negligible amount of background impurities. The NTD crystals of a known neutron dose are most suitable for this purpose. If the experimental conditions are somewhat different from ours, the data points deviate systematically from our calibration curves. In that case, one should shift the calibration curves so that the data points lie on the curves.

The specifications of the PL method in the case of using our particular experimental system are as follows: The reproducibility of the PL data is within a 20% deviation. The detection limits of B and P are 1×10^{11} and 5×10^{10} cm^{-3}, respectively. We can not discuss the accuracy of the PL method at present, since there are no other methods which can determine individual impurities in the concentration range between 10^{11} and 10^{15} cm^{-3}. Therefore, we compare the $|N_P - N_B|$ value obtained from the PL analysis with the $|N_D - N_A|$ value from the resistivity measurements. The difference between the two values is less than 30% in the concentration range between 10^{11} and 10^{15} cm^{-3}.

The errors in the previous calibration curves are due mainly to the errors in resistivity measurements and partly to the lack of the base-line consideration in the spectral analysis.[17)] The PL intensity ratio depends linearly on the impurity concentration in the range lower than 10^{13} cm^{-3} in the revised calibration curves. This simple relationship seems to be correlated with the fact that the calibration curves hold good not only for dominant impurities but for compensating impurities.

4. CHARACTERIZATION OF RADIATION-INDUCED DEFECTS

The NTD crystals should be subjected to heat-treatment in order to anneal out various radiation-induced defects. Since the remaining defects detrimentally affect the properties of NTD crystals, the analysis of the defects is quite important. We investigate the behavior of the defects under various conditions of neutron-irradiation and subsequent heat-treatment.

4.1. Sample Preparation

The four FZ ingots of similar specifications were used as a starting material for NTD. The resistivity of the ingots was higher than 2000 Ω·cm (n-type). The diameter was 63 mm. The O concentration was below our IR detection limit (< 1×10^{16} cm^{-3}) and the C concentration was less than 5×10^{16} cm^{-3}. The ingots were separately irradiated by four different reactors, #1 - #4. The thermal to fast

Table 1. Characteristics of NTD samples.

sample	#1	#2	#3	#4
target ρ (Ω·cm)	27	12	50	120
Φ_{th} / Φ_f	5-10	10-20	400	1000
$\Phi_{th}\cdot t$ ($\times10^{14}cm^{-2}$)	9500	22000	5200	2200
$\Phi_f\cdot t$ ($\times10^{14}cm^{-2}$)	950-1900	1100-2200	13	2.2

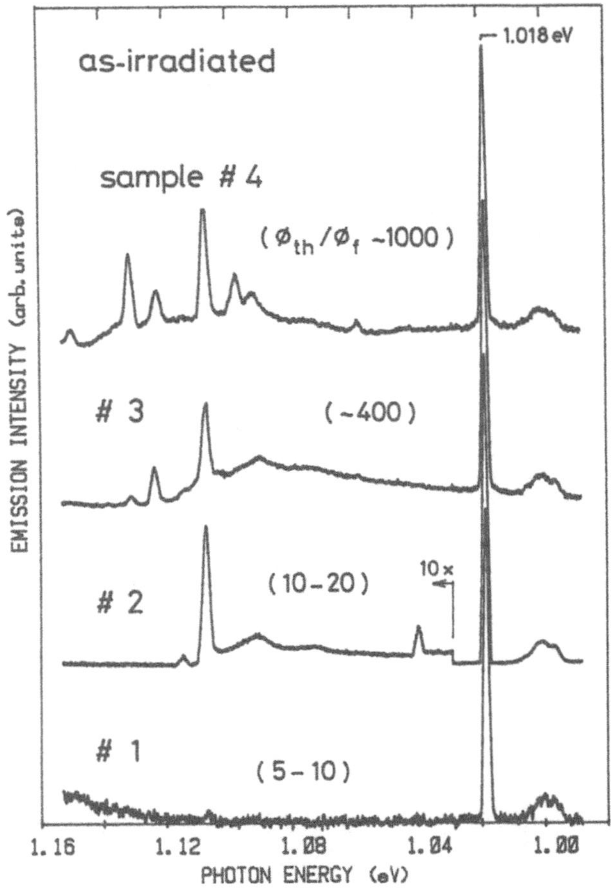

Fig. 4. PL spectra from as-irradiated samples #1 - #4. The characteristics of the samples are summarized in Table 1.

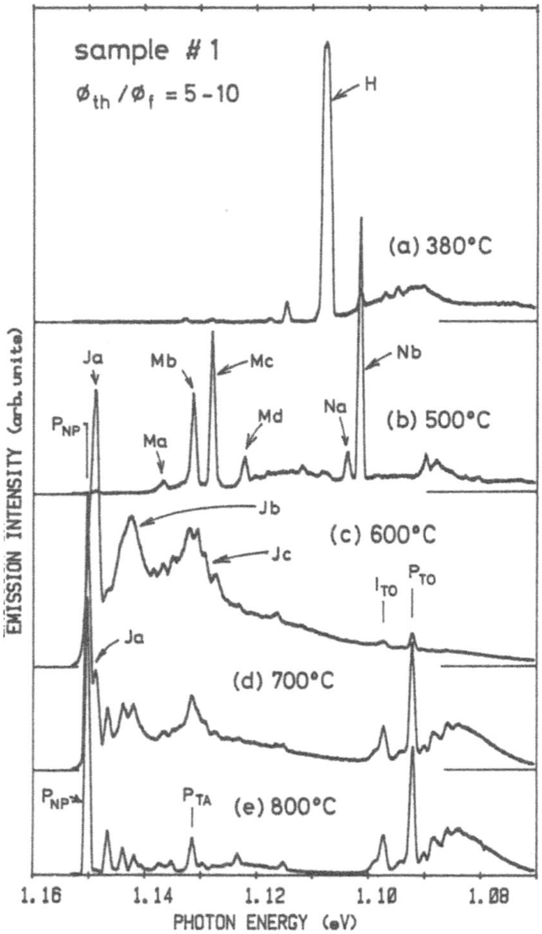

Fig. 5. PL spectra from sample #1 (Φ_{th} / Φ_f = 5-10) observed at various stages during the 1-h isochronal annealing. (a) 380°C, (b) 500°C, (c) 600°C, (d) 700°C. (e) 800°C. (submitted to Jpn. J. Appl. Phys.)

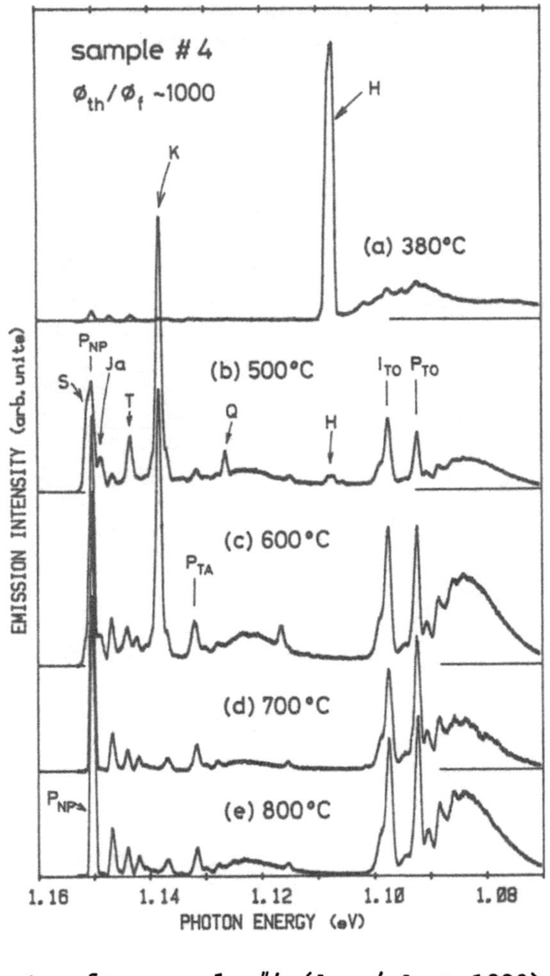

Fig. 6. PL spectra from sample #4 (Φ_{th} / $\Phi_f \sim 1000$) observed at various stages during the 1-h isochronal annealing. (a) 380°C, (b) 500°C, (c) 600°C, (d) 700°C, (e) 800°C. (submitted to Jpn. J. Appl. Phys.)

neutron flux density ratio, Φ_{th} / Φ_f, of #1, #2, #3, and #4 reactors is 5-10, 10-20, ∿400, and ∿1000, respectively. We distinguish the ingots by the reactor number. The characteristics of the ingots are summarized in Table 1. After irradiation, wafers were sliced from the ingots and subjected to various kinds of heat-treatments at 380 - 1200°C for 10 min - 30 h in an N_2 atmosphere. The resistivity of the wafers was measured by the four-probe method. The PL spectra of the center part of the wafers were measured at 4.2 K.

4.2. As-Irradiated Samples

The resistivity of the wafers sliced from the as-irradiated ingots was higher than 10^5 Ω·cm. The PL spectra of the as-irradiated wafers are shown in Fig. 4. These spectra were completely different from those of ordinary crystals. Neither free exciton nor bound exciton luminescence was detected. The 1.018-eV line was commonly detected for all samples. This line is always observed as the strongest peak in an as-irradiated crystal,[8] and is accompanied by TA-phonon wing at lower energy side. The 1.018-eV line was reported to appear in the PL spectra from a silicon crystal irradiated by high-energy electron,[9] neutron[10] or ion[10,11] followed by room-temperature or low-temperature ($\lesssim$ 400°C) annealing. This line has been identified[11] as radiative recombination at a five-vacancy-cluster, since the thermal stability of the 1.018-eV line is quite similar to that of the five-vacancy-cluster observed in ESR studies (P1 center).[12] Other than the 1.018-eV line, many lines and bands appear in samples #2, #3 and #4, although their origins are unidentified at present. Some components are commonly observed in the two or three samples.

4.3. Isochronal Annealing

The four groups of neutron-irradiated wafers were subjected to the isochronal annealing at temperatures ranging from 380 to 800°C for 1 h. Annealing under these conditions introduced new spectra, which could never be observed in unirradiated crystals. For comparison purpose, we measured such wafers as are annealed at 1200°C for 3 h in order to anneal out the radiation-induced defects completely.

Figure 5 shows the spectra of sample #1 (Φ_{th} / Φ_f = 5-10) observed at various stages during the 1-h isochronal annealing sequence. The intense peak H appears after a 380°C anneal. Highly resolved spectra show that this peak consists of two sharp lines at 1.1078 and 1.1069 eV. This peak is still observed at 500°C, but vanishes above 550°C. The H peak is interpreted to be the same as the reported 1.108-eV peak from the coincidence of the peak position and the thermal stability. According to the reported data,[10,13] this peak appears only in B-doped FZ crystals and show no splitting.

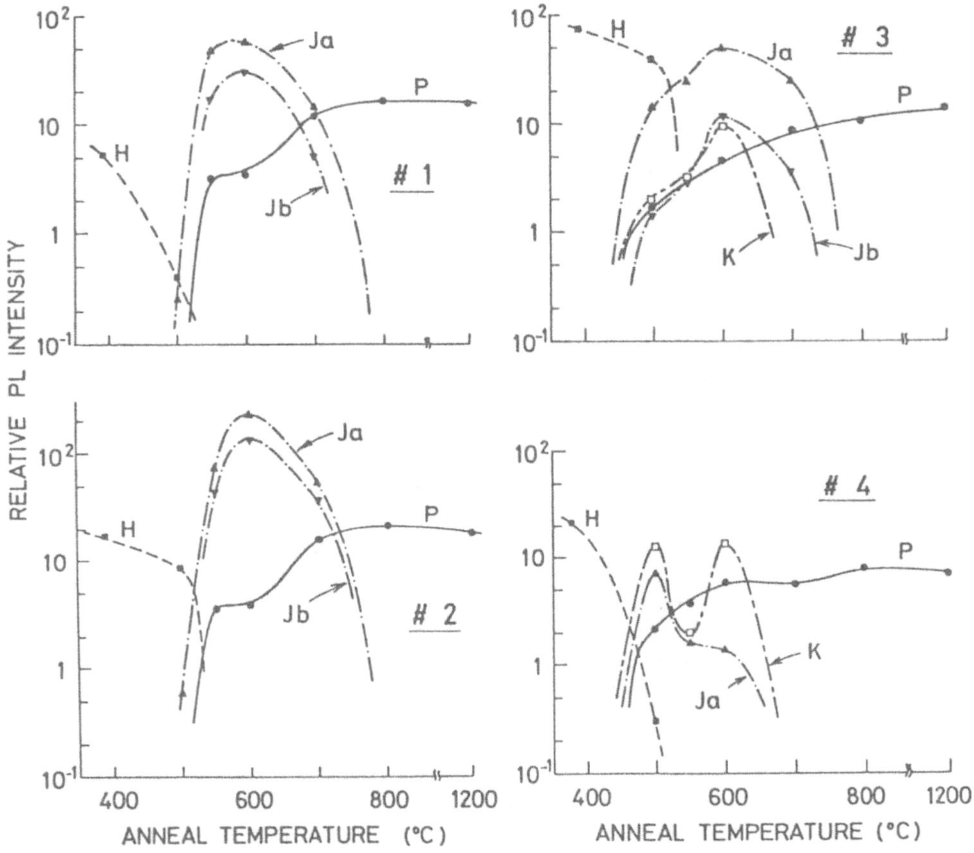

Fig. 7. Isochronal annealing curves for PL intensity for samples #1 - #4 (for 1 h). The intensities of the H, Ja, Jb, K, and P bands are plotted against annealing temperature.

The intrinsic luminescence, labeled I, and the luminescence cm^{-3}) and show splitting.

Annealing at 500°C introduced the M and N bands, labeled Ma - Md, Na, Nb. The M bands consist of two or three sharp lines. These components appear only after a 500°C anneal, and do not appear after annealing at 400 or 600°C. There have been no reports on these M and N bands. The characteristic J bands, labeled Ja - Jc, appear above 500°C. The Ja band consists of three sharp lines at 1.1492, 1.1486, and 1.1483 eV. The J bands dominate the spectra at annealing temperatures between 550 and 700°C. We have detected these J bands for the first time and investigated their heat-treatment behavior as discussed later.

However, our H peak appears in non-doped FZ crystals ($N_B < 5x10^{12}$ associated with P, labeled P, can be detected weakly at 550°C. They grow gradually, becoming the dominant peak above 700°C, and their intensities become saturated above 700°C. The thermal stability of these bands is summarized in Fig. 7 , where the peak intensities of main components are plotted against annealing temperature.

The radiation-induced luminescence depends strongly on the irradiation condition. Figure 6 shows the spectra of sample #4 ($\Phi_{th} / \Phi_f \sim 1000$) observed at various stages of the isochronal annealing. A fast neutron dose for sample #4 is about 1/1000 of that for sample #1, as shown in Table 1. There are two large differences between Figs. 5 and 6. First, the P component is recovered at lower temperatures in sample #4. Second, the Ja component is quite weak and the Jb and Jc components are undetectable in sample #4, while these components appear as dominant peaks in sample #1. The PL data of the isochronal annealing for other samples are summarized in Fig. 7.

4.4. Isothermal Annealing

In order to investigate the time-dependence of annealing behavior of the radiation-induced defects, the isothermal annealing was performed at temperatures of 600 and 800°C.

Figures 8(a) and (b) show the resistivity change due to annealing at 600 and 800°C, respectively. In these figures, the resistivity is normalized by the resistivity obtained for each wafer annealed at 1200°C for 3 h. The profile of neutron flux density in the reactor was taken into account in these values.

For a 600°C anneal, the resistivity is decreased and approaches the target resistivity with increasing annealing time. The deviation of the resistivity from the target value becomes larger in the order of samples #2, #1, #3, and #4. This order corresponds with that of the fast neutron dose, as shown in Table 1. The PL data for the isothermal annealing at 600°C are summarized in Fig. 9, where the peak intensities of the Ja, Jb, K and P components are plotted against annealing time. The J components due to radiation-induced defects are dominant and the P component due to transmuted P^{31} is weak for samples #1, #2, and #3. The J components grow gradually, reaching an intensity maximum after a 4-h anneal, and then their intensities decrease. In contrast with this, the P component grows continuously with annealing time. The recovery of the P component is more rapid for lower fast neutron dose. This corresponds with the resistivity results. It should be noted that the J components are still dominant after 30-h annealing for samples #1, #2, and #3, although the resistivity is only 20% higher than the target resistivity.

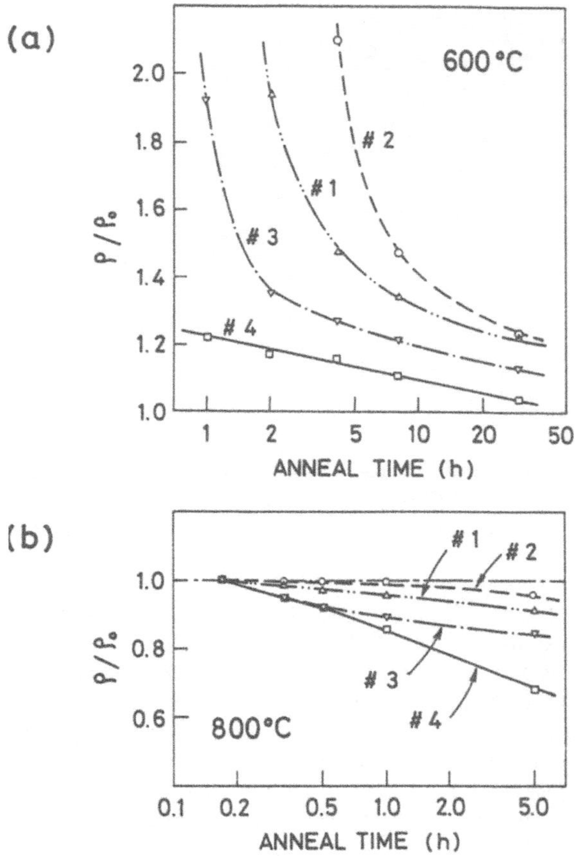

Fig. 8. Isothermal annealing curves for resistivity for samples #1 - #4. (a) 600°C, (b) 800°C. The resistivity is normalized by the target resistivity, ρ_0, and plotted against annealing time. (submitted to Jpn. J. Appl. Phys.)

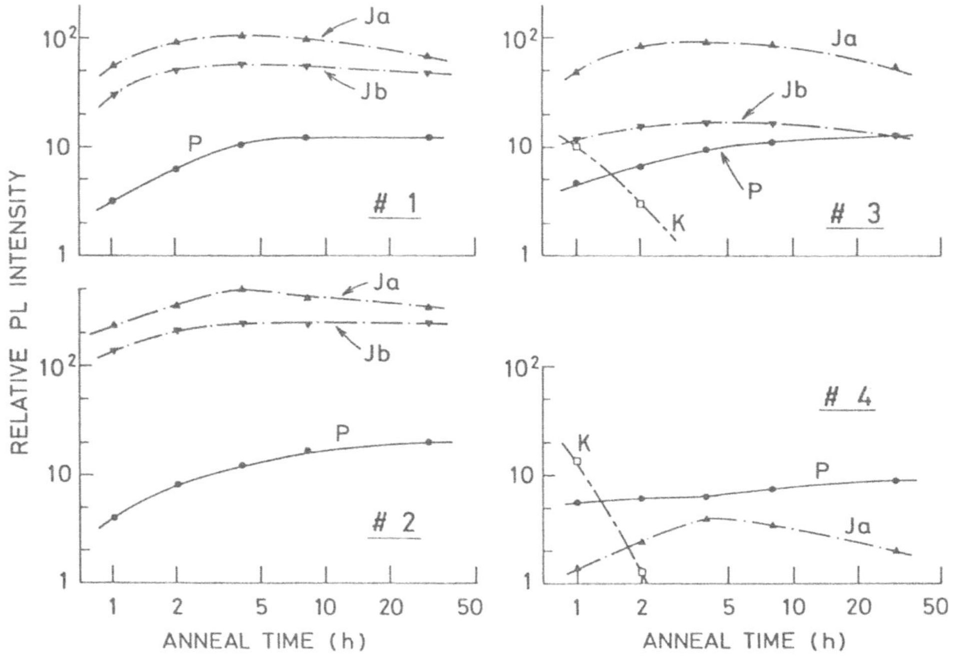

Fig. 9. Isothermal annealing curves for PL intensity for samples #1 - #4 (at 600°C). The PL intensities of the Ja, Jb, K, and P bands are plotted against annealing time.

The resistivity data of the isothermal annealing at 800°C, shown in Fig. 8(b), present a striking contrast with Fig. 8(a). The resistivity of each wafer after short (10 min) annealing is the same as the target value within the experimental error. However, the resistivity is decreased with increasing the annealing time, i.e., these samples undergo a reverse anneal. The resistivity approaches again to the target value after long (30 h) annealing. In spite of the striking change in the resistivity, no significant difference was observed in the PL spectra from the reverse-annealed samples. The intensities of the P and I components are nearly the same as those of 1200°C annealed samples within the experimental errors.

4.5. Discussion

The various types of radiation-induced defects are grown and annihilated by heat-treatment. Some of them act as efficient luminescence centers. The most characteristic luminescence in this experiment is the Ja band, which appears in all samples and becomes most intense after annealing at 600°C. As for the Jb and Jc bands, the thermal stability of these bands is nearly the same as that of the Ja band. Therefore, the Jb and Jc bands are considered to be associated with the Ja band.

The whole spectral shape of the J bands have some similarities with that of the 0.79- and 0.97-ev bands, which were observed in an silicon crystal irradiated by high-energy-electron,[14] fast neutron,[10] or ion.[10,11] The 0.79- and 0.97-eV bands were reported to be due to an O-modified divacancy or C-O-vacancy complex and a C-related center, respectively, from the comparison with ESR and IR results.[14] These bands consist of a narrow no-phonon band accompanied by two broad phonon wings, where the associated phonons are TA- and TO-modes. From the similarity of the band shape,[15] it is speculated that the Ja band is the no-phonon band and the Jb and Jc bands are its phonon sidebands. However, the spacing between the Ja and Jb bands does not coincide with the energy of any normal-mode phonon. Therefore, the associated phonon is considered to be the local-mode phonon. The Jc band may be identified as the transition involving the TA-phonon.

We can not discuss the origin of the J band at present, since neither ESR nor IR data are available concerning the defects responsible for the J band. However, the J band is considered to be due to complexes consisting of vacancies and impurities from analogy of the spectral shape and the thermal stability of the J band to those of the 0.79- and 0.97-eV bands.

5. SUMMARY

The NTD Si crystals with a controlled neutron dose have been used as standard samples for the calibration of the PL method for the quantitative impurity analysis. The obtained calibration curves make it possible to analyze the P and B concentrations simultaneously from the PL analysis. The difference between the $|N_P - N_B|$ value obtained from the PL analysis and the $|N_D - N_A|$ value from the resistivity measurement is less than 30% in the concentration range between 10^{11} - 10^{15} cm^{-3}. The reproducibility of the PL data is within a 20% deviation. The detection limits of B and P are $1x10^{11}$ and $5x10^{10}$ cm^{-3}, respectively.

In order to investigate the properties of radiation-induced defects, neutron irradiation has been performed by four different reactors with the thermal to fast neutron density ratio varying from 5 to 1000, followed by heat-treatment at 380 - 1200°C for 10 min - 30 h. The characteristic J band has been observed for the first time in all samples annealed at temperatures between 500 and 700°C. The J band consists of three sharp no-phonon lines at 1.1492, 1.1486 and 1.1483 eV and two broad phonon wings. The intensity of the J band becomes more intense for higher fast-neutron-dose. The J band still dominates the spectra after a 30-h anneal at 600°C, although the resistivity is only about 20% higher than the target resistivity. The J band is considered to be due to the radiative recombination at the impurity-vacancy-complexes.

ACKNOWLEDGMENTS

The authors would like to thank T. Iizuka, T. Ishiguro and Y. Tokumaru for helpful discussions. The authors are also grateful to T. Abe of Shin-Etsu Handotai Co., Ltd. for supplying several crystals together with their accurate resistivity measurements.

REFERENCES

1) H. M. Janus and O. Malmros, IEEE Trans. ED-23, 797 (1976).
2) H. Herzer, Semiconductor Silicon 1977, edited by H. R. Huff and E. Sirtl (Electrochem. Soc., Princeton, 1977) p. 106.
3) J. M. Meese (ed.), Neutron Transmutation Doping in Semiconductor (Plenum, New York, 1979) Chap. 5.
4) M. Tajima, Appl. Phys. Lett. 32, 719 (1978).
5) M. Tajima, A. Yusa and T. Abe, Proc. 1979 Internat. Conf. Solid State Devices; Jpn. J. Appl. Phys. Supplement 19-1, 631 (1980).
6) M. Tajima, S. Kishino, M. Kanamori, and T. Iizuka, J. Appl. Phys. 51, 2247 (1980).
7) J. C. Irvin, Bell Syst. Tech. J. 41, 387 (1962).
8) M. Tajima, Appl. Phys. Lett. 35, 242 (1979).
9) A. V. Mudryi and A. V. Yukhnevich, Sov. Phys.-Semicond. 8, 875 (1976).
10) V. D. Tkachev and A. V. Mudryi, Radiation Effects in Semiconductors 1976, edited by N. B. Urli and J. W. Corbett (Inst. Phys., London, 1977) Conf. Ser. No. 31, p.231.
11) C. G. Kirkpatrick, J. R. Noonan and B. G. Streetman, Radiat. Eff. 30, 97 (1976).
12) Y.-H. Lee and J. W. Corbett, Phys. Rev. B8, 2810 (1973).
13) C. E. Jones, E. S. Johnson, W. D. Compton, J. R. Noonan, and B. G. Streetman, J. Appl. Phys. 44, 5402 (1973).
14) J. R. Noonan, C. G. Kirkpatrick and B. G. Streetman, J. Appl. Phys. 47, 3010 (1976).
15) A. P. G. Hara, G. Davies and A. T. Collins, J. Phys. C, 5, 1265 (1972).
16) M. Tajima, unpublished.
17) M. Tajima, submitted to Appl. Phys. Lett.

PRECISION RESISTIVITY MEASUREMENTS ON NTD-SILICON

O. Malmros

TOPSIL A/S

Frederikssund, DENMARK

The major reason for producing NTD-silicon is the resistivity homogeneity, which may be obtained. This puts special demands on the resistivity measurement equipment and procedures, demands which did not exist in the pre-NTD era.

Conciderations are given to the best measurement procedure and geometric correction factors are reviewed. Special attention is paid to the correlation between ingot and wafer resistivities and it is shown that the widely accepted procedure for wafer measurement gives resistivities, which are 1-10 % too high. A possible explanation for this phenomenon is given.

1. INTRODUCTION.

Resistivity measurement with a linear four point probe is a well established technique, which has been applied to silicon characterization for many years. However, with the large scale introduction of neutron transmutation doped silicon 5 years ago, new demands were placed on exactness and accuracy.

Individual NTD-orders are in some cases specified with a +/- 5 % center resistivity tolerance, one spec. being for instance 90-100 Ohmcm. In special cases even +/- 3 % has been requested.

In a round robin experiment conducted in 1965,[1] a standard deviation of 2.7 % was found between various laboratories, and that is rather high compared with the above +/- 3 %. In this experiment no geometric or temperature correction factors were used, and certainly, the later introduction of such factors has helped appreciably. However, a number of details remain regarding the proper procedure,

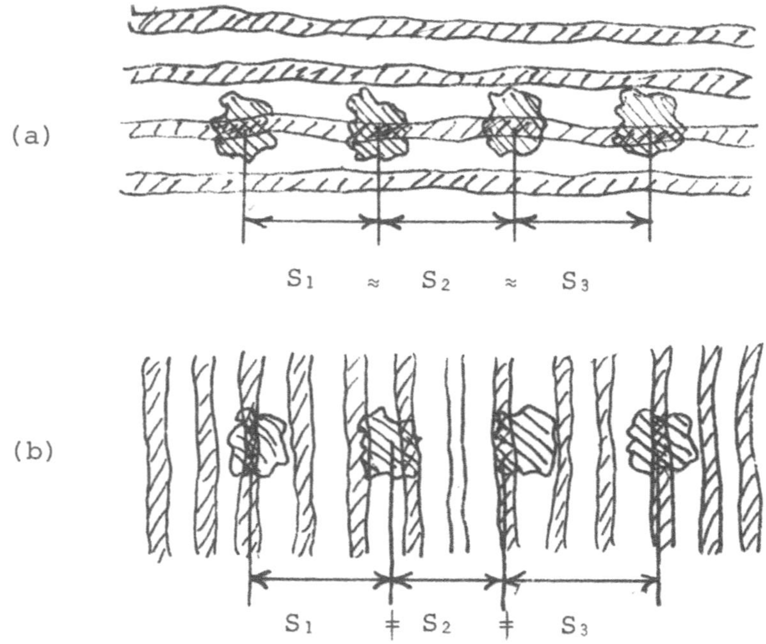

Fig. 1 Resistivity measurements on a surface with saw marks. Saw marks and probe line are parallel in (a), hence the 3 probe distances S_1, S_2 and S_3 are almost equal.

Saw marks and probe line stand perpendicular to each other in (b), hence S_1, S_2 and S_3 are different.

details which according to my experience are not generally known, and which should not be overlooked.

Four independent topics are dealt with in the present paper, all of which contribute significantly towards making the measurements more reliable. These are:

1. Silicon surface preparation.
2. Temperature control during measurement.
3. Geometric correction factors by measuring on the end face of cylinder shaped ingots.
4. Correlation between ingot and wafer resistivity.

Details about the measurement equipment are given in Appendix I.

2. EXPERIMENTAL WORK.

2.1 Silicon surface preparation.

In ASTM F43-78 the following surface preparation of the test specimen is recommended:

"The surface on which the resistivity is determined with either the two- or four-probe method shall be abraded by lapping with No. 600 alumina or by sandblasting with No. 280 Carborundum, and the measurement shall be made thereafter without unnecessary delay. The specimen shall show no surface cracks or other defects observable with normal vision and it shall be free of surface oxide. The abrasive grain sizes specified for this purpose shall comply with the appropriate specifications of Product Standard PS8-67."

These are very stringent demands, which of course are suitable for standardization purposes, but rather difficult to work with in practice. Fortunately less will do.

First of all, we have not been able to measure any difference in resistivity as a function of the age of the surface. The natural 10-20 Ångstrøm thick oxide layer, which develops on the surface, does not interfere with the measurements.

Secondly, other surface preparations can be used, and most important, it is possible to measure on a surface from an I.D. saw if the saw marks are held parallel with the probe line.

Each probe tip has a certain contact area and it is the precise position of this along the probe line, which determines the measured resistivity. When the probe line and the saw marks are perpendicular to each other the effective points of indentation will change slightly in the direction of the probe line (Fig. 1a). This creates a certain

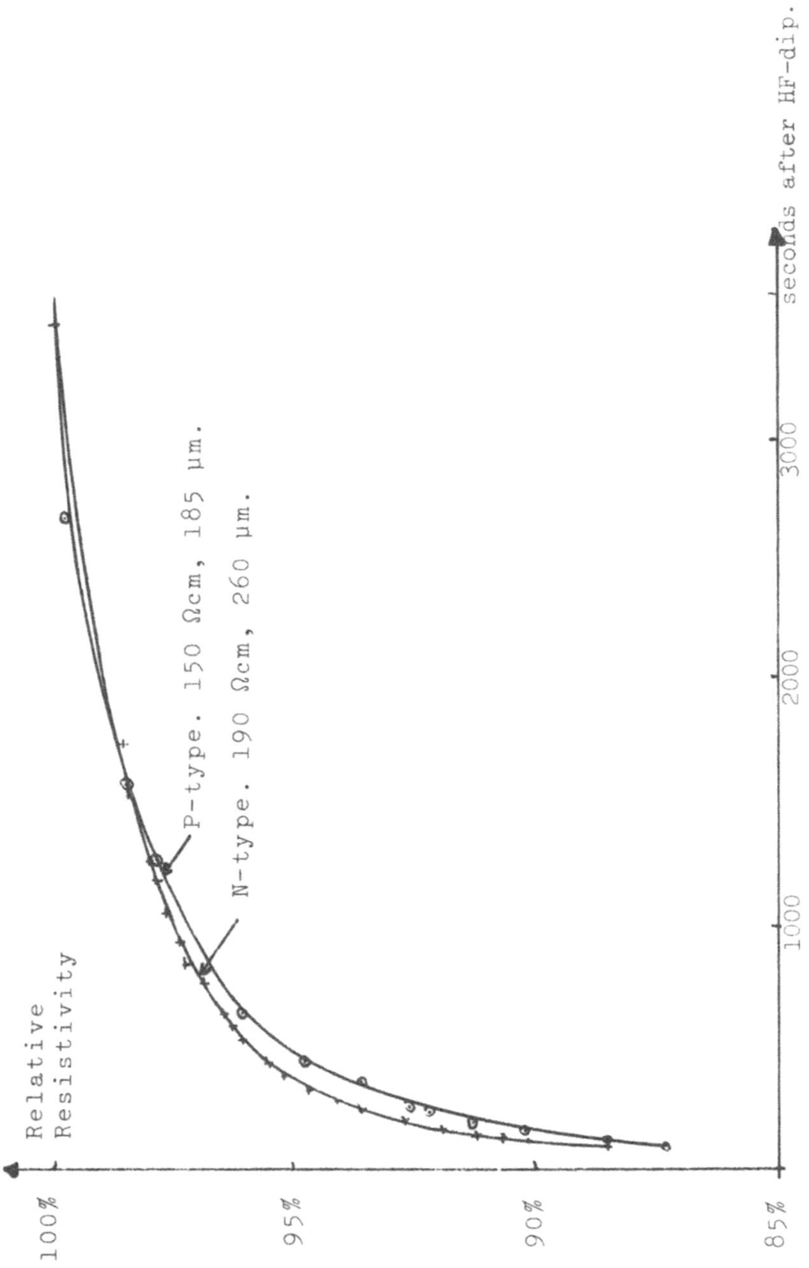

Fig. 2 Relative resistivity as a function of elapsed time after HF-dip. Resistivities normalized with the value after 1 hour.

amount of scatter in the resistivity reading. Conversely, when the saw marks and the probe line are parallel to each other the indentation points will change in the direction perpendicular to the probe line, a change which is practically without importance for the resistivity reading (Fig. 1b). Therefore, a smaller scatter is experienced in this case.

The above effect is demonstrated by the following experiment:

Center resistivity was measured several times on one end of a 35 Ωcm NTD ingot with the probe parallel or perpendicular to the saw marks. For comparison, the roughness was measured with a Perth-o-Meter in the direction of the probe line. The following results were obtained:

	Rel.Std.dev. on resistivity	Roughness RMS
Probe and saw marks parallel	0.5%	0.65 μm
Probe and saw marks perpendicular	1.1%	1.06 μm

For comparison, a surface lapped according to the ASTM-procedure gave an 0.4 % relative standard deviation and an 0.40 μm roughness. This is slightly better than the parallel case, however, the latter is considered quite adequate.

The correlation between standard deviation and roughness, which can be seen from the above figures, has been noticed earlier[2].

One further note regarding the proper surface condition has to be added. Vieweg-Gutberlet et al[3] have found that resistivity was not measured correctly on polished P-type wafers when they were dipped in HF. Upon this treatment the resistivity of a 16.5 Ωcm wafer increased to 30.0 Ωcm.

We do not observe this phenomenon on lapped wafers, although there is an effect from HF. For both conductivity types the resistivity drops (Fig. 2) and then drifts back towards the original value. After 1 hour the resistivity is within 3 % of this. The effect disappears if the wafers are dipped in water after the HF-treatment, but prior to the measurement.

2.2 Temperature stability.

On silicon the temperature dependence of the resistivity is rather high, for N-type resistivities between 10 and 300 Ωcm, the typical NTD-interval, there is an 0.82 % increase in resistivity for each 1° C increase in temperature.

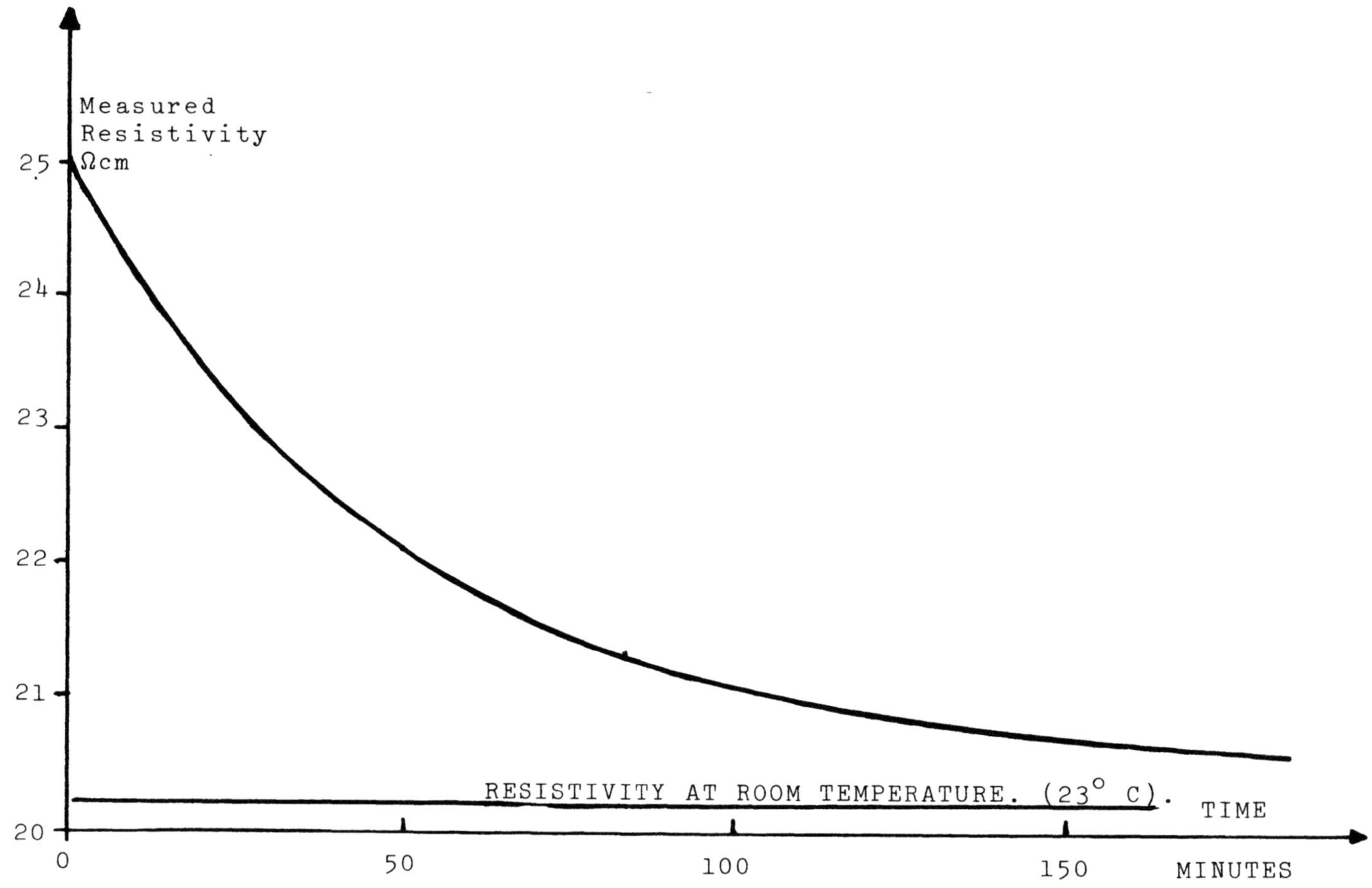

Fig. 3 Resistivity measurements on a 3" ingot, 285 mm long heated to 51.3° C and left at room temperature, 23° C. Resistivity is still not stable after 3 hours.

When measuring on NTD-Silicon it is therefore important that temperature is well defined. The room must be thermostatically controlled and the silicon has to be stationed in it for sufficient time to reach the temperature of the air.

It is surprising to see how long time it takes an ingot to establish equilibrium with the surroundings.

In Fig. 3 the resistivity measured in the center of a 285 mm long, 3 inch crystal is shown. The crystal was held in hot water with a temperature of 51.3° C and removed at time zero. As can be seen from the figure, the resistivity still changes after 3 hours.

In production the last process before ingot resistivity measurement is normally a sawing or grinding operation. These are performed much closer to room temperature than the above 51° C, hence a shorter quarantine time will suffice. One hour has been chosen as the minimum.

2.3 Geometric correction factors on the end face of cylinder shaped ingots.

It is well known that geometric factors have to be applied in case the measurements are not performed in semiinfinite volumes, and fortunately all important correction factors have been calculated from pure mathematics.

The reason for dealing with this topics in spite of the above is, that the correction factors on the end face of an ingot are not generally known, and hence certainly not used.

The correction factors applicable to this geometry have been calculated by Polyakov et al.[4] and they are shown in Fig. 4a. For comparison, the corresponding factors for thin wafers, calculated by Schwarzendrüber[5] are shown in Fig. 4b.

Fig. 5 shows the center values derived by Polyakov et al. together with experimental values from an ingot measured after grinding to various diameters. The agreement between the calculated curve and the measured values is obvious.

An extensive experimental program to check the factors outside the center confirmed the calculations of Polyakov et al. in all cases.

Also shown in Fig. 5 is the center correction factors for wafers. It can be seen that it is better not to correct ingot measurements at all, than utilizing the wafer factors for lack of other tabulated values.

The following is worth emphasizing regarding ingot measurements:

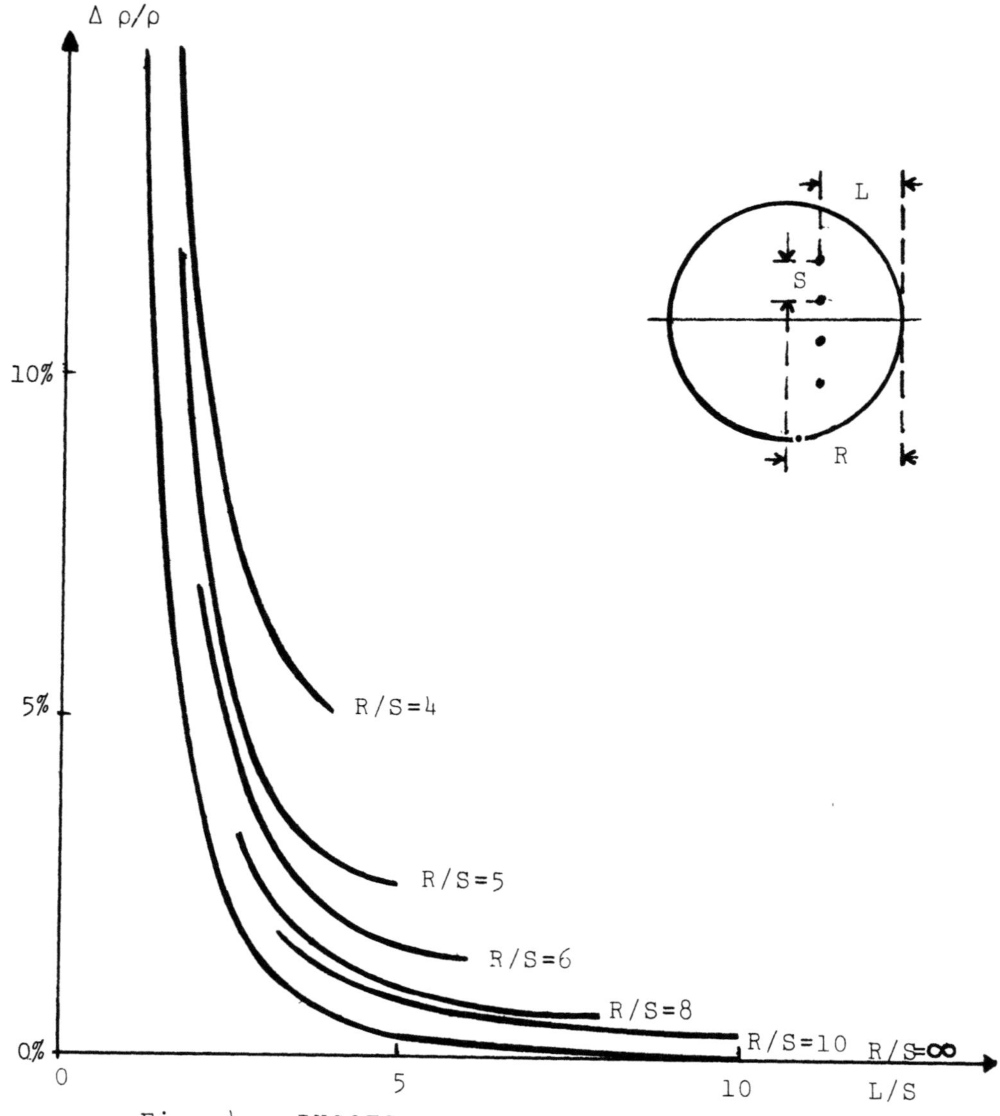

Fig. 4a. INGOTS.

Geometric correction factors by resistivity measurements on circular plane surfaces. Excess resistivity reading in % of true value.

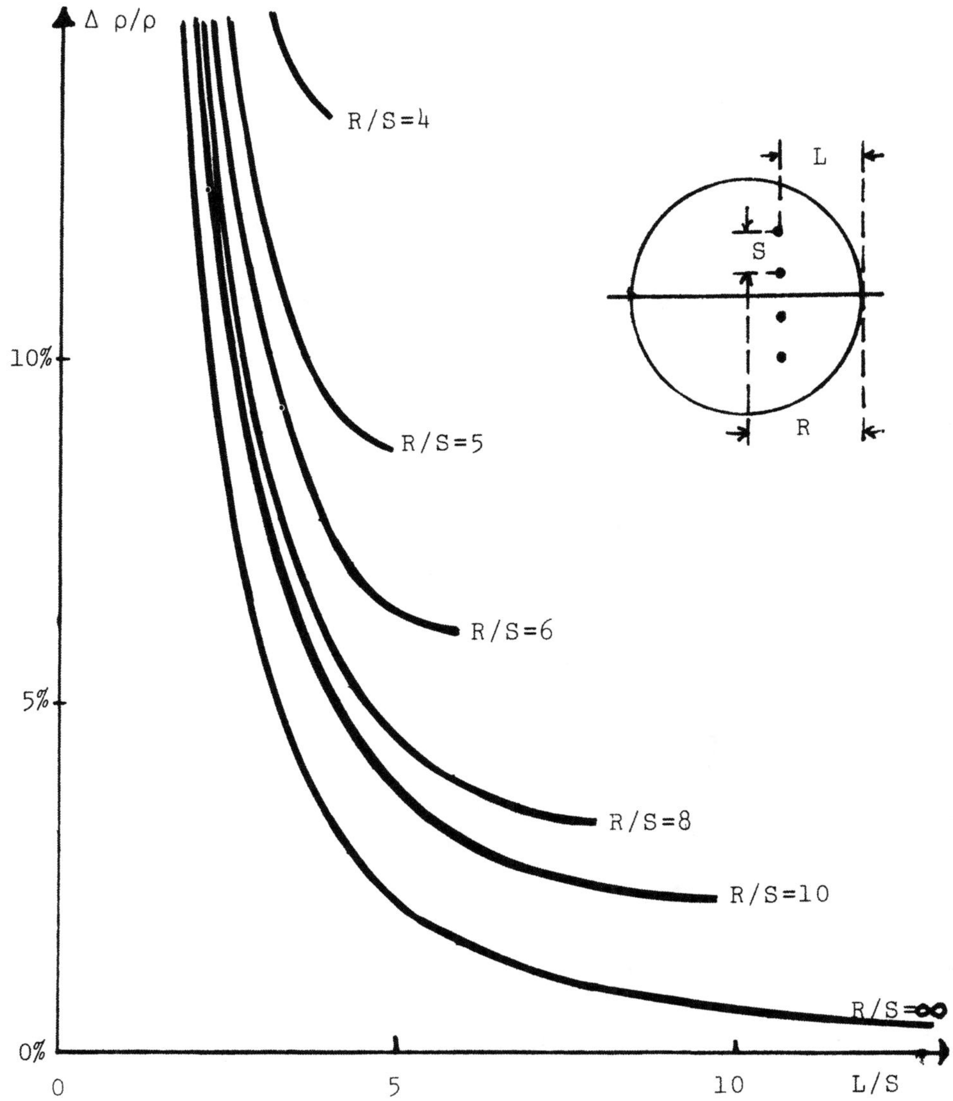

Fig. 4b. WAFERS.

Geometric correction factors for resistivity measurements on circular plane surfaces. Excess resistivity reading in % of true value.

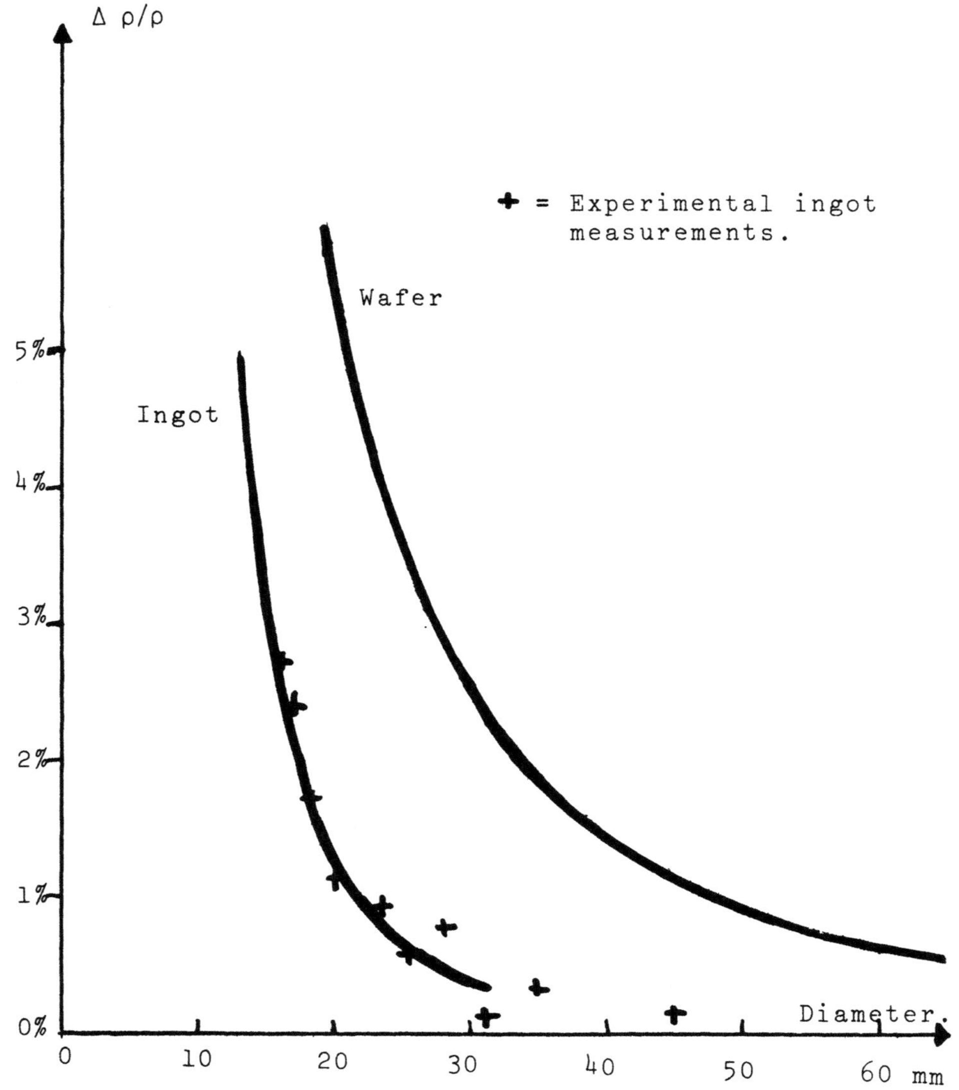

Fig. 5 Comparison of Geometric correction factors in center of wafers and ingots. Probespacing 1.592 mm.

1) On a one inch ingot the correction factor will decrease the measured resistivity with 0.6 % in the center, this is not in all cases without importance. On a two inch ingot the corresponding factor is around 0.1 %. This can for all practical purposes be disregarded.

2) By measuring radial variation on ingots according to ASTM F81-77 (which actually is only valid for wafers) sampling plan B or C, resistivity has to be measured 6 mm from the edge. Assuming a perfectly flat NTD-resistivity profile and a diameter of 2" or more, it can be seen that the resistivity in this position will be measured 0.7 % too high. This is not insignificant as the radial variation in some cases is guaranteed to be less than 2 %.

2.4 Correlation between ingot and wafer resistivity.

When originally introducing NTD-Silicon, we did not expect to encounter any problems with correlating the resistivity of an ingot and the wafers cut from it.

However, we consistently experienced higher resistivity on the wafers than the ingots. This finding could not be explained by resistivity variations along the crystal length, as the known neutron flux profile should give lower resistivities in the middle part of the crystal and not higher.

As a result of preliminary experiments we came up with a theory saying, that the thickness of the wafer, which actively carries the current, is smaller than the measured value.

This led to the following conciderations:

By measuring the resistivity of a thin wafer with a linear four point probe, we have according to Smits'[6] :

$$\rho = \frac{\pi}{\ln 2} \cdot T \cdot \frac{V}{I} \qquad (1)$$

where

ρ(Ohmcm) = Resistivity.
T(cm) = Wafer thickness.
V(Volt) = Voltage drop between inner probes.
I(Ampere) = Current through outer probes.

Assuming now that there is an inactive thickness, which does not carry any current, we measure in stead an apparent resistivity:

$$\rho_T = \frac{\pi}{\ln 2} \cdot (T - \Delta T) \cdot \frac{V}{I} \qquad (2)$$

Table I. The materials used for comparison of ingot and wafer resistivities.

Material type	Crystal	Resistivity Ohmcm	Inactive Thickness T, m	Comments
N-type	I	15	10	
Phosporous	II	100	13	
NTD	III	125	15	
	IV	190	13	
	V	210	12	
N-type Phosporous Conv.	VI	1000	-	Too large scatter in data.
P-type	VII	9	12	Rather large scatter in data.
Boron	VIII	100	(11)	
Conv.	IX	150	16	
	X	600	15	

where

ρ_T(Ohmcm) = Apparent resistivity.
ΔT(cm) = Inactive thickness.

Dividing equation (2) by (1) and rearranging the terms we obtain:

$$\rho_T = \Delta T \cdot \rho_T/T + \rho \quad (3)$$

Defining ρ_{rel} as ρ_T/ρ we obtain finally:

$$\rho_{rel} = \Delta T \cdot \rho_{rel}/T + 1 \quad (4)$$

Concequently, if we for a number of wafers with different thicknesses plotted ρ_{rel} versus ρ_{rel}/T we should obtain a straight line with the slope ΔT.

An experiment was designed in order to check this model. 10 float zone crystals with varying resistivity and type (Table 1) were chosen. Each crystal was cut into wafers with thicknesses from 200 to 1500 µm. The knowledge of the position of each wafer along the crystal was kept, and at short intervals reference wafers were cut with a thickness around 500 µm. These were needed for determining the local resistivity. From each crystal 50-80 wafers were cut.

The wafers were cleaned in isopropylalcohol and their thickness were measured twice in center of each wafer.

The thickness measurement equipment was calibrated with standard thickness elements of stainless steel and care was made to ensure that the adjustment did not change during the measurements.

Finally the resistivity was measured 5 times in the center of each wafer in the same order as they were cut. In this way we could compensate for any drift in temperature, measurement current etc.

For each wafer the apparent resistivity was calculated according to (1), however, for wafers with thicknesses above 650 µm an additional thickness correction factor was applied according to Uhlir'[7].

The 500 µm wafers were used for obtaining the local resistivities, and all resistivities were normalized with respect to these. The 500 µm values were taken for convenience rather than the true ingot resistivities.

Plotting the relative resistivities against ρ_{rel}/T a straight line was obtained from which a first approximation of ΔT was obtained. This was fed back into the original data so that better local

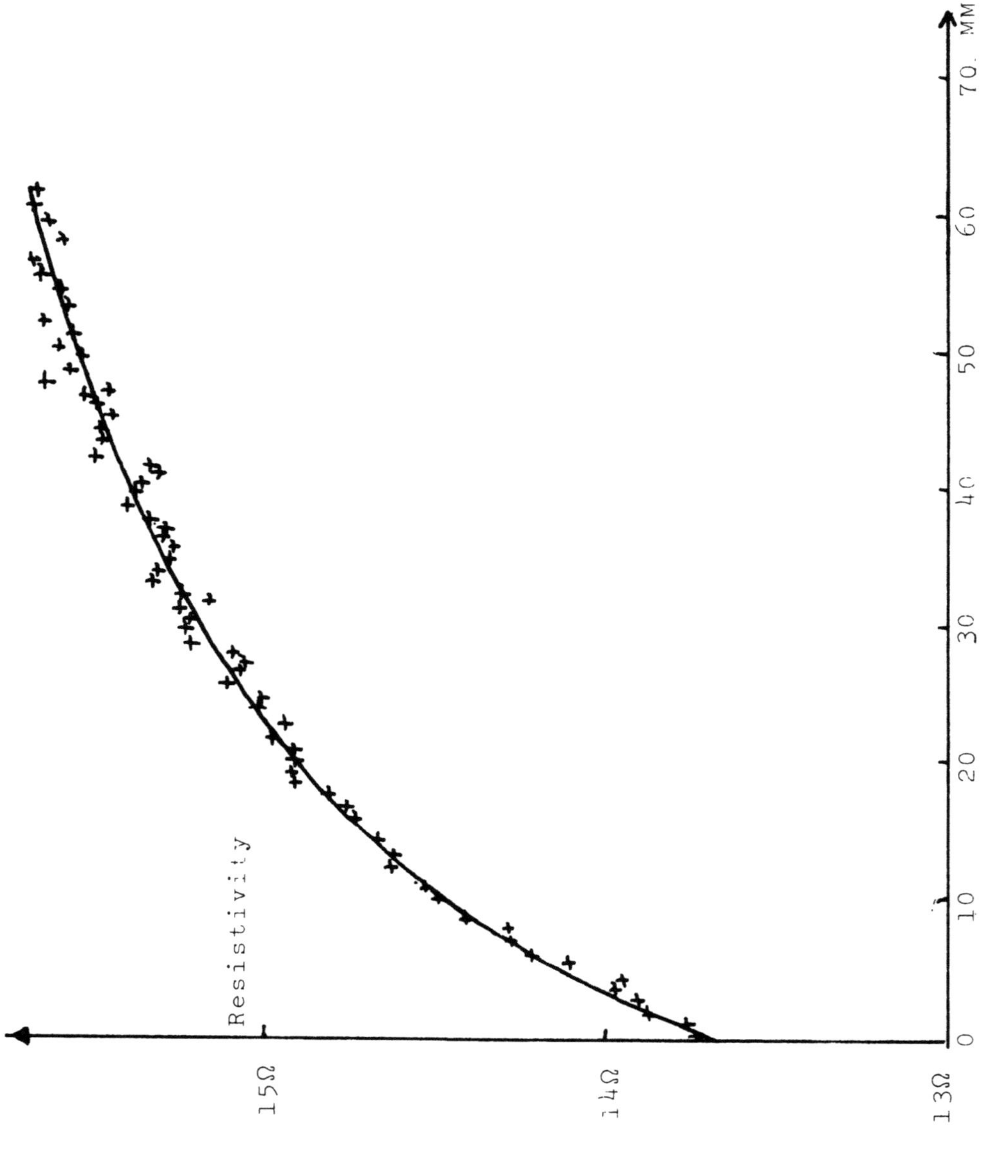

Fig. 6 Local resistivity variation on crystal I.

resistivities were found and hence a better value for the inactive thickness. In Fig. 6 an example of the local resistivity variation is shown, and in Fig. 7 and 8 two examples of the determination of ΔT can be seen. The results for all 10 crystals are given in Table I, which gives rise to the following comments:

1) It was more difficult to measure ΔT on P-type crystals than on NTD, at it was impossible to measure on conventional N-type. This is probably caused by the dopant striations, which let the average resistivity of a wafer fluctuate as compared with the ingot average resistivity.

2) All thicknesses are in the same 10-16 μm range without any visible correlation with resistivity or conductivity type. However, the individual values are clearly different.

3) The inactive thickness is not caused by an electric charge at the surface. If that were the case the effect should go in opposite directions for N- and P-type, furthermore, there should be a much more pronounced effect at high resistivities, than low.

An additional experiment was performed in order to further clarify the reason for the inactive thickness. The wafers from crystal I were etched 60 μm (30 μm from each side) in an acid mixture, boiled in nitric acid and measured again.

The result is shown in Fig. 9. It is evident that the slope of the straight line is now zero, hence, in this case it is the directly measured thickness, which goes into the resistivity calculation.

The exact reason for the excess thickness reading is not known. There are at least 2 possibilities, which maybe work together:

a) The mobility is reduced at the surface due to heavy work damage.

b) The surface can be considered as a landscape with mountains and vallies. The thickness sensor guage touches down on the tops, while the current has to run underneath the vallies. From roughness measurements it is known that the distance from top to bottom is 3-4 μm, and as the 2 sides of the wafers are identical this gives 6-8 μm. That is not enough to explain the size of ΔT, however, there is probably a number of micro-cracks extending below the vallies, which will go undetected by the roughness measurements, but not by the current.

Measuring damage depth by bevel polishing and Sirtl etch gives a one side damage of 10-13 μm, i.e. 20-26 μm on 2 sides. This is larger than all ΔT values.

A correlation between roughness and ΔT was sought, but could

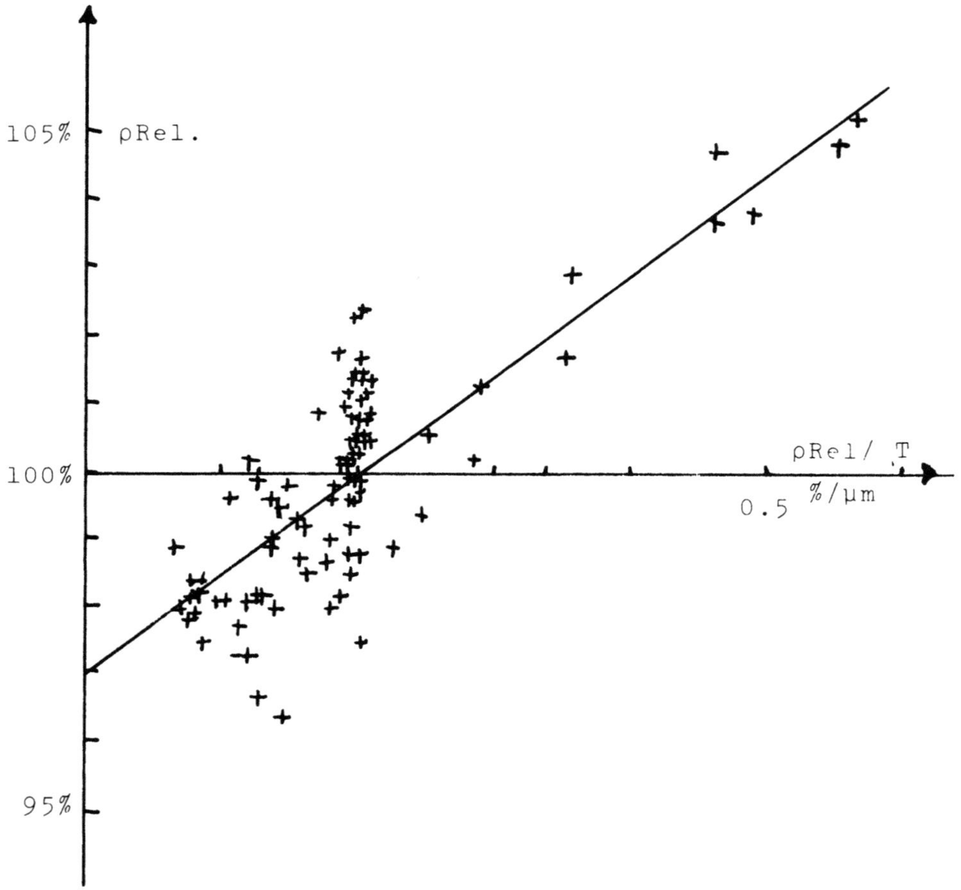

Fig. 7. Crystal X, 600 Ωcm, P-type. The slope of the straight line gives an inactive thickness of 15 μm.

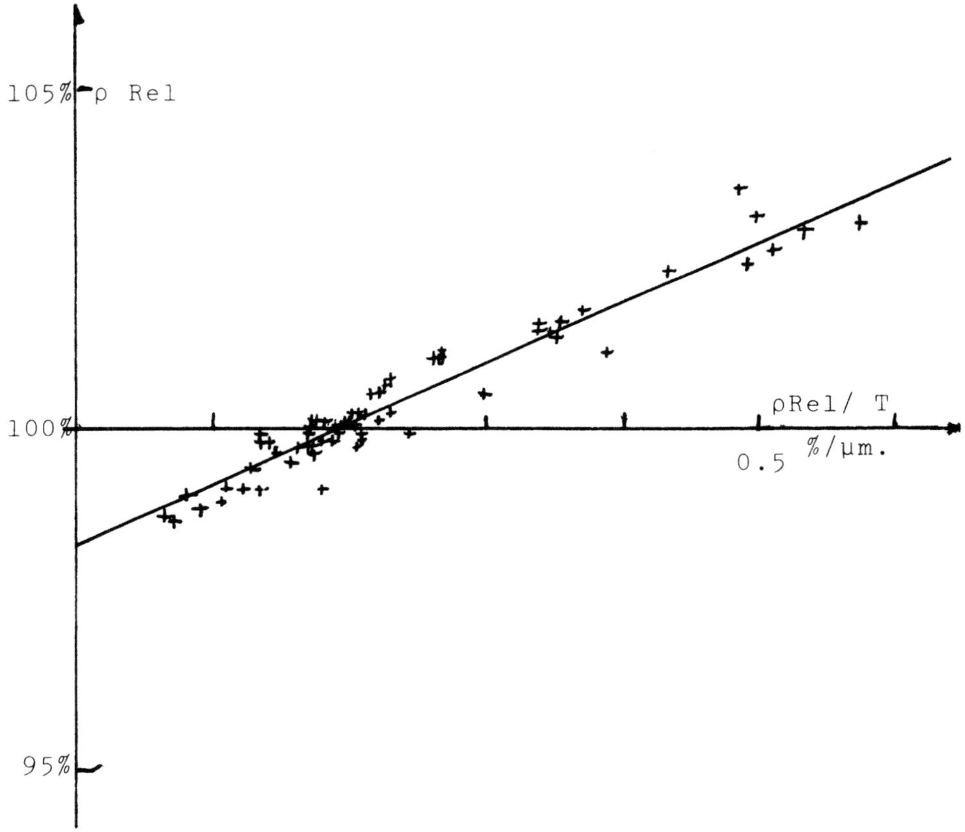

Fig. 8 Crystal I. 15 Ωcm, N-type. The slope of the straight line gives an inactive thickness of 10 μm.

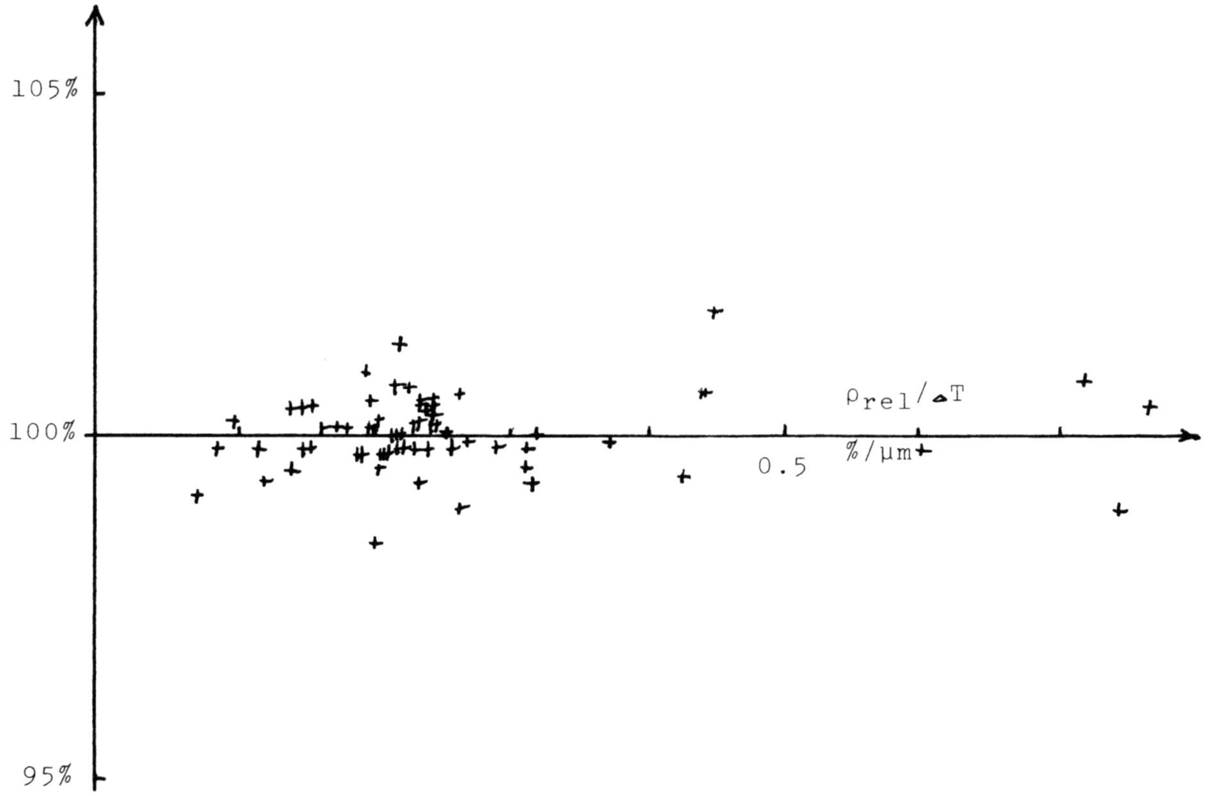

Fig. 9. Crystal I. Etched 60 μm. The resistivities are now measured correctly independent on wafer thickness.

not be found. Nevertheless, the difference between individual crystals is probably related to the sawing step anyway.

It has not in this investigation been established whether the same ΔT effect can be seen on lapped wafers as well. However, if it is a surface roughness problem it should be present on this type of wafers too.

It is customary that the resistivity guaranteed on wafers is the one, which can be measured without taking ΔT into account. This means that the wafer supplier must know his average ΔT value in order to avoid resistivity losses after cutting. In device fabrication there will, for each company, be a number of empirical correlations between resistivity and the parameters derived from it, e.g. break down voltage, therefore, if the device maker has always used the apparent resistivity he can continue to do so. Only if he changes wafer supplier or thickness may he experience a slight change.

Contrary to this, it is not possible to disregard the thickness effect in scientific work. It is important whether resistivity is 10 % higher or lower.

3. CONCLUSIONS.

In conclusion, the four aspects of high precision resistivity measurements on NTD-Silicon can be summarized in the following manner:

1) Ideally the surface on which measurements are to be made shall be lapped. However, it is possible to use an I.D. cut surface if the probe line and the saw marks are held parallel.

2) Measurements must be performed in a room with thermostatically controlled temperature. When handling ingots, these must be left in the room for one hour before being measured.

3) The geometric correction factors on the end face of an ingot are given by Polyakov et al. In the center, the correction factor can be disregarded for ingots with a diameter of two inches or more. On a one inch ingot there is an 0.6 % correction.

When measuring 6 mm from the edge on ingots, two inches or greater in diameter in order to determine radial variation, there will be an 0.7 % correction.

4) Resistivity of as cut wafers will be measured too high because the thickness, which carries the current is 10-16 μm less than the measured wafer thickness.

Dependent on wafer thickness this will give a resistivity error between 1 and 10 %.

If all the above recommendations are followed, it is my belief that resistivity on NTD-Silicon can be measured with a standard deviation less than 1 %.

ACKNOWLEDGEMENTS:

I wish to thank those of the staff of Topsil, who have helped me with the experimental work, notably Mobin Ul Haq Khan. I would further like to thank Jens Guldberg, Kaj Lund and Noel de Leon for reading the manuscript and suggesting valuable corrections.

APPENDIX I

Measurement equipment.

Measurement of resistivity (ρ) on wafers requires a resistance and a thickness measurement.

For the measurements we use the following equipment:

A Resistance:

Probe spacing : 1.592 mm

Current : 10 μA for $\rho \geq$ 100 Ωcm.
100 - - 100 Ωcm $> \rho \geq$ 10 Ωcm.
1000 - - 10 Ωcm $> \rho \geq$ 1 Ωcm.

Temperature : 20-25° C.

Probe head : Fell

Voltmeter : Datron DC-Voltmeter 1045.

Current source : Home built.

Electric isolation resistance : 10^{13} Ω.

B. Thickness:

Apparatus: Tesatronic digital, type 44.10001.

Measurement Principle: The wafer is placed between a ball with a 4 mm diameter and a flat plate with 2.25 mm diameter.

Precision: +/- 1 μm from 0-2000 μm.

REFERENCES:

1. ASTM F43-78.

2. I.K. Hargreaves: Brit. J. Appl. Phys 13, (1962), 231.

3. F.G. Vieweg-Gutberlet: J. Electrochem. Soc. 126, (1979), 1792.

4. N.N. Polyakov: Istv. Akad. Nauk SSR 35, (1971), 538 (in Russian).

5. L.J. Swartzendrüber: Solid State Electronics 7, (1964), 412.

6. F.M. Smits: Bell System Techn. J. 37, (1958), 711.

7. A. Uhlir: Bell System Techn. J. 34, (1955), 105.

PHOTOLUMINESCENCE ANALYSIS OF NTD-SILICON

R. B Hammond and J. M. Meese

Los Alamos Scientific Laboratory, Los Alamos, NM 87545
Missouri University Research Reactor, Columbia, MO 65201

INTRODUCTION

Crystalline Si, which is optically excited at low-temperature (~4 K), exhibits luminescence characteristic of several states for nonequilibrium electrons and holes. A single electron-hole pair can bind into an exciton that is chargeless but free to move in the lattice. Also, such a **free exciton** (FE) can become bound to a neutral impurity atom and become a **bound exciton** (BE). The energy of the BE state is lower than that of the FE by its binding energy to the impurity. An exciton will be bound to every type of neutral impurity atom with a different binding energy. Thus, when an electron-hole pair in a FE or BE state recombines radiatively, it emits a photon whose energy uniquely identifies the type of impurity atom to which it was bound or indicates that it was unbound. Recent work[1] has shown that an exciton luminescence spectrum from optically excited Si at ~4 K is a very sensitive indicator of the various types of dopant impurity atoms present in the crystal. More recent work[2] explained the kinetics of exciton capture and emission from impurity atoms and laid the ground work for making exciton luminescence a quantitative measure of impurity concentrations in Si. In this paper we, first, use the kinetics model of Hammond and Silver[2] to calculate the impurity concentration dependence of exciton luminescence intensities in Si and, second, report detailed measurements on NTD Si with known substitutional P concentrations. We show, first, the correctness of the kinetics model and, second, the practicality of exciton luminescence as a precise and extremely sensitive impurity diagnostic for Si.

Theory

Using the model of Hammond and Silver,[2] we can write

$$\frac{df_i}{dt} = \frac{f_i}{\tau_i} + \sigma_i \nu_{FE} n_{FE} - f_i \sigma_i \nu_{FE} \frac{AT^{3/2}}{g_i} \left[1 + \exp(-\Delta E/T)\right] \exp(-E_i/T)$$

where f_i is the fraction of type i impurity atoms that hold a BE, τ_i is the recombination lifetime of an exciton bound to a type i impurity, σ_i is the capture cross section for an exciton at a type i impurity, ν_{FE} is the FE thermal velocity, n_{FE} is the FE density, A is the density of states factor for FE's, T is the absolute temperature, g_i is a BE degeneracy factor, ΔE is the FE ground-state splitting 0.31 meV (Ref. 3), and E_i is the binding energy of an exciton bound to a type i impurity. This equation holds when $f_i \ll 1$.

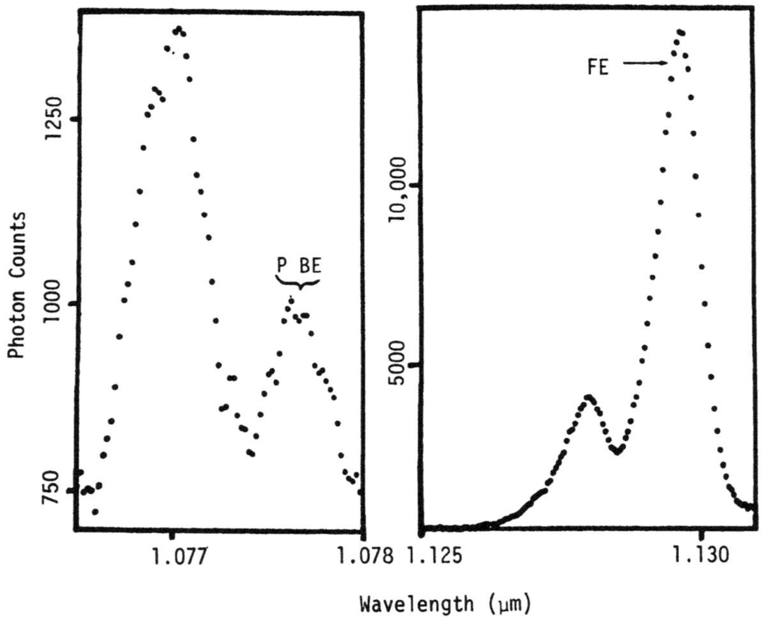

Fig. 1. Photoluminescence spectra taken at 3.96 K with an optical excitation of 9 mW/cm². The P no-phonon-assisted BE luminescence at 1.0776 μm and the FE to-phonon-assisted luminescence are shown. The P content of this crystal was 9×10^{10} cm^{-3}.

If we now assume that the Si crystal is excited with a continuous optical source (dc excitation), $df_i/dt = 0$, and

$$\frac{f_i}{n_{FE}} = \frac{\tau_i \sigma_i \nu_{FE}}{1 + \tau_i \sigma_i \frac{AT^{3/2}}{g_i} [1 + \exp(-\Delta E/T)] \exp(-E_i/T)}$$

The important aspect of this equation is that the right-hand side depends only on the type of impurity atom, i, and on the absolute temperature. We draw three major conclusions from this fact. First, the ratio $(n_i f_i / n_{FE})$ of excitons bound to type i impurities to FE's is **independent** of other types of impurities in the crystal. Second, this ratio is **independent** of the FE lifetime (a poorly understood and highly variable quantity). Third, this ratio is **directly proportional** to the concentration of type i impurity atoms if we assume a constant temperature.

Experimentally we can measure the ratio of BE to FE luminescence intensity. This measurable ratio is proportional to the ratio $n_i f_i / n_{FE}$. Thus, if we measure BE to FE luminescence intensity ratios at low excitation (that is, $f_i \ll 1$) and at fixed temperature we should find the results depend **only** on impurity concentration n_i and are **directly proportional** to impurity concentration.

Experimental

In this study we used 15 NTD Si samples doped in the range from 2×10^{11} cm^{-3} to 2×10^{15} cm^{-3}. Each of the samples contained $\sim 1 \times 10^{12}$ cm^{-3} B. The P doping concentrations were either calculated from neutron dose, from measured resistivity, or from Hall effect results. The crystals were chemically etched on all surfaces. They were immersed in LHe at 3.96 K and irradiated with a calibrated GaAs LED. The luminescence was analyzed with a SPEX 1702 3/4-m grating spectrometer and detected with an RCA 7102 S-1 photomultiplier operated at 77 K in the photon counting mode. The spectrometer broadening was 0.43 meV at 1.1503 eV and 0.39 meV at 1.0972 eV FWHM. Ratios were taken of the peak intensities of the P no-phonon-assisted BE luminescence at 1.1503 eV to the TO-phonon-assisted FE luminescence at 1.0972 eV.

Results

With optical excitation intensity below ~50 mW/cm² we found, first, that no P-bound multiexciton complex lines were observed and, second, that the ratio of BE-to-FE luminescence intensities was independent of excitation intensity. This confirmed that below ~50 mW/cm² at 3.96 K, our assumption $f_i \ll 1$ was valid. In order to have a safe margin for error we took all our data for this work at 3.6 mW/cm². Figure 1 shows representative FE and BE spectra taken under these conditions. Figure 2 shows the results of our measurements of P BE-to-FE luminescence intensity ratios plotted versus substitutional P concentration. The results show the proportionality predicted by the kinetics model over 4 orders of magnitude in impurity concentration. The standard deviation of these data from the proportionality is 7%. We found that the samples, which were more carefully characterized for P content, have less deviation from the precise proportionality. In general, the deviations reflect the accuracy with which the actual P concentrations were known for these crystals.

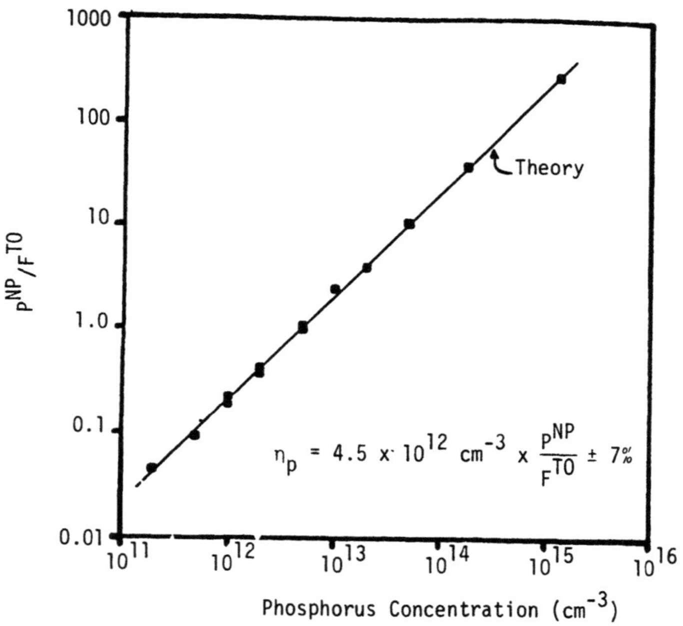

Fig. 2. Ratios of the intensities of the P NO-phonon BE luminescence peak at 1.1503 eV to the FE TO-phonon luminescence peak at 1.0972 eV are plotted vs P concentration. Results from 15 NTD Si samples are shown. Data were taken in the low-excitation regime at 3.96 K. Spectrometer broadening was 4 Å FWHM. Optical excitation intensity was 3.6 mW/cm². All 15 samples contain ~1 x 10^{12} cm^{-3} B. The line through the data corresponds to $n_p = 4.5 \times 10^{12}$ cm^{-3} (p^{NP}/F^{TO}). The data fit the line with a standard deviation of 7%.

Conclusions

First, we have shown that at a fixed temperature of 3.96 K, BE/FE ratios are directly proportional to P concentration in NTD Si. Second, we have also shown that a variation from 0.75 μs to 30 μs for FE lifetime in our samples did not affect this result. Thus, we have confirmed the major predictions of the theory and established an extremely sensitive and accurate impurity diagnostic for NTD Si.

Our experimental arrangement gives us a sensitivity to ~1×10^9 cm^{-3} P. Also, once the experiment is operating, a single sample may be measured and the P concentration calculated in approximately 15 minutes. We normally use 1-cm^2 samples about 0.05 cm thick. The extreme sensitivity of this method suggests its use for very small samples and for impurity profiling of samples.

Acknowledgments

The authors would like to thank R. N. Silver for discussions and encouragement. We also thank Jens Guldberg of Topsil and Jim Baukas and Mary Young of Hughes Research Labs for some of the samples and sample characterization. Finally, many thanks to Gary Bernstein for help with the data taking.

References

1. R. B. Hammond, T. C. McGill, and J. W. Mayer, Temperature Dependence of the Electron-Hole Liquid Luminescence, Phys. Rev. B13:3566 (1976).
2. R. B. Hammond and R. N. Silver, Temperature Dependence of the Exciton Lifetime in High-Purity Si, Appl. Phys. Lett. 36:68 (1980).
3. R. B. Hammond and R. N. Silver, Analysis of LO and TO Phonon Assisted Free Exciton Luminescence in Si, Solid State Commun. 28:993 (1978).

QUANTITATIVE DETERMINATION OF B AND P IN SILICON BY IR SPECTROSCOPY*

Bernard Pajot and Dominique Débarre

Groupe de Physique des Solides de l'E.N.S.
Université Paris 7, Tour 23
75221 Paris Cedex 05, France

INTRODUCTION

Many of the applications of silicon share the need for a high-purity homogeneous material with a controlled concentration of residual shallow and deep impurities. One of its utilization is the fabrication of coupled charge devices (CCD) linked to extrinsic Si detectors arrays. This has given a new impetus to the study of impurities and defects in this material as well as to the need for accurate compensation techniques since spurious effects arise from the presence of uncompensated and unwanted impurities, even at low concentration. Besides atomic impurities, electrical, optical and photoconductive studies have shown the existence in the band gap of silicon of shallow acceptor levels due to impurity complexes whose nature is not yet wholly elucidated[1-3].

The optical characterization of shallow-impurities in silicon seems to be a well-known story; then, why to spend time with this problem ? It is because if many papers have been published on the position of the shallow impurity lines in silicon and on their variation under external and internal perturbations[4], not so many deal with a quantitative correlation between the intensity of the lines and the actual impurity concentration[5]. IR absorption is a non-destructive characterization technique with good sensitivity for the detection of B and P in silicon, but the theoretical prediction of the intensity of the impurity lines is not reliable enough to use it as a primary standard. One has then to calibrate

* Work supported in part by D.R.E.T. contract n° 78-077.

the intensities against an absolute method giving the impurity concentration. The measurement of the Hall effect *vs* temperature, C-V and junction breakdown techniques[6,7], all techniques which are used also for their own can be used for this goal. A simpler and widely employed technique is the room temperature resistivity measurement with the four-point probe. We thought that for P impurity, the accuracy of the transmutation doping technique was high enough to try to use N.T.D. samples as primary standard instead of an electrical standard, taking also opportunity of the homogeneity of the P distribution in N.T.D. silicon.

Once made quantitative, IR spectroscopy allows the determination of the B and P concentrations, whether they are the main impurities or the residual ones, as well as the measurement of the compensation ratio. We report in this communication on our attempt to obtain reliable absorption cross sections and integrated intensities of the strongest lines of B and P in silicon.

SPECTROSCOPIC CHARACTERIZATION METHOD

In a semiconductor the intensity of an electronic impurity line depends of the neutral impurity concentration. Now, under thermal equilibrium conditions at low temperature, when impurities are minority ones, they are ionized, and an equal amount of the shallowest majority impurities is also ionized. This results from the minimization of the electronic energy of the system; it is the compensation phenomenon. As a consequence, the minority impurities are never detected under equilibrium and, to be detected, the shallowest majority impurity must be in concentration greater than $\sum_i N_i min$.

For instance, a residual majority center like Sb or Li in n-type silicon, whose ionization energy is less than that of P remains undetected if its concentration is less than N_A, standing for the sum of the compensating acceptor impurities. Hence in P-doped silicon, the net P concentration is $N_P + N_{D_r} - N_A$ where N_{D_r} is the total concentration of residual donors shallower than P, as long as $N_A > N_{D_r}$. Similarly in n-type silicon doped with an impurity deeper than P (Bi, for instance), the net concentration of the deeper impurity is $N_{Bi} + N_{D_r} - N_A$; here, the P concentration is included in N_{D_r}. When $N_{D_r} > N_A$, the material is termed as undercompensated and some residual impurities are still neutral; when $N_{D_r} < N_A$, all the residual donors are ionized and the material is said to be over-compensated.

The compensating majority and minority impurities can be detected under non equilibrium conditions, by injection of excess carriers, which are trapped by the ionized centers which become neutral and for some experimental conditions, both typesof impurities can be detected spectroscopically at their actual concentration.

When a steady-state pairs concentration is produced by continuous illumination of the sample with band gap light ($\sim$ 1.2 eV), the photo-carriers are trapped by the ionized impurities. The excitons are also trapped by neutral impurities and this indeed is used to detect these impurities by the luminescence technique[8]. The position of the B and P lines in Si is given in Table 1. For B, the lines were initially numbered by order of increasing energy, but when using higher resolution new lines were observed. All the P lines have been attributed to specific transitions and they are labelled by the excited state of the transition.

Table 1. Position (cm^{-1}) of the electronic excitation lines of B and P in silicon[9]

Line	Boron	Line	Phosphorus
1	244.99[a]	2 p_o	275.09
2	278.27	2 $p_\pm$	315.95
3	309.47	3 p_o	323.42
4	319.35	3 d_o	{ 336.8
4-A	319.96		337.6
4-B	321.90	4 p_o	340.84
5	334.52	3 $p_\pm$	342.42
	338.05	4 f_o	348.8
	339.41	5 p_o	349.50
6	340.13	4 $p_\pm$	349.92
7	344.75	4 $f_\pm$	352.31
	346.13	5 f_o	354.33
8	348.3	5 $p_\pm$	355.81
	349.1	5 $f_\pm$	357.47
9	352.8	6 $p_\pm$	358.85
	354.1	6 $h_\pm$	360.44
10	357.4		
2 p'	668.5		
3 p'	693.0		
4 p'	701.3		

[a] These lines belong to the $p_{3/2}$ series while lines 2 p', 3 p' and 4 p' belong to the $p_{1/2}$ series[4].

In the P spectrum, the most intense line is 2 $p_\pm$. In the B spectrum lines 4-4A and 2 are of about equal intensities, but line 2 is $\sim$ 40 cm^{-1} lower in energy than 4-4A and necessitates a longer scan time if recorded with 2 $p_\pm$ (P).

In a silicon sample of thickness d, the absorption coefficient of an impurity is obtained from the relative transmission T_r through the relationship (1), which takes into account the multiple reflections in the sample, with a reflectivity R = .3

$$(1) \quad \alpha = -\left[\ell n \, (5.056 \, T_r^{-1} \, ((1 + .4347 \, T_r^2)^{1/2} - 1))\right] /d$$

Under the assumption that the peak absorption α_{ij} of a line corresponding to a transition i → j of the impurity atom is proportional to the (neutral) impurity concentration, N_{imp}, an optical cross section $\sigma_{ij} = \alpha_{ij}/N_{imp}$ can be obtained for i → j from calibration samples and its inverse used as a "detectability factor" or "calibration ratio" for line i → j.

Alternatively, the integrated intensity of the line i → j,

$$(2) \quad I_{ij} = \int_{\tilde{\nu}_1}^{\tilde{\nu}_2} \alpha \, (\tilde{\nu}) \, d \, \tilde{\nu}$$

where $\alpha(\tilde{\nu})$ describes the spectral dependence* of the absorption coefficient can be used. The measurement of I_{ij} is less accurate than that of α_{ij}, but I_{ij} is not sensitive to instrumental effects and concentration effects.

The instrumental effect arise from the finite spectral band-pass (s.b.p.) of the spectrometer used. The observed profile of the line from which α_{ij} is derived is obtained from the convolution of the true profile f $(\tilde{\nu})$ with the instrumental function g $(\tilde{\nu})$. Two limits can be foreseen :

1) $g\,(\tilde{\nu} - \tilde{\nu}_o) \simeq \delta(\tilde{\nu} - \tilde{\nu}_o)$, i.e. a s.b.p. negligible with respect to the true width of the line. The observed profile is then the true profile

2) $f\,(\tilde{\nu} - \tilde{\nu}_o) \simeq \delta(\tilde{\nu} - \tilde{\nu}_o)$, i.e. the true width of the line much less than the s.b.p.. In this case, the observed width is the s.b.p..

When g $(\tilde{\nu})$ and f $(\tilde{\nu})$ are both gaussian or lorentzian, it is easy to interpolate between these limits, but when g $(\tilde{\nu})$ is gaussian and f $(\tilde{\nu})$ is lorentzian, it is difficult to obtain explicitly the dependence of σ_{ij} on the s.b.p.. The evolution from case 2 to case 1 is shown in Figure 1 for line 2 $p_\pm$ (P).

As the easiest quantity to measure is α_{ij}, the comparison between σ_{ij} and α_{ij} for an unknown sample is to be made with the

*The symbol $\tilde{\nu}$ is used for the wave number (per unit length). The common unit is the cm^{-1}. To convert cm^{-1} into meV, multiply by the factor .124.

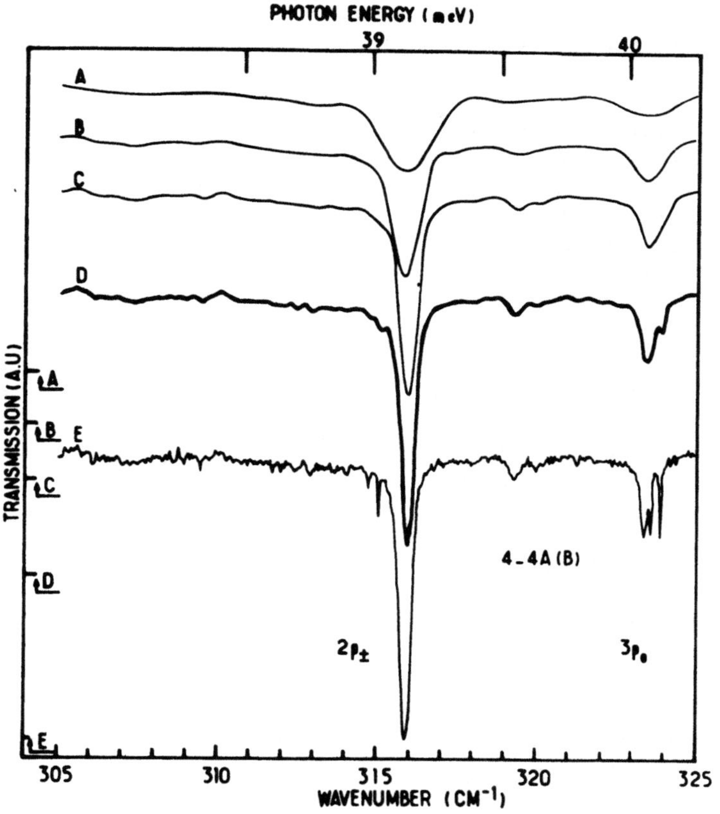

Fig. 1. Transmission spectrum of a 4 mm thick P-doped Si sample (100 Ωcm) as a function of the s.b.p.. The s.b.p. are 2, 1, .53, .26 and .07 cm^{-1} from A to E. The relative peak transmission for 2 $p_{\pm}$ decreases from ∿ .72 for a s.b.p. of 2 cm^{-1} to about zero for .07 cm^{-1}.

same s.b.p.. If α_{ij} is rigorously proportional to N_{imp}, a log-log plot of α_{ij} *vs* N_{imp} should exhibit a slope unity and a lowering of this slope would indicate concentration broadering. High resolution spectra of P in silicon have shown[9,10] that for a concentration less than 10^{15} P atoms/cm^3, the width of the P lines is under 0.2 cm^{-1}. For the B lines, we have found widths ∿ 0.4 cm^{-1} for lines 4 and 4A, and slightly lower values for lines pertaining to shallower levels[9].

RESULTS AND DISCUSSION

The Si samples used for this study were all cut in F.Z. crystals. The net carrier concentration $N_{maj} - N_{min}$ was obtained from careful room temperature resistivity measurements. Some of the test P-doped samples were obtained by nuclear transmutation at the Centre d'Etudes Nucléaires de Grenoble (CENG).

The irradiated test samples (Cd ratio ∿ 10) were cut into parallelipipeds 15 mm x 8 mm x d and optically polished on the four large sides. They were annealed for 1 hour in vacuum at 900° C with a slow rise and cool down. With these samples were also analyzed reference samples, 8 mm thick, cut in the same slice of the intrinsic Si crystal.

The samples are mounted laterally on the cold finger of a liquid helium optical cryostat with two perpendicular pairs of windows. The samples can be illuminated on the side by band-gap light, obtained by suitable filtering of a quartz halogen lamp. The temperature of the samples is 8 K. The cryostat is placed behind a double-pass grating monochromator equipped with a 51 line/mm grating blazed at 417 cm^{-1} (24 μm) in the first order. The first order is selected by the calcium fluoride source mirror and by a filter blocking wave numbers greater than 450 cm^{-1}.

The transmission of the sample can be measured :

1) with monochromatic light (the cryostat is carefully blinded to avoid band-gap light to reach the sample). In this case, the majority impurity is detected and its absorption coefficient is proportional to $N_{maj} - N_{min}$;

2) with monochromatic light plus lateral band-gap light pumping. In this case, both majority and minority impurities are detected and the absorption coefficient should be proportional to N_{maj} and N_{min}, respectively.

The difference between this geometry and that used by Kolbesen (a room temperature Si filter inserted between the source and the cryostat in a commercial spectrometer) is that in the first case :

1) The energy on the detector remains the same for the two measurements (same setting of the amplifier).

2) The Si lattice is not heated directly by the IR source.

3) Only the transition observed is excited during the measurement.

The output of the lock-in amplifier is digitized by a A/D converter trigerred by the controller of the stepping motor which rotates the grating and recorded on one of the tape cartridges of a HP 9845 desktop computer where it is subsequently processed.

The peak absorption coefficient $\alpha_{2p\pm}(P)$ was measured with a s.b.p. of .45 cm^{-1} (56 μeV) and a S/N ratio of ∿ 50. For the samples whose concentration was deduced from resistivity measurements, the absorption was measured with monochromatic light. For the N.T.D. samples, the absorption was measured under pumping and the absorption of the reference sample was substracted to obtain only the contribution of the transmutation*. The P concentration introduced by transmutation was calculated from the thermal neutron dose using the factor 1.7 x 10^{-4} P at/thermal neutron.

Table 2. Characteristics of the intrinsic reference samples

Sample	ρ (k Ωcm)	$\alpha_{2p\pm}$ (cm^{-1})	α_{4-4A} (cm^{-1})
P_1 P_2	∿ 20	.08 ± .01	.03 ± .01
P_2 P_3	∿ 40	.09 ± .01	.02 ± .01
P_3 P_4	∿ 36	.08 ± .01	.03 ± .01
P_4 P_5	∿ 36	.08 ± .01	.03 ± .01
P_5 P_6	∿ 40	.08 ± .01	.02 ± .01

The B and P concentrations are homogeneous in the series. A resistivity of 40 k Ω cm corresponds to 3.5 x 10^{11} net acceptors/cm^3 or to 1.3 x 10^{11} net donors/cm^3. In sample P_4 P_5 neutral P at equilibrium was detected with an absorption coefficient of .02 cm^{-1}, indicating that the sample was n-type.

The results for the N.T.D. samples are summarized in table 3.

Table 3. Experimental data on the N.T.D. samples

Sample	therm.n_o ($10^{17}/cm^2$)	P conc. (10^{12}at/cm^3)	calc. ρ[a] (Ωcm)	2p±(P) (cm^{-1}) pump.	M.L.
P_1 P_2 T	10.1	172	29.7	15.1±.1	14.4 ±.1
P_2 P_3 T	4.09	69.5	73.4	5.84±.05	5.42±.05
P_3 P_4 T	1.15	19.6	260	1.81±.03	1.64±.03
P_4 P_5 T	.404	6.9	740	(.35±.01)	(.26±.01)
P_5 P_6 T	.123	2.1	2430	.37±.01	.28±.01

[a]We assume that ρ(Ωcm) ≃ 5.1 x $10^{15}/N_p$ (at/cm^3).

*For the routine experiments, we use samples with a maximum thickness of 4 mm. For the calibration experiment, the maximum thickness was 8 mm for the N.T.D. samples and 4.8 mm for the other samples.

One can note the increase of the difference between the absorption coefficients obtained under pumping and with monochromatic light. There has been some confusion on the identification of sample P_4 P_5 T containing nominally 6.9 x 10^{12} P atoms/cm^3, which was found to contain actually $\sim$ 2 x 10^{12} P atoms/cm^3. This was confirmed by resistivity measurements. The graph between $\alpha_{2p\pm(P)}$ and the P concentration is given in Figure 2.

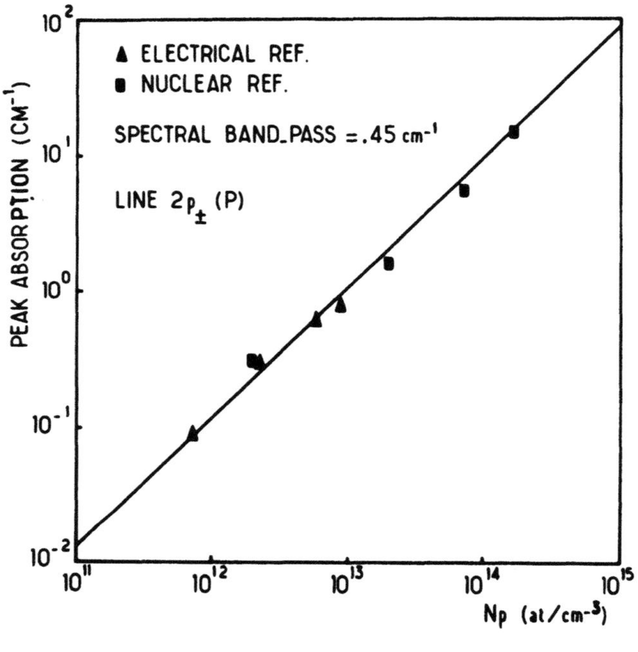

Fig. 2. Peak absorption coefficient of 2 $p_{\pm}$(P) *vs* P concentration at 8 K with a s.b.p. of .45 cm^{-1}.

It can be expressed numerically as :

$$(3) \quad N_p \; (at/cm^3) \; = \; 9.6 \times 10^{12} \left[\alpha_{2p_{\pm}} \; (cm^{-1})\right]^{1.04}$$

for a s.b.p. of .45 cm^{-1}. The true width of 2 $p_{\pm}$ (P) has been found to be .15 cm^{-1} at low concentration. The s.b.p. used is three times this value and in our spectra, we cannot distinguish the width of the residual water vapor lines from that of the 2 $p_{\pm}$ (P) line. From this, we infer that we already are in the resolution region where $\alpha_{2p\pm}$ is unversely proportional to the s.b.p.. We can then derive a relation including the s.b.p. $\Delta\tilde{\nu}_s$ and it reads :

$$(4) \quad N_p\ (at/cm^3) = 2.13 \times 10^{13} \times \Delta\tilde{\nu}_s\ (cm^{-1})\ \left[\alpha_{2p\pm}\ (cm^{-1})\right]^{1.04}$$

This relation should be valid for $\Delta\tilde{\nu}_s > .5\ cm^{-1}$ and for resistivities greater than 5 Ωcm. For low resistivities, $\alpha_{2p\pm}$ becomes high and thin samples mounted in a strain free manner must be used. We have found that the N.T.D. samples investigated were unusually sensitive to strains despite careful mounting and a new sequence of smoothing and polishing after the annealing. For $\Delta\tilde{\nu}_S = 1\ cm^{-1}$, (4) compares with the calibration factor of 2.3×10^{13} P atoms/cm^3 obtained by Kolbesen at a higher temperature (15 K). For lower s.b.p. we anticipate $\sigma_{2p\pm(P)}$ to increase accordingly, and from the value of the observed width at high resolution, we expect a gain in sensitivity by comparison with the present measurement by a factor 2.5 - 2.8 using a s.b.p. less than .1 cm^{-1}.

We have also measured the absorption coefficient of the $3\,p_\pm$(P) line in samples P_1 P_2 T and P_2 P_3 T; they are 3.83 and 1.57 cm^{-1} respectively for a s.b.p. of .58 cm^{-1}, in good agreement with the ratio of the P concentration (2.47).

The same experimental conditions are used to establish a calibration factor for line 4.4A (B). We have also computed a calibration factor for line 3 (B) at 309.5 cm^{-1}, which is about nine times less intense than line 4.4A (B). The exponents found are nearly the same as that for $2\,p_\pm$(P).

$$(5) \quad N_B\ (at/cm^3) = 2.54 \times 10^{13} \times \left[\alpha_{4.4A(B)}\ (cm^{-1})\right]^{1.05}$$

$$(6) \quad N_B\ (at/cm^3) = 2.35 \times 10^{14} \times \left[\alpha_{3(B)}\ (cm^{-1})\right]^{1.04}$$

Relation (6) can be used to measure the B content in p-type samples with 150 Ωcm > ρ > 2 Ωcm.

It is difficult to extrapolate (5) to higher spectral band pass because of the doublet structure of the line (Figure 3).

Relations (3) - (6) allow not only the optical determination of the net impurity concentration, but also the measurement of the compensation, as N_P and N_B can be measured separately, and this was pointed out by Kolbesen. An obvious check of the above-described technique was to verify that all the ionized minority impurities are neutralized by optical pumping. This was done for n-type Si as follows : We first measured spectroscopically the net and total B content in a p-type sample with $N_B - N_D = 3.7 \times 10^{13}$ at/cm^3 and $N_B = 4.1 \times 10^{13}$ at/cm^3. This sample was then converted into n-type by $(3.9 \pm .9) \times 10^{17}$ therm. n_o/cm^2. When remeasured, we found a net P concentration of 3.1×10^{13} at/cm^3 and $N_B = 3.7 \times 10^{13}$ at/cm^3,

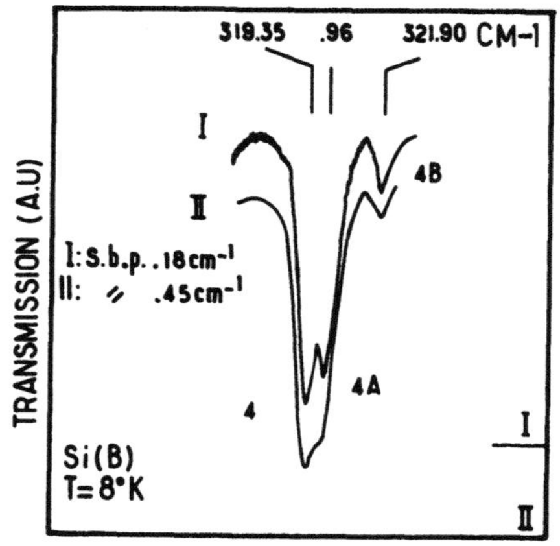

Fig. 3. Group of lines 4-4A-4B of "residual" boron in Al doped Si (10 Ωcm). The detector used to obtain these spectra was a Si bolometer cooled at 1.8 K.

in reasonable agreement with the initial B content. For p-type Si, we plan to transmute two samples : one intrinsic and the second p-type with $\rho \sim 200$ Ωcm with $\sim 10^{17}$ therm. n_o/cm^2. After annealing, the first sample must be converted into n-type with $\rho \sim 300$ Ωcm and the second into p-type with $\rho \sim 264$ Ωcm.

The relations we have derived have been used to gain additional information on the high purity samples investigated. For P_4 P_5 sample, we obtain total and net P concentrations of $(7.1 \pm 1) \times 10^{11}$ at/cm^3 and $(1.6 \pm 1) \times 10^{11}$ at/cm^3, respectively and a total B concentration of $(6.4 \pm 2) \times 10^{11}$ at/cm^3. Within the experimental accuracy, a reasonable agreement exists between the compensation measurements and the net P concentration. As a comparison, $N_D - N_A$ from the four point probe measurement is 1.4×10^{11} cm^{-3}.

The reliability of the compensation technique for the p-type material can be estimated from Figure 4, showing the difference between the "pump" and monochromatic light spectra of a p-type Si sample with $N_A - N_d = 9.3 \times 10^{12}$ at/cm^3.

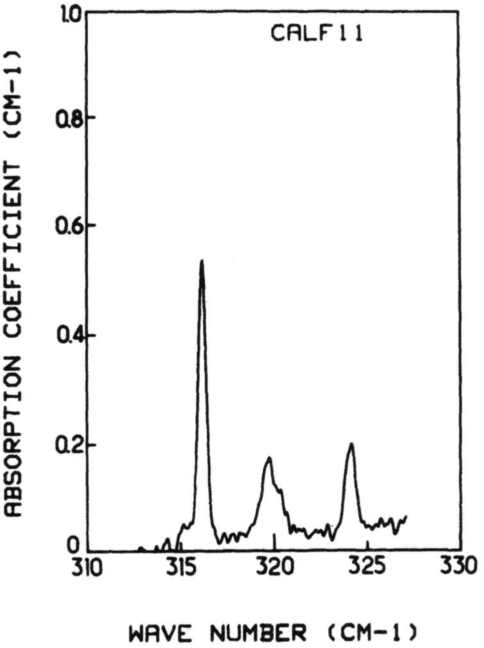

Fig. 4. Net effective neutralization of B and P in a 1500 Ωcm p-type Si sample at 8 K. The line near 325 cm^{-1} is a residual water vapor line.

We deduce from this figure $N_{P^+} = 5.0 \times 10^{12}$ at/cm^3 and $N_{B^-} = 4.2 \times 10^{12}$ at/cm^3, *i.e.* a compensation ratio of ∿ 40 %.

For highly compensated n-type material, we have found a noticeable difference between N_{P^+} and N_{B^-} with $N_{P^+} < N_{B^-}$. This could be explained by assuming the presence of a donor impurity with an ionization energy lower than that of P, and participating to the compensation mechanism, but we were unable to detect the 2 $p_\pm$ (Li) or 2 $p_\pm$ (Li-O) lines at 214.5 and 268.6 cm^{-1} respectively and the question remains open.

In the N.T.D. samples investigated here, we note an increase of the compensation with the dose and we find that the sequence (irradiation + annealing) has introduced acceptors at a nearly constant rate. We would expect the annealing at 900° C to have destroyed the acceptor complexes, but it does not seem to be the case.

The integrated intensity of the 2 $p_{\pm}$ (P) line has been found to be proportional to the concentration between 2×10^{12} and 2×10^{14} at/cm^3. We obtain an integrated absorption cross section, *i.e.* the ratio of the integrated intensity over concentration of $(5.9 \pm 0.5) \times 10^{-14}$ cm for 2 $p_{\pm}$ (P). For boron, we give results for the group 4-4A-4B which cannot be separated for a s.b.p. greater than 1 cm^{-1}. The integrated capture cross section for the B lines is $(5.4 \pm .4) \times 10^{-14}$ cm. These quantities are independent of the s.b.p. and should be independent of concentration as long as pairs absorption is negligible[11]. They can then be used to determine the impurity concentration once the integrated intensity has been measured unambiguously. In the case where measurements are made with white light, caution must be exercized because of the overlap of the B and P lines at high s.b.p.[5].

SUMMARY AND CONCLUSION

To summarize, we have shown that P introduced in Si by transmutation doping could be used to obtain optical calibration points independent from the resistivity measurements. We derive calibration factors for P and B in silicon for a lower s.b.p. than that used previously[5], hence a gain in sensitivity when a high S/N ratio is achieved. The set up for measuring the compensation is also different and we give results on high resistivity material obtained with it. The actual sensitivity is $\sim 3 \times 10^{11}$ P at/cm^3 and $\sim 6 \times 10^{11}$ B at/cm^3 for 4 mm thick samples.

Photoluminescence is certainly more a sensitive technique than absorption to measure the B content in high purity silicon and it allows a more local analysis, but it does not compare directly with resistivity measurements. Optical measurements can give a reliable value of the net concentration of a given impurity (this is also true for other impurities than B and P, but with lower sensitivities), and also a value of the compensation of the material. We would like to emphasize the interest of absorption techniques for a precise knowledge of the compensation in N.T.D. silicon for an indirect detection of residual impurities introduced during the processing or simply to know if residual shallow acceptors are under or overcompensated by P in p-type Si doped with a deep acceptor.

ACKNOWLEDGEMENTS

We wish to thank D. Roche and F. Blanc from C.E.N.G. for supplying the N.T.D. samples and we acknowledge the assistance of S. Squelard for the annealing. We are also indebted to J. Gulberg from Topsil A/S for providing some of the calibration samples used in this study.

REFERENCES

1. R. N. Thomas, T. T. Braggins, H. M. Hobgood, and W. J. Takei, Compensation of residual B impurities in extrinsic In-doped Si by neutron transmutation of Si, J. Appl. Phys. 49:2811 (1978).
2. M. H. Young, O. J. Marsh, and R. Baron, Shallow defect levels in neutron irradiated p-type extrinsic Si, J. Appl. Phys. 50: 3755 (1979).
3. W. Scott, and C. E. Jones, IR spectra of new acceptor levels in B- and Ga-doped Si, J. Appl. Phys. 50:7258 (1979).
4. F. Merlet, B. Pajot, Ph. Arcas, and A.M. Jean-Louis, Experimental study of the Zeeman splitting of B-levels in Si, Phys. Rev. B 12:3297 (1975).
5. B. O. Kolbesen, Simultaneous determination of the total content of B and P in high resistivity Si by IR spectroscopy at low temperature, Appl. Phys. Lett. 27:353 (1975).
6. R. Baron, M. H. Young, J. K. Neeland, and O. J. Marsh, Characterization of high resistivity Si by Hall measurements, in "Semiconductor Silicon 1977", H. R. Huff, and E. Sirtl, ed., The Electrochemical Society, p. 367 (1977).
7. D. K. Schroder, T. T. Braggins, and H. M. Hobgood, The doping concentrations of In-doped Si measured by Hall, C-V, and junction breakdown techniques, J. Appl. Phys. 49:5256 (1978).
8. M. Tajima, Determination of B and P concentrations in Si by photoluminescence analysis, Appl. Phys. Lett. 32:719 (1978).
9. B. Pajot, J. Kauppinen, and R. Anttila, High resolution study of the group V impurities absorption in Si, Solid State Commun. 31:759 (1979).
10. C. Jagannath, Z. W. Grabowski, and A. K. Ramdas, A high resolution study of the excitation spectrum of the P donors introduced in Si by neutron transmutation, Solid State Commun. 29:355 (1979).
11. M. Capizzi, G. A. Thomas, F. De Rosa, R. N. Bhatt, and T. M. Rice, Observation of donor exciton band in silicon, Solid State Commun. 31:611 (1979).

EXTRINSIC NTD SILICON FOR INFRARED APPLICATIONS*

T. T. Braggins and R. N. Thomas

Westinghouse Research and Development Center

Pittsburgh, Pennsylvania 15235

ABSTRACT

All solid state imaging became a reality in the early seventies with the advent of low noise, charge transfer devices. More recently, the spectral range of these silicon MOS-based sensors has been extended to the infrared region through the use of extrinsically doped silicon. Using indium or gallium-doped silicon which covers the 3-5 and 8-12 μm atmospheric windows, respectively, sensors or focal plane arrays which combine high resolution detection with on-focal-plane CCD signal processing have been developed and are of great interest for infrared imaging systems. Uniformly doped p-extrinsic silicon of the highest purity is needed for infrared detection. A particular consideration is the need to compensate any residual acceptor impurities which are shallower than the IR sensitive dopant in order to achieve the highest operating temperature. Unfortunately, however, overcompensation results in a loss of sensitivity so that the ability to precisely compensate residual acceptors, such as boron, is highly desirable.

The preparation of high detector quality indium-doped silicon by combining *very high purity crystal growth* (by float zoning and modified Czochralski pulling) with *precision neutron transmutation* compensation is reviewed. It is shown that high purity indium-doped silicon substrates with net compensation densities in the low 10^{12} cm^{-3} range can be produced by this technology. Detectors compensated by neutron transmutation offer improved uniformity and peak infrared

*Supported in part by DARPA under Contract DAAG53-76-C-0170 and monitored by Night Vision and Electro-Optics Laboratories, Fort Belvoir, VA 22060.

responsivities of up to 100 A/W compared to sensitivities of <20 A/W obtained by conventional compensation techniques. The effect of focal plane array processing on high sensitivity extrinsic silicon substrates will also be discussed.

INTRODUCTION

The development of charge coupled devices (CCDs) during the 1970s has generated great interest in solid state imaging systems.[1] The powerful signal processing capabilities of CCDs when combined with large numbers of infrared detectors can lead to improved performance for infrared imaging.[2] One approach which is currently being investigated is the monolithic one, in which CCD signal processors are fabricated directly on an infrared sensitive substrate.[2,3] Due to the well-developed silicon integrated circuit capability, infrared sensitive substrates of extrinsically doped silicon are a candidate for monolithic focal plane arrays. A monolithic test array consisting of 32 x 16 detector elements, which is currently under development at the Westinghouse R&D Center, is shown in Fig. 1. Employing indium-doped silicon substrates, infrared sensitivity over the 3-5 μm atmospheric transmission band is obtained while gallium-doped silicon is utilized for the 8-14 μm range.

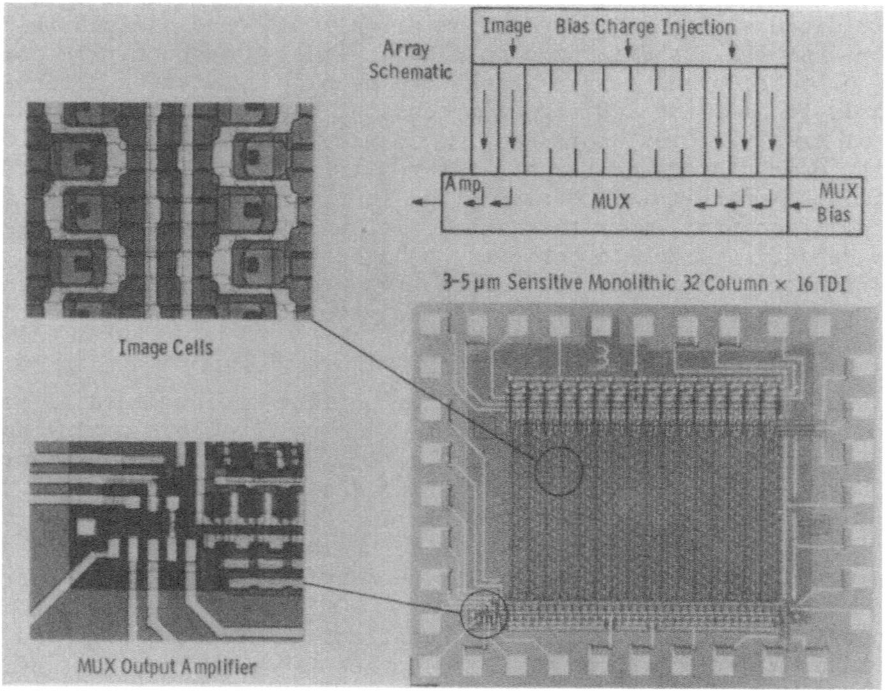

Fig. 1. Monolithic 32 x 16 extrinsic silicon focal plane array.

For extrinsic silicon monolithic arrays, the materials properties of the extrinsically doped substrate are of critical importance to the infrared performance. In particular, to maximize the photocurrent in the extrinsic substrate, one must maximize the product $\eta\mu\tau$, where η is the quantum efficiency, μ is the mobility and τ is the majority carrier lifetime. To obtain high quantum efficiency, high concentrations of the extrinsic dopant are required consistent with high crystalline perfection so as not to degrade the mobility. For extrinsic photoconductors, the majority carrier lifetime is determined by recombination at ionized impurities. With indium or gallium doped silicon at their maximum operating temperatures, the recombination center density is equal to the net compensation density, N_D-N_A, and the lifetime is given by

$$\tau = 1/B(N_D - N_A) \quad (1)$$

where N_D is the donor density, N_A is the shallow acceptor density, and B is the recombination coefficient. High carrier lifetime is therefore obtained by compensating as closely as possible, residual shallow acceptor impurities, such as boron, which are always present in melt-grown silicon, by donor impurities. Undercompensation ($N_D < N_A$) imposes serious penalties in that thermal generation from shallow acceptors requires lower operating temperatures to obtain background limited infrared performance. On the other hand, excessive overcompensation ($N_D \gg N_A$) is equally deleterious, reducing the carrier lifetime and resulting in lower responsivity detector material.

NTD EXTRINSIC SILICON

The use of neutron transmutation doping has been investigated for the controlled introduction of phosphorus donors into indium-doped silicon infrared detector material.[4] Neutron transmutation doping is widely used in the manufacture of power devices primarily due to the improved radial uniformity of the phosphorus dopant and the accuracy of the transmuted phosphorus concentration (i.e. resistivity) which can be specified.[5] These advantages are of great importance to the neutron transmutation doping of extrinsic silicon. The uniformity of the transmuted phosphorus can lead to improvements in the photoresponse uniformity of infrared detector arrays while the accuracy with which a given concentration can be obtained allows precise compensation of residual acceptor impurities leading to improvements in responsivity.

Neutron transmutation doping of both Czochralski and float-zone grown Si-In detector material has been investigated. From initial studies using Czochralski Si:In, it was demonstrated that infrared detector properties were unaffected by neutron irradiation.[4] For example, following an anneal at 850 to 900°C in an inert atmosphere to remove radiation damage, the spectral response was shown to follow the photoionization cross section of indium and background limited detectivities were observed at operating temperatures below about 45 to 50K, as predicted theoretically. Due to the approximately 10^{14} cm^{-3} boron contamination arising from the dissolution of the natural quartz crucibles which are utilized industry-wide in silicon Czochralski growth, infrared responsivities of 1 A/W were achieved consistent with the low - 10^{14} cm^{-3} net compensation densities obtained by transmutation.

Czochralski Si:In material of significantly improved purity (low - 10^{13} cm^{-3} boron concentrations) can, however, be prepared by utilizing certain higher purity synthetic quartz crucibles.[6] The lower shallow acceptor content of this material when combined with precision neutron transmutation doping (in which the neutron fluence is continuously monitored during irradiation[7]) has enabled net compensation densities of 2.7 x 10^{12} cm^{-3} to be obtained, as shown by the Hall-effect measurements of Fig. 2. Also shown in Fig. 2 are the Hall-effect measurements of the as-grown Czochralski material prior to irradiation together with the impurity parameters determined from analysis of the Hall data, as has been described previously.[4] Thermal carrier generation from the shallow acceptors is clearly evident at low temperatures for the as-grown sample. Following neutron irradiation and annealing, compensation of the shallow acceptors by the addition of phosphorus donors is indicated by the exponential decrease of the carrier concentration with a slope characteristic of the indium-X level activation energy.[4,8]

Substantial improvements in the infrared responsivity are observed for precisely compensated Czochralski Si:In material, as shown in Fig. 3, where the responsivity of samples prepared by high purity Czochralski pulling and neutron transmutation, and conventional Czochralski growth and compensation (where phosphorus was added directly to the Czochralski melt) are compared. The neutron compensated sample exhibits a responsivity of 50 A/W (λ = 5.9 μm, T = 50K, E ≃ 1000 V/cm), a significant increase over conventionally compensated samples for which responsivities generally lie between 1 and 10 A/W for the net compensation densities in the 10^{13} cm^{-3} range obtained from synthetic crucible growth.[6] The temperature dependence of the responsivity in precisely compensated Si:In detector material (see Fig. 3) is complex due to carrier recombination at compensated shallow acceptors. A detailed discussion of these effects is beyond the scope of this work and has been reported elsewhere.[9,10]

Neutron transmutation doping of extrinsic silicon can be most advantageously used with float-zone grown material. The inherently low residual acceptor concentrations (low - 10^{12} cm^{-3}) obtained by float-zoning enable comparable net compensation densities to be readily obtained by transmutation. Additionally, improvements in

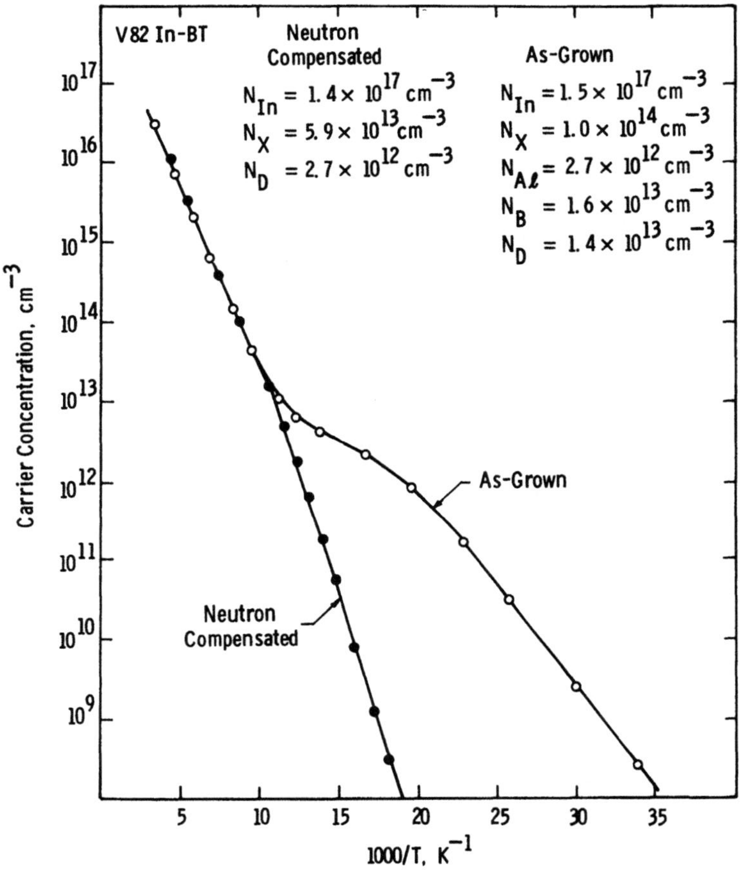

Fig. 2. Carrier concentration versus reciprocal temperature for as-grown and neutron compensated samples of Czochralski Si:In.

the uniformity of infrared response are possible as higher ratios of transmuted phosphorus to residual impurities can be obtained without unduly sacrificing responsivity as the magnitude of the net compensation density is increased.[4]

With float-zone growth Si:In, exceptionally high infrared responsivities (100 A/W, λ = 5.9 μm, T = 50K, E ≃ 1000 V/cm) have been obtained by neutron transmutation.[9,11] Many samples have been neutron irradiated and net compensation densities less than 5×10^{12} cm^{-3} can be routinely obtained. This is illustrated in Fig. 4 where the net compensation density is shown for samples of neutron-compensated indium-doped silicon prepared by Czochralski pulling and float-zoning.

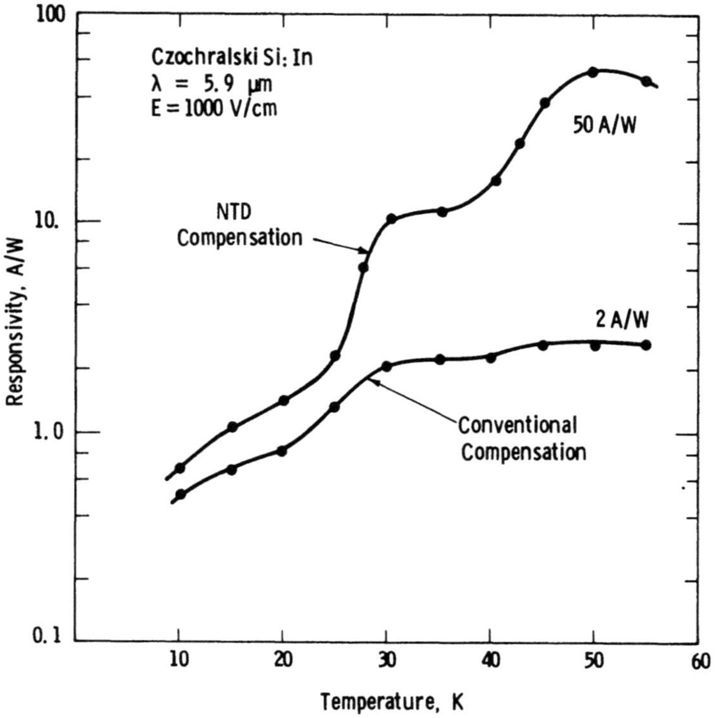

Fig. 3. Comparison of responsivity for neutron compensated and conventionally compensated Czochralski Si:In.

The high infrared responsivities observed in neutron compensated Si:In result directly from the long photocarrier lifetimes obtained for low net compensation densities. From a knowledge of the material parameters, the photocarrier lifetime can be determined from the measured responsivity as has been described previously.[9] The lifetime as a function of the net compensation density is shown in Fig. 5 exhibiting the inverse dependence which is observed in accordance with Eq. 1. From these data, an average value of the recombination coefficient, $B = (1.6 \pm 0.6) \times 10^{-6}$ cm^3/sec, was determined for a temperature of 50K and a bias field of 1000 V/cm. These data clearly indicate that long photocarrier lifetimes, and hence high infrared responsivities, can be obtained through control of the net compensation density.

EFFECTS OF HIGH TEMPERATURE PROCESSING

Of particular interest in the development of monolithic focal plane arrays is the effect of the high temperature (1000-1200°C) processing employed in CCD fabrication on the responsivity of infrared sensitive substrates. It has been shown previously that responsivity is drastically reduced following high temperature processing due to

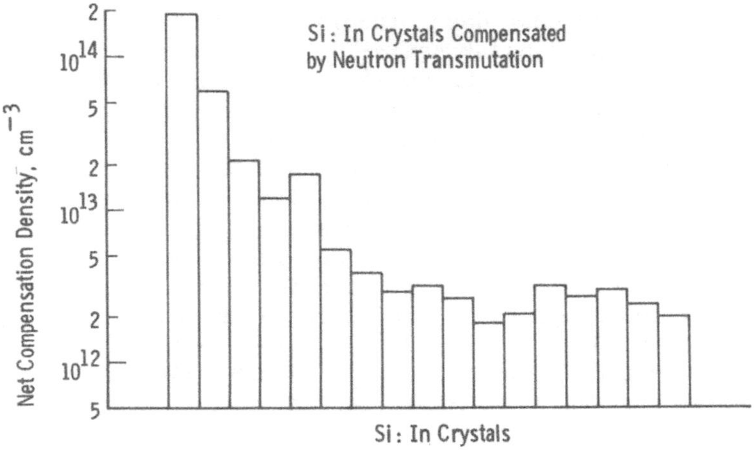

Fig. 4. Net compensation densities obtained by neutron transmutation in Si:In crystals. Historically, earlier crystals are to the left and latter toward the right.

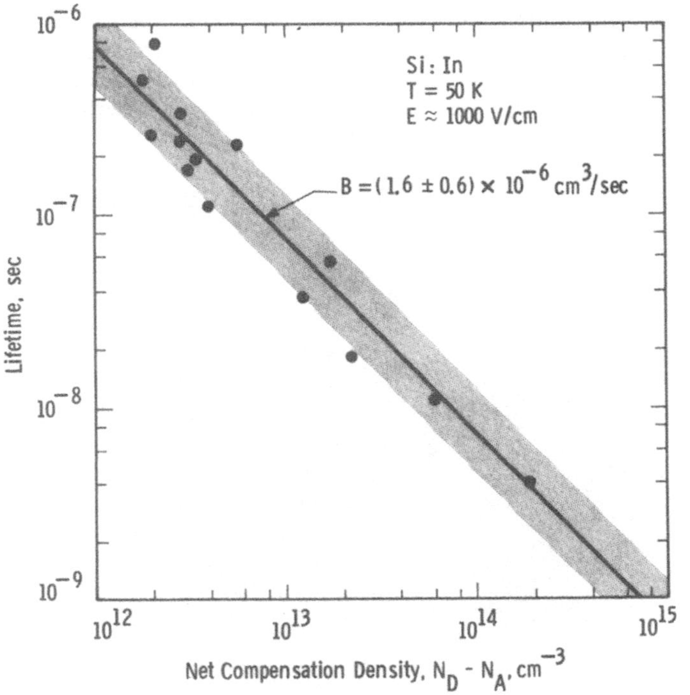

Fig. 5. Lifetime as a function of net compensation density. Solid line is calculated lifetime for a recombination coefficient of (1.6 ± 0.6) x 10^{-6} cm^3/sec.

an increase in the net compensation density presumably arising from the introduction of deep donor levels due to heavy metal contamination.[9] Although phosphorus gettering has been shown to be effective in reducing the donor contamination and restoring the responsivity in simple detector test structures, the use of phosphorus and other well-known gettering treatments in complex CCD processing on indium and gallium-doped silicon substrates, has been found to be less effective at restoring infrared sensitivity. An alternative approach is to fabricate hybrid extrinsic detector/CCD arrays in which the CCD fabrication is optimized for a separate bulk-silicon substrate and the extrinsic detector array is also fabricated separately employing only low-temperature processing techniques. The detector array is then bonded to the CCD on a pixel-to-pixel basis using a suitable interconnect metallization. A SEM photograph of a hybrid extrinsic silicon array is shown in Fig. 6.

CONCLUSIONS

Neutron transmutation doping can be advantageously employed with both Czochralski and float-zone grown extrinsic silicon infrared detector material to precisely compensate residual shallow acceptors present in extrinsic silicon. Precise compensation is shown to result in the long photocarrier lifetimes required to obtain improved infrared responsivity of extrinsic silicon detectors. Infrared responsivities of 50 A/W and 100 A/W have been obtained for neutron compensated indium-doped silicon grown by the Czochralski and float-zone growth methods, respectively.

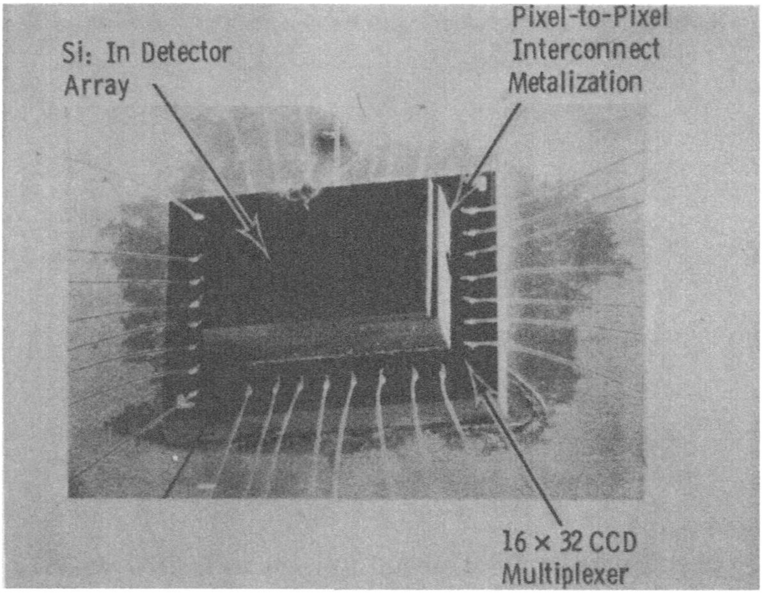

Fig. 6. Extrinsic silicon hybrid detector/CCD array.

The use of hybrid extrinsic silicon/CCD arrays is proposed to overcome the problems of impurity contamination arising from the high temperatures required for monolithic CCD/detector array fabrication.

ACKNOWLEDGMENTS

The authors would particularly like to thank Dr. H. M. Hobgood and Mr. M. M. Sopira for their contributions to this work in providing the extrinsic silicon crystals and the CCD fabrication results, respectively. The technical assistance of W. E. Bing and D. P. Nebel for Czochralski and float-zone crystal growth, respectively, and W. F. Valek and R. S. Nye for CCD fabrication and detector characterization, respectively, is gratefully acknowledged.

REFERENCES

1. D. F. Barbe, Imaging Devices Using the Charge Coupled Concept, Proc. IEEE, 63:38 (1975).
2. R. D. Nelson, Infrared Charge Transfer Devices — The Silicon Approach, Optical Engr. 16:275 (1977).
3. A. J. Steckl, R. D. Nelson, B. T. French, R. A. Gudmundson and D. Schechter, Application of Charge Coupled Devices to Infrared Detection and Imaging, Proc. IEEE, 63:67 (1975).
4. R. N. Thomas, T. T. Braggins, H. M. Hobgood and W. J. Takei, Compensation of Residual Boron Impurities in Extrinsic Indium-Doped Silicon by Neutron Transmutation of Silicon, J. Appl. Phys. 49:2820 (1978).
5. Special Issue on High Power Semiconductor Devices, IEEE Trans. Electron Devices ED-23:795 (1976).
6. H. M. Hobgood, T. T. Braggins, M. M. Sopira, J. C. Swartz and R. N. Thomas, Growth and Characterization of Indium-Doped Silicon for Extrinsic IR Detectors, IEEE Trans. Electron Devices ED-27:14 (1980).
7. S. Gunn, J. M. Meese and D. M. Alger, High Precision Irradiation Techniques for NTD Processing at the University of Missouri Research Reactor, in: "Neutron Transmutation Doping of Semiconductors," J. M. Meese, ed., Plenum, New York (1979).
8. R. Baron, M. H. Young, J. K. Neeland and O. J. Marsh, A New Acceptor Level in Indium-Doped Silicon, Appl. Phys. Lett. 30: 594 (1977).
9. T. T. Braggins, H. M. Hobgood, J. C. Swartz and R. N. Thomas, High Infrared Responsivity Indium-Doped Silicon Detector Material Compensated by Neutron Transmutation, IEEE Trans. Electron Devices ED-27:2 (1980).
10. D. H. Alexander, R. Baron and O. M. Stafsudd, Temperature Dependence of Responsivity in Closely Compensated Extrinsic Infrared Detectors, IEEE Trans. Electron Devices ED-27:71 (1980).
11. H. M. Hobgood, T. T. Braggins, J. C. Swartz and R. N. Thomas,

Role of Neutron Transmutation in the Development of High Sensitivity Extrinsic Silicon IR Detector Material, in: "Neutron Transmutation Doping of Semiconductors," J. M. Meese, ed., Plenum, New York (1979).

IMPURITY DOPING AND ISOLATION PROCESSING

BY HIGH ENERGY ELECTRON BEAM

Takao Wada

Faculty of Engineering
Mie University
Kamihama, Tsu, 514, Japan

ABSTRACT

The present work describes the experimental results for the introduction of aluminum (indium, gallium) impurity into Si from an Al (In, GaP) sheet in contact with its surface by high energy electron bombardment. The surfaces of the Al (In, GaP) sheet ($4\times4mm^2$ area, t~0.3~1.5mm), which was put on B-doped Si wafer (~4×4 mm^2 area) were exposed with a dose of $(1\sim3)\times10^{17}$ electrons/cm^2 at 9 and 10 MeV. They were put in a running water bath in order to prevent them from temperature rise. The intensity ratios of $^{27}Al^+$ ($^{115}In^+$, $^{69}Ga^+$) ions to $^{28}Si^+$ ions for the specimen were obtained from $\sim6\times10^{-4}$ to $\sim10^{-4}$ as a function of depth (0~900Å) from the Si surface by using secondary ion mass spectrometry. This technique is useful for the impurity doping of semiconductors.

High energy electron bombardment produced high resistivity layers in Si, GaP and GaAsP, and varied from an original orange colour of GaP crystal to black. After bombardment, the Laue pattern of the wafers was the same normal pattern as that before bombardment. The spatial distributions of high resistivity layer and red luminescence in a linelike region irradiated by scanned 1MeV electron beams were investigated using the scanning electron microscope (SEM). An electron beam induced current and cathodoluminescent images by SEM presented a very marked line due to bombardment. This new microstructure technique by focussed beam is useful for isolating GaP, GaAs and GaAsP junction devices.

1. INTRODUCTION

The technique of "Recoil Implantation" is the introduction of impurities into a solid by atomic recoil form. The number of recoils as a function of recoil energy passing from an evaporated layer into the substrate by irradiation with energetic ions has been calculated using cross section for atom displacement by Nelson [1]. However, in practice, the fraction of atoms which penetrated greater than just a few tens of atomic distances was very small.

Recently, 50 and 300KeV Krypton ion beams have been used to introduce Sb atoms in silicon from thin evaporated layers deposited on surface. These experimental results [2] have been good agreement with transmission sputtering theory of Sigmund. [3] Also, the experimental and theoretical investigation of recoil implantation at elevated temperatures has been presented. [4] It has been founded that a radiation-induced solute flux away from the substrate/overlayer interface was effective in carrying implanted atoms out of the shallow surface layer. [4]

Proton bombardment has been used to convert both p-and n-type GaAs into high resistivity layers which can be used to isolate GaAs junction devices. [5]

In the present paper a new method of recoil implantation in Si using high energy electron bombardment is investigated. Aluminum (Indium, Gallium) impurities are introduced into Si from an Al (In, GaP) sheet in contact with its surface by 10MeV electron bombardment. The distributions of Al (In, Ga) atoms implanted into Si in depth are measured using secondary ion mass spectrometry. [6] Also, high energy (1~10MeV) electron bombardment produces high resistivity layers in semiconductors and varies from an original orange colour of GaP crystal to black. This new technique is useful for electrically and optically isolating GaP junction diodes. [7~10]

The method is excellent for fabricating the deeper isolation with narrower width.

2. EXPERIMENTAL PROCEDURES

2.1 Impurity doping

The samples used in the experiments were B doped p-Si wafers with an area of $4\times4mm^2$ ($t\simeq300\mu m$), whose resistivity was 10~20Ω-cm. The Al (In) materials were 99.99 (99.999)% pure sheet and GaP wafers were LEC n-type crystals with a dimension of $4\times4\times0.3\sim1.5(mm)^3$. The surfaces of the Al (In and GaP) sheet

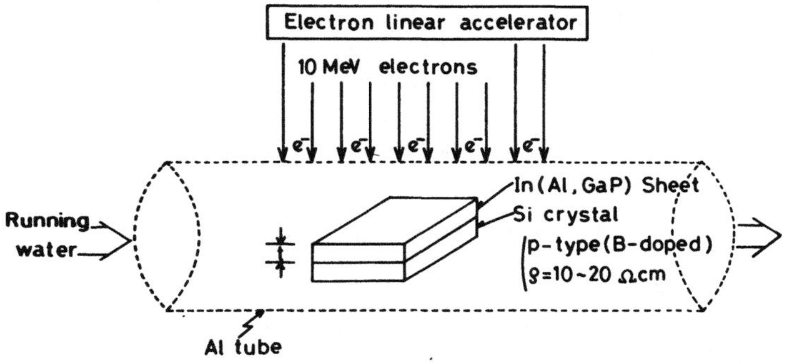

Fig. 1. Schematic diagram for the experiments of impurity doping by electron beam.

in contact with the Si wafer were bombarded with electrons of 9 and 10MeV from an electron linear accelerator with a pulse width of 4μs, ~80Hz duty cycle, and an average electron beam current density of 1.6μA/cm^2. The electron beam was scanned over a wide enough range to provide uniform bombardment. The samples were put in a running water bath to prevent temperature rise. Fig.1 shows a schematic diagram for the bombardments. After bombardment, the samples were left at room temperature for a few days to remove the annealing effect sufficiently.

The concentration profiles of aluminum (indium, gallium) atoms into silicon were measured with a Hitachi secondary ion mass spectrometer type IMA-2(IMA). A focussed positive oxygen ion beams with an ion energy of 7 and 12KeV were used for the measurements of IMA.

2.2 Isolation Processing

Two kinds of GaP samples were prepared. The first was fabricated by epitaxial growth of an (S, N)-doped n-type (N_D~6~7× 10^{16}cm^{-3}, 130~140μm) on an LEC S -doped n-type (N_D~5×10^{17}cm^{-3}) substrate (~400μm). The second p (epi)-n(epi)-n(LEC) was prepared by a double epitaxial growth of a (Zn,O)-doped p-type (N_D~5×10^{17} cm^{-3}) and a Te-doped n-type (N_D~5×10^{17}cm^{-3}) on an LEC n-type (~5×10^{17}cm^{-3}) substrate.

The samples were irradiated with electron beams of 7, 9MeV from an electron linear accelerator, and 1, 2MeV from an ultra-high voltage electron microscope. In the case of 7, 9MeV, they were cooled with running water in order to prevent the temperature from rising, and at 1,2MeV they were exposed to a spot size electron beam with 5~30μm in diameter at room temperature.

3. EXPERIMENTAL RESULTS

3.1. Impurity doping

The typical IMA signals by a recorder with two pens are shown in Fig.2 for the bombarded samples. In the figure, the lower trace (mark ① (a)) indicates the Si^+ and Al^+ peaks. In a mark ① (b), the only $^{27}Al^+$ peaks are traced as a function of time (corresponds to depth) with ten times increased meter sensitivity. The samples were irradiated with a total dose of ~1.3×10^{17} electrons/cm^2 at 10MeV. The IMA measurements were done by using the primary ion (O_2^+) beam (diameter 1.0mm) with ion energy of 7KeV and in the vacuum of 1.1×10^{-6}Torr. For both an unbombarded silicon and Si samples bombarded from a surface of Al foils

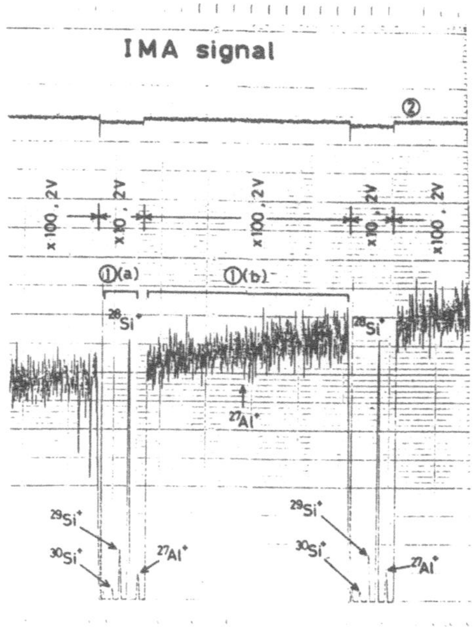

Fig. 2. Typical IMA signals for the recoiled Al atoms in silicon.

(t~60μm) in contact with its surface, the $^{27}Al^+$ peak disappeard. The thick Al sheet is necessary for Al impurity doping. The intensity ratios of $^{27}Al^+$ ions to $^{28}Si^+$ ions for the bombarded specimen are shown by marks ○ and Δ in Fig.3 as a function of depth from the Si surface at different thicknesses t of an Al sheet. At a given depth, the concentration of aluminum implanted into Si in the case of t~1.0mm is higher than that in t~1.5mm. Fig.4 shows the ratio of $^{27}Al^+/^{28}Si^+$ as a function of depth in the different experimental conditions (IMA measurements in the vacuum of 2.6×10^{-7}Torr, electron dose ~1.8×10^{17} electrons/cm^2.) for the bombarded sample. The concentration of Al atoms distributes up to a depth of ~900Å below the surface and the ratio of $^{27}Al^+/^{28}Si^+$ varies from ~6×10^{-4} to ~10^{-4}. These distribution profiles follow an exponential decay curve.

In the case of Ga doping, the surfaces of GaP crystal (~4×4mm^{2}area, t~0.4mm), which was put on B-doped Si wafer were exposed to 9MeV electron beams. Fig.5 shows the intensity ratios of $^{69}Ga^+/^{28}Si^+$ as a function of depth x for the silicon sample, which follow exp(-x/L) law. For the case of In doping experiment by using In sheet (t~0.3mm), Fig. 6(a) and (b) shows the typical IMA signals of $^{115}In^+$ peaks and the intensity ratios of $^{115}In^+/^{28}Si^+$ as a function of depth for the silicon sample, respectively.

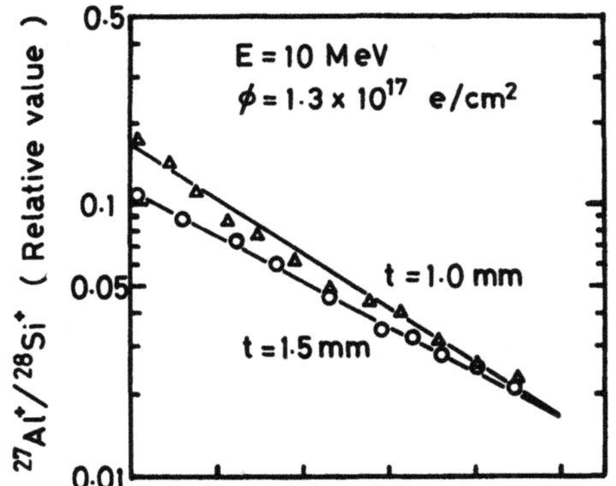

Fig. 3. Intensity ratios of $^{27}Al^+/^{28}Si^+$ in Si as a function of depth at different thicknesses of Al sheet.

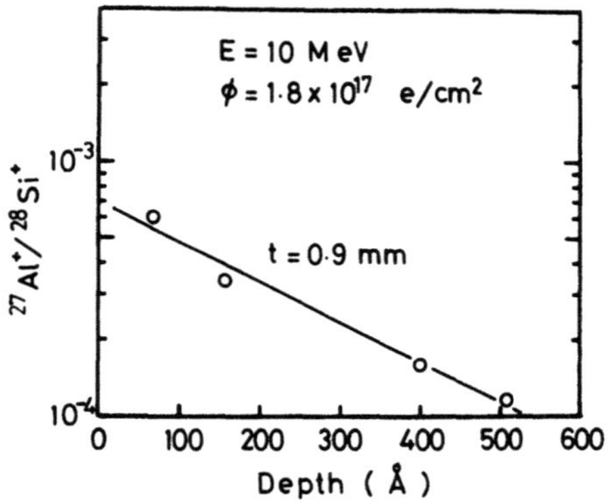

Fig. 4. Intensity ratios of $^{27}Al^{+}/\ ^{28}Si^{+}$ in Si as a function of depth.

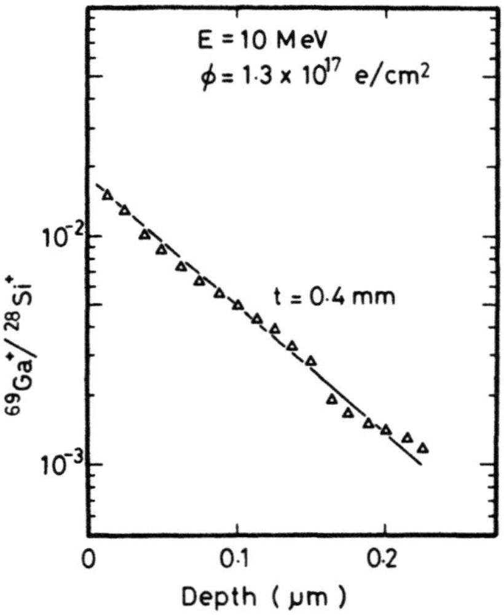

Fig. 5. Intensity ratios of $^{69}Ga^{+}/^{28}Si^{+}$ as a function of depth.

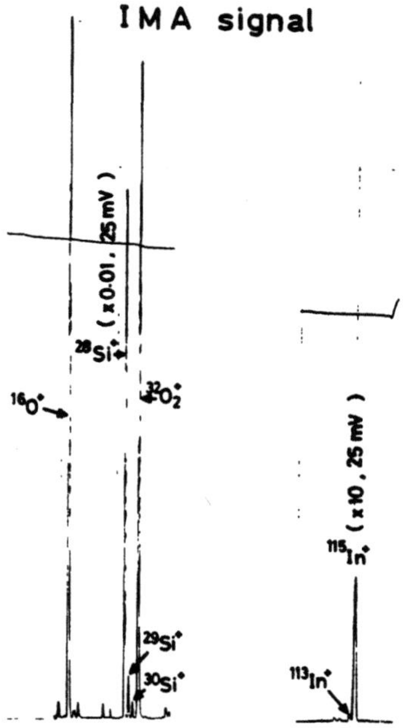

Fig. 6.a) Typical IMA signals for the recoiled In atoms in Si.

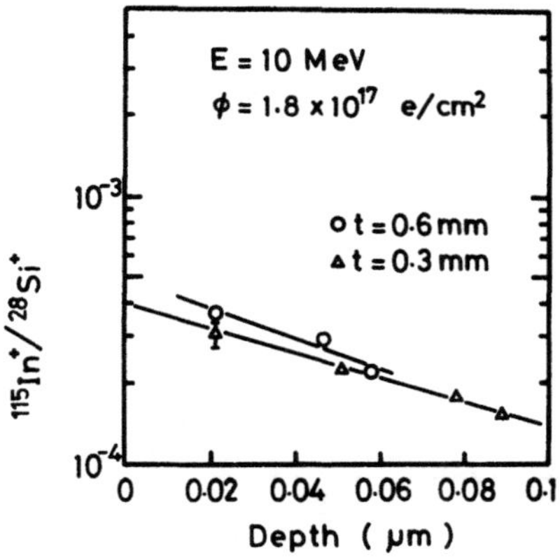

Fig. 6.b) Intensity ratios of $^{115}In^+/^{28}Si^+$ in Si as a function of depth.

a)

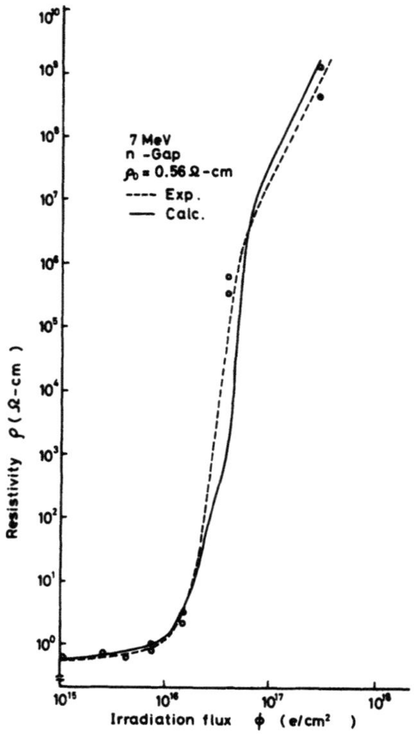

b)

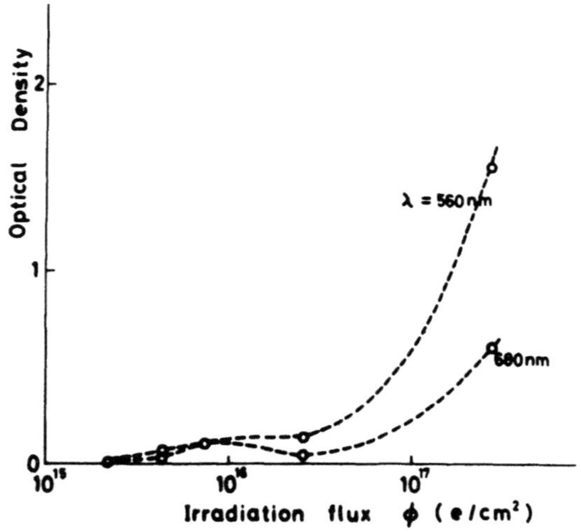

Fig. 7. Variation of the resistivity a), and the optical density b) as a function of irradiation flux for the epitaxial layer of n-GaP.

3.2. Isolation Processing

The resistivities ρ of the epitaxial layer of the GaP sample irradiated with 7MeV electrons are shown in Fig.7.a) as a function of total flux ϕ. Fig.7.b) indicates the optical density (=$\log_{10} I_0/I$, where I_0 and I are incident and transmitted light intensities respectively) for the sample as a function of electron flux at different wavelengths. Both the resistivity and the optical density similarly increase with increasing the flux.

The electron irradiation introduces the six deep-lying energy states in the forbidden band of n-GaP. [11] The resistivity of the crystal increases with increasing the irradiation flux due to the trapping of electrons at deep defect levels. ρ versus ϕ curve can

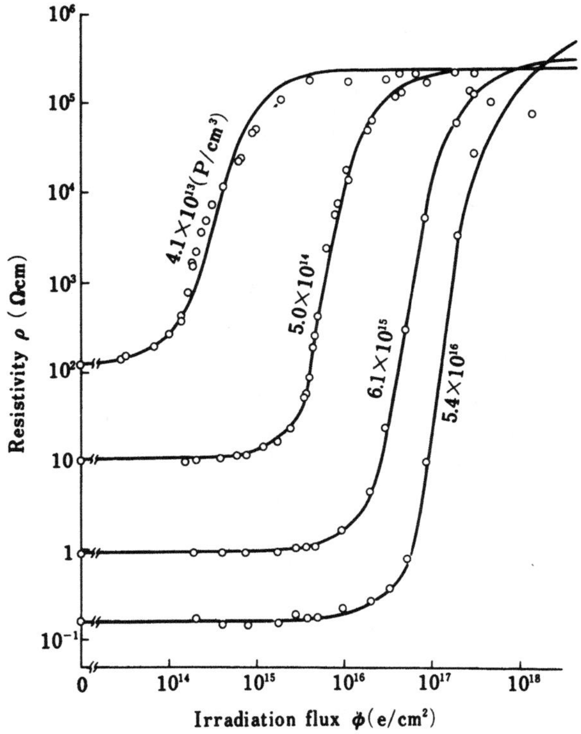

Fig. 8.a) Variation of the resistivity as a function of irradiation flux for the wafers of n-Si at different phosphorus impurity concentrations.

be quantitatively interpreted by taking account of the equation of charge neutrality condition and the statistics of the charge distribution.[12] Also, the various optical wavelengths are absorbed in the irradiated crystal and then the colour of the crystal changes to black.

The resistivities of Si (GaAsP) sample irradiated with 9 (7) MeV electrons are shown in Fig.8.a) (8.b)) as a function of irradiation flux with different phosphorus impurity densities for Si. In the case of Si, ρ finally approached the intrinsic value even at different impurity densities.

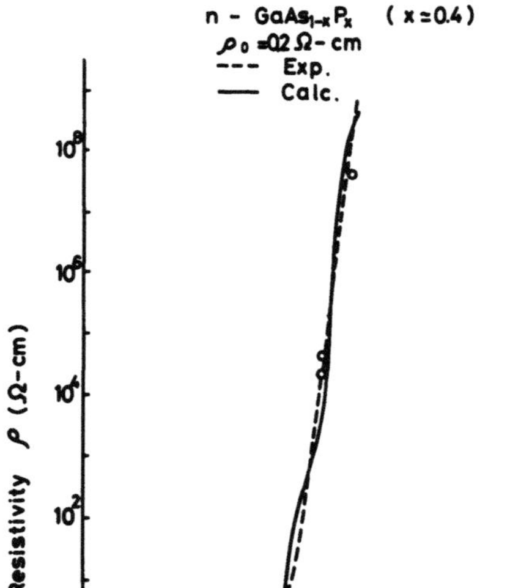

Fig. 8.b) Variation of the resistivity as a function of irradiation flux for the wafers of n-GaAsP.

The electron diffraction pattern of the materials irradiated with the total flux of ~2×10^{17}electrons/cm^2 at 7MeV represents sharp Laue spots, as shown in Fig.9.

Fig. 9. Electron diffraction pattern of the sample irradiated with a flux of ~2×10^{17}e/cm^2 at 7MeV.

Fig.10.a) and b) shows a schematic diagram for the experiments of n-GaP crystal exposed to focussed electron beams at 2MeV from a high voltage electron microscope and a photomicrograph of a cleaved cross section of the sample irradiated with spot size (~7μmφ) electron beams (current~0.15μA), respectively. The profile of the brack colour corresponds to the distribution of the damage defect density.[13] The optical densities and spacings of the changed colour of the crystal in the photographic films were measured by using a microdensitometer.

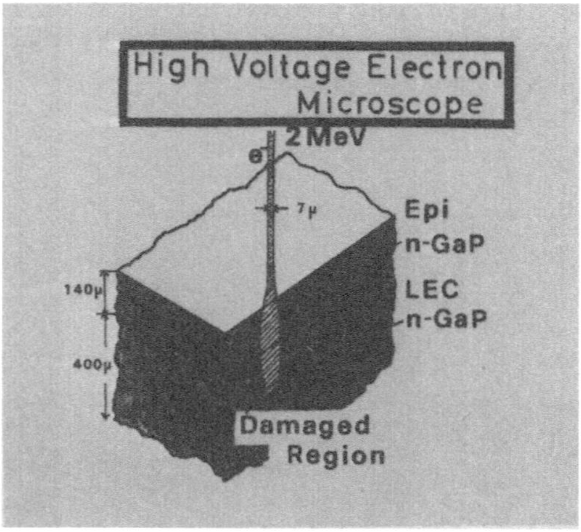

Fig. 10. a) Schematic diagram for the irradiation from a high voltage electron microscope.

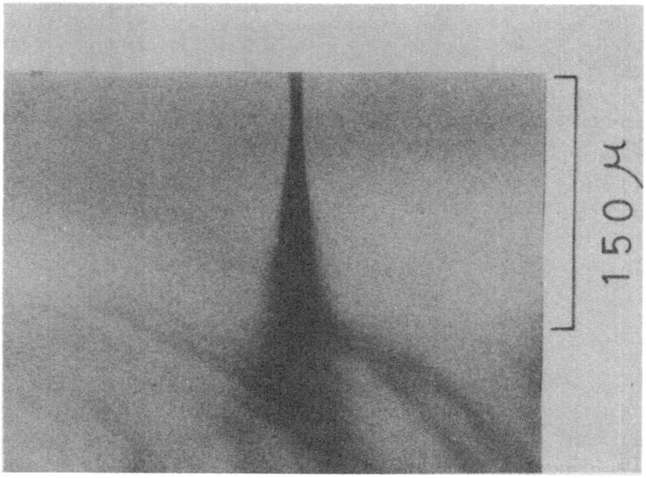

Fig. 10. b) Photomicrograph of a cleaved cross section of the GaP wafers exposed to a focussed beam with a flux of ~6.5×10^{20}e/cm^2 at 2MeV.

Fig.11 indicates the typical two-dimensional profiles of equi-optical density for the wafer irradiated with electrons and isochronally annealed in a hydrogen gas atmosphere for 10 min at different temperatures. The profiles become a smaller distance off axis and depth with increasing the annealing temperature.

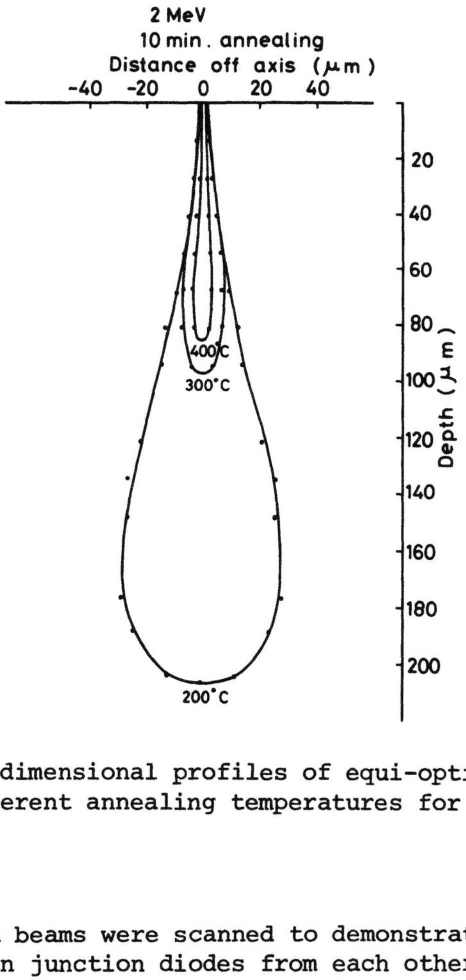

Fig. 11. Two-dimensional profiles of equi-optical density at different annealing temperatures for the irradiated GaP.

The electron beams were scanned to demonstrate the separation of an array of p-n junction diodes from each other, as shown in Fig.12.a). Fig.12.b) represents a photomicrograph of the surface, which has several black lines. They are produced by scanned electron bombardment. A size of the matrix is about 100μm square and the beam spot size is 5μm. The total flux was $\sim 2\times 10^{20}$ electrons/cm^2 at 1MeV and the irradiated sample was annealed at 300°C. The surface, which has the bombarded line is chemically etched by the aqua regia. An etch rate of the line is larger than

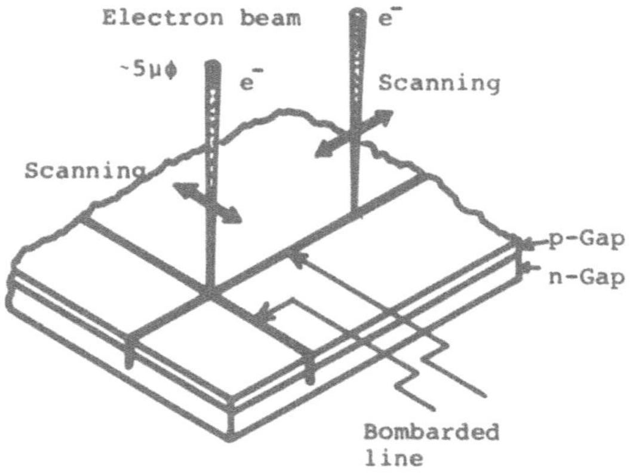

Fig. 12. a) Schematic diagram for the experiments of separation

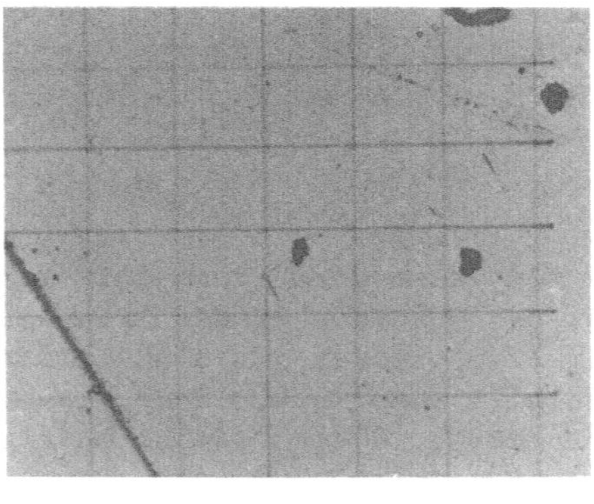

b) Separation (~100μm square) of an array of GaP p-n junction diodes by using electron beam.

that of an unirradiated region, because of a high defect density in the irradiated line. A SEM micrograph of the GaP surface after treatment with the etch indicates a deeper removed layer, which has a steep edge along the line with a 5μm width, as shown in Fig.13.

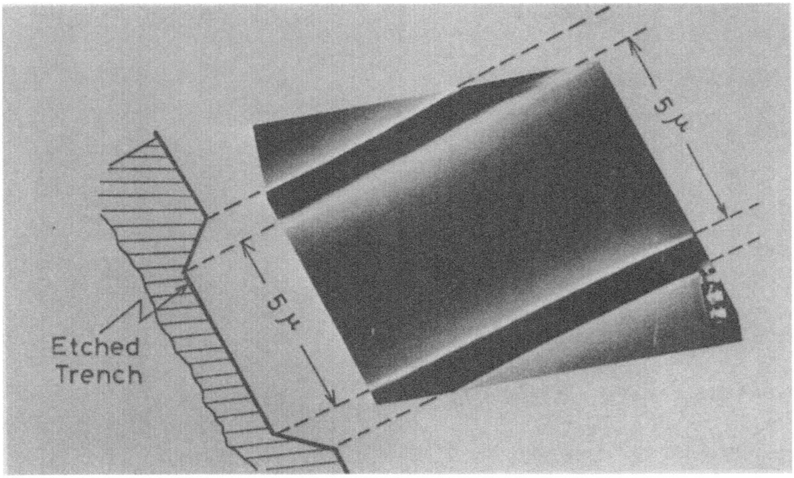

Fig. 13. SEM micrograph of the GaP surface exposed to scanned electron beam after treatment with the etch.

Fig.14.a) shows an electron beam induced current image by SEM for the surface of GaP LED, which is irradiated at a total flux of $\sim 2\times 10^{20}$ electrons/cm^2 by scanned electron beam (1MeV) with ~32μ width. It indicates a very marked line (~32μ width) produced by high resistivity layer. An example of cathodoluminescent image for the surface of GaAsP LED irradiated by scanned electron beam is shown in Fig.14.b). It presents a marked line resulting from that the red luminescence disappears in the irradiated region.

The structure, indicated in Fig.15.a), was fabricated to demonstrate the simple separation of p-n junction diodes from each other by scanned electron bombardment. Fig.15.b) shows that a luminescence of a forward-biased diode appears, but that of another no-biased diode, which is separated from the biased diode disappears. An area of diodes is $\sim 1\times 2.5 mm^2$. The luminescence area is optically and electrically isolated by the bombarded line.

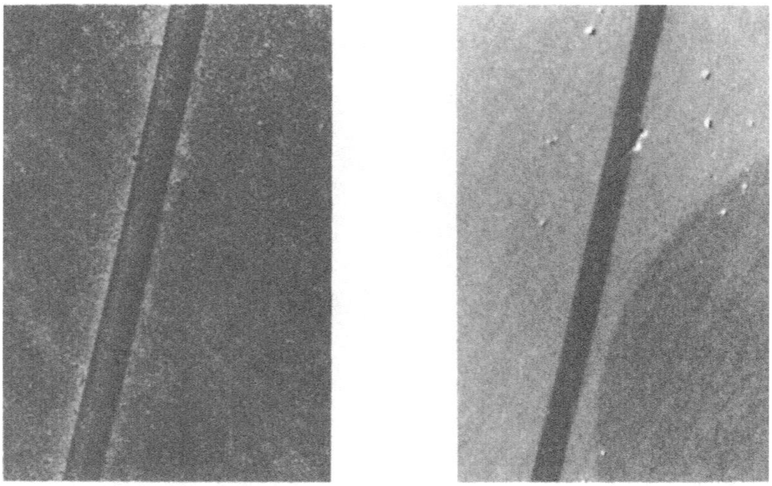

a) b)

Fig. 14. a) Electron beam induced current image of the irradiated GaP LED.
b) Cathodo-luminescent image of the irradiated GaAsP LED.

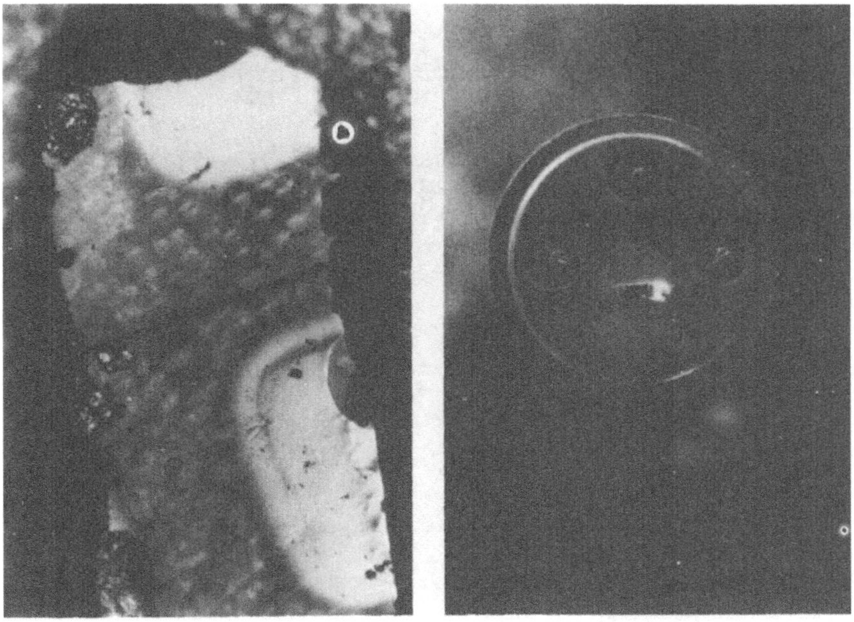

a) b)

Fig. 15. a) Simple separation of p-n junction diodes from each other by scanned electron beam.
b) Luminescence of the forward-biased diode, which is separated from another diode.

4. DISCUSSION

4.1 Impurity doping

The range of electron with 10MeV in Si (or Al) is about 21mm and in this case the production rate of defects is ~9cm^{-1}. Then, the Al sheet layer is sufficiently thin to allow the bombarding electrons to penetrate into the substrate without a significant loss in kinetic energy.

i) Recoil process by electron bombardment

When the flux and energy spectrum of recoil atoms crossing

the interface between the Al layer and the substrate are calculated, the depth distribution of the recoil impurity atoms within the substrate can be estimated. The energy spectrum of recoil atoms during electron bombardment can be obtained by McKinley and Feshbach expression[14], giving

$$d\sigma_{McF}=\pi\left(\frac{Ze^2}{mc^2}\right)^2\left(\frac{1}{\beta^4\gamma^2}\right)T_m\left[1-\beta^2\frac{T}{T_m}+\pi\frac{Z}{137}\beta\left\{\left(\frac{T}{T_m}\right)^{1/2}-\frac{T}{T_m}\right\}\right]\frac{dT}{T^2} \quad (1)$$

where $d\sigma_{McF}$ is a differential cross section, $\gamma^2=\left(\frac{E}{mc^2}+1\right)^2$, $\beta^2=1-\frac{1}{\gamma^2}$, T_m the maximum recoil energy $T_m=2\frac{m}{M}\frac{(E+2mc^2)E}{mc^2}$, E electron energy, T recoil kinetic energy and M mass of atom.

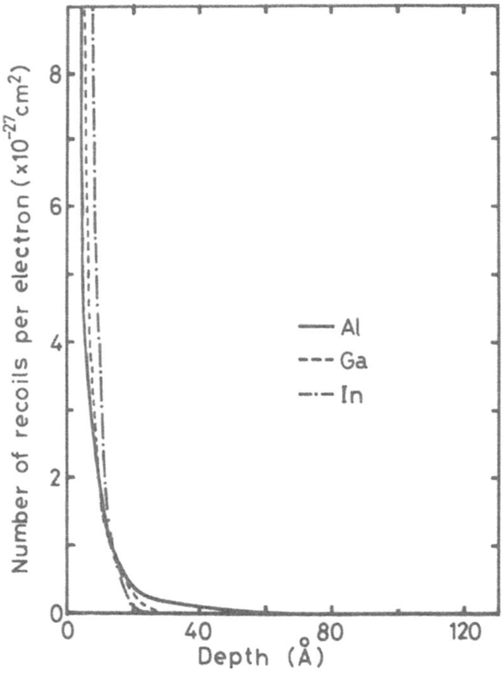

Fig. 16. Theoretical concentration profiles of recoil atoms into Si by ~10MeV electron beams as a function of depth.

The recoil spectrum, i.e. the number of recoils crossing an interface as a function of recoil energy are calculated. The maximum recoil energy becomes to be ~8.8KeV for 10MeV electrons. While, the range of implanted ion in Si has been calculated as a function of ion energy[15]. By using the both results, the penetration distribution during the recoil implantation of Al into Si by 10MeV electron bombardment is theoretically obtained as a function of depth, as shown in Fig. 16. The implanted impurity atoms within Si crystal distribute up to a depth of ~100Å below the surface.

However, in this experiments the concentration profiles of Al atoms into Si are obtained up to a depth of ~1300Å below the surface. As the incident electrons pass through the sheet into silicon, the defects in Si are introduced from the surface up to the range of electrons. Then, mechanisms for such profiles of Al atoms are not only the recoil process of Al atoms, but also enhanced diffusion of impurity atoms in Si.

ii) Enhanced diffusion of Al atoms by electron bombardment.

The high-energy electron bombardment introduces deep-lying energy states in the forbidden gap of Si.

The defect density of $N_D = \phi\Sigma\eta[\sim 9(e\cdot cm)^{-1}\times 1.3\times 10^{17}(e/cm^2)] \sim 10^{18}$ defects/cm^3 distributes almost uniformly in the layer ($t \sim 300\mu m$) of Si crystal[16], where η is the production rates of various defects.

Under these conditions the mobility-enhanced and/or the concentration-enhanced diffusion of the recoil atoms due to the increase in the vacancy concentration and an interstitial component may occur[17].

It is experimentally the most special feature that the concentration profiles of Al impurity follow a simple exp(-x/L) law, as shown in Figs.3,4,5 and 6. If a diffusion process is composed of two stream with different diffusivities and if there is an exchange of flow between them, the solutions of the equation are given as exp(-x/L)[18]. One of the physical interpretations for a double stream process is a trapping mechanism. When the diffusion interstitial is trapped by vacancy in the crystal or other trapping complexes, this type of mechanism gives exponential rather than complementary error function profiles[18]. In the present experiments it is suggested that the anomalous diffusion may be due to a double stream process which is composed of free species (probably Al interstitials) and trapped ones (Al substitutionals).

The concentration $C_{Al}(0)$ of Al impurity at the Si surface is estimated to be $\sim 10^{18}$Al/cm^3 from the results in Fig.4 by assuming that the secondary ion yields of Si and Al are ~0.02

and ~0.53 respectively. This value agrees well with the defect density in Al layer ($N_D=\eta\cdot\phi$).

Fig.17. shows the relative number of Al atoms N_{Al} at a given depth in Si as a function of thickness of Al sheet. The value of N_{Al} becomes a maximum value at the proper thickness. The increasing region in Fig.17 is due to the production rate of defects ($\eta\simeq 9$ defects/cm for 10MeV electrons). In a range of smaller thicknesses, N_{Al} increases with increasing the thickness. While, in the decreasing region, grossly Θ_ω^2 is directly proportional to the sample thickness according to the theory of electron multiple scattering,[14] where Θ_ω is the angle at which the distribution has a value 1/e of its maximum value. Thus, N_{Al} may decrease at larger thicknesses.

4.2 Isolation processing

The theoretical profiles of damage defects N_D as a function of space coordinates in the crystal is given by[9]

$$N_D(E_0,x,y,z)=\int_{E_d}^{E_0} n(E_0,E_B,x,y,z)\,\eta(E_B)\,dE \qquad (2)$$

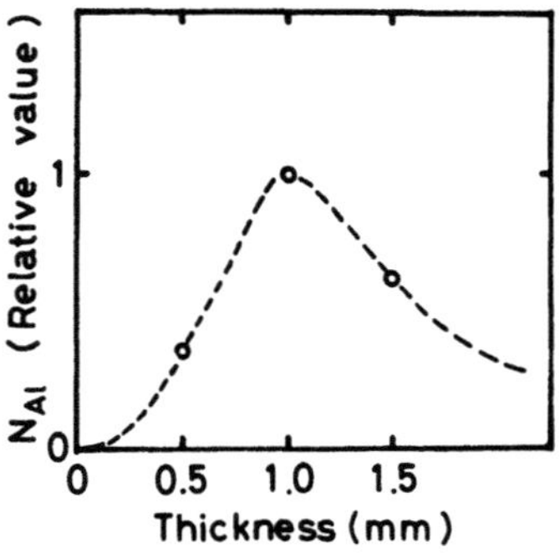

Fig. 17. The relative number of Al atoms at a given depth in Si as a function of thickness of Al sheet.

where E_0 is the initial electron energy and E_d the threshold energy for atomic displacements. $n(E_0,E_B,x,y,z)$ is the energy spectra of the electrons as a function of space coordinates for E_0 and electron bombardment energy E_B. The theoretical profiles are obtained using the production rates of each defect[12)] and the energy spectra[13)] calculated by a Monte Carlo simulation for the theory of fast electron penetration in matter. The theoretical results agree well with the experimental profiles.[19,20)]

Fig. 18 a) and b) shows how to fabricate the X-Y leads to the LED displays in the both case of the usual type and the present work. In the usual type, both leads of X and Y are connected to the same surface side. While, in our work, the row and column leads can be connected to the face and back sides of the wafer respectively, as shown in Fig. 18.b). Then, the higher-density matrix LED displays are available using this technique.

Fig. 19 indicates the value of Θ_ω^2 as a function of electron energy E[14)]. With increasing the electron energy, the distribution at the angle in materials due to electron multiple scattering largely decreases. Then, the present processing by higher energy electron beam is well for fabricating the deeper isolation with narrower width.

5. CONCLUSION

A new recoil implantation in Si by using high energy electron bombardment is investigated. The surface of an Al (In, GaP) sheet ($4\times4mm^2$ area) in contact with Si wafers ($4\times4mm^2$ area) were exposed with 10MeV electron beam. After bombardment, the intensity ratios of $^{27}Al^+$($^{115}In^+$, $^{69}Ga^+$) ions to $^{28}Si^+$ ions in Si crystal were obtained from $\sim6\times10^{-4}$ to 10^{-4} as a function of depth (0~900Å) below the surface using secondary ion mass spectrometry. The concentration profiles of Al impurity follow an exp(-x/L) law.

The mechanism for the introduction of impurity atoms into Si are both the recoil process and the enhanced diffusion of Al impurity atoms during high energy electron bombardment. The concentration varying as exp(-x/L) may be due to a trapping mechanism.

This technique is useful for the impurity doping of semiconductors. Also, the isolation processing by electron beam is demonstrated by using an ultra-high voltage electron microscope.

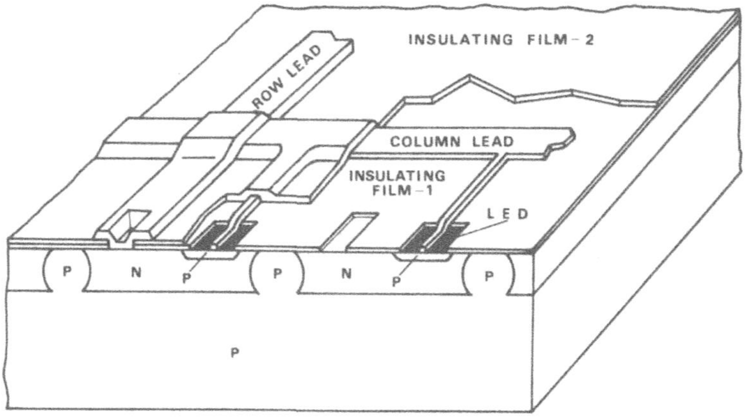

a)

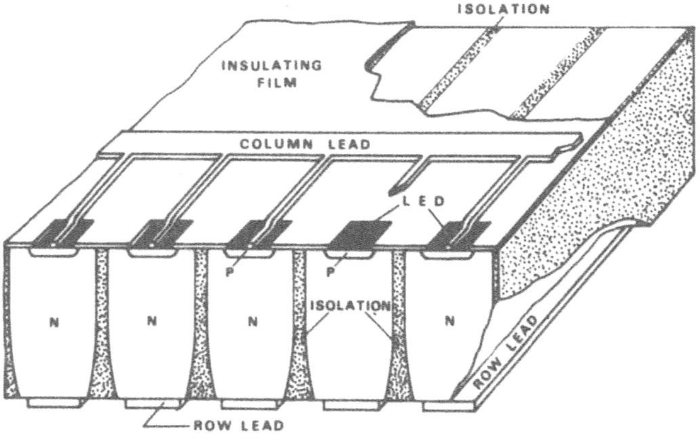

b)

Fig. 18. Structures of X-Y leads to the LED displays in the usual type (a), and the present work (b).

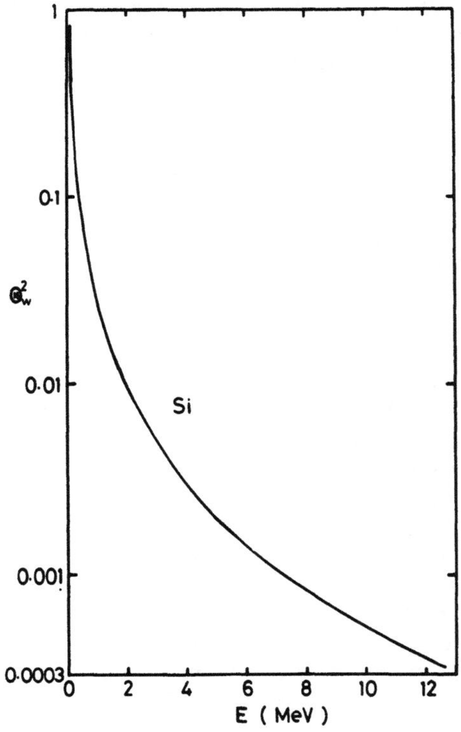

Fig. 19. Θ_{ω}^{2} as a function of electron energy.

ACKNOWLEDGEMENT

Several helpful discussions with Profs. A.Seeger, W.Frank and Dr.U.Gösele of Max-Planck-Institute, Prof.H.Y.Fan of Purdue University, Prof.J.W.Corbett of State University of New York, and Prof. T.Arizumi of Daido Institute of Technology are gratefully acknowledged. The author wishes to thank Drs. S.Okabe, T.Tabata, H.Fukuda et al. of the Radiation Center group of Osaka Prefecture for their help in the irradiation of the samples, and Drs. T. Suzuki, N.Ishizuka, M.Kakehi, Mrs. N.Kitamura, S.Uemura, M.Kaneiwa T.Nagasako et al. for their help in IMA measurements and experiments. The work was partly supported by a Grant-in-Aid for Scientific Research from the Ministry of Education, Science and Culture.

REFERENCES

1) R.S.Nelson, Rad.Effects, 2. 47 (1969)
2) A.Grob, J.J.Grob, N.Mesli and P.Siffert, 2nd Int.Conf. on Ion Beam Mod. of Materials, Albany, A-9, (1980)
3) P.Sigmund, Phys. Rev. 184, no2 383 (1969)
4) A.D.Marwick and R.C.Piller, 2nd Int.Conf. on Ion Beam Mod. of Materials, Albany, A-15 (1980)
5) A.G.Foyt, W.T.Lindley, C.M.Wolfe and J.P.Donnelly., Solid State Electro. 12. 209 (1969)
6) T.Wada and M.Kaneiwa, 2nd Int. Conf. on Ion Beam Mod. of Materials, Albany, A-16 (1980)
7) T.Wada and S.Uemura, Tech.Digest of Int.Electron Devices Meet. Washington,D.C. 192 (1975)
8) T.Wada and S.Uemura, Tech.Digest of Int.Electron Devices Meet.Washington,D.C.486 (1977)
9) T.Wada, S.Uemura, M.Kakehi and N.Kitamura, Tech.Digest of Int.Electron Devices Meet.Washington,D.C. 638 (1978)
10) T.Wada, S.Uemura and N.Kitamura, Proc.of Int.Conf. on Semi-Insulating III-V Materials, Nottingham, 14th April (1980)
11) D.V.Lang and L.C.Kimerling, Appl.Phys.Letters. 28, 248 (1976)
12) T.Wada, K.Yasuda, S.Ikuta, M.Takeda and H.Masuda, J.Appl. Phys. 48 2145 (1977)
13) T.Wada, S.Uemura and A.Ishikawa, Proc. of 5th Int.Conf. on High Voltage Electron Microscopy. 497 (1977)
14) J.W.Corbett, Electron Radiation Damage in Semiconductors and Metals, (Academic Press, New York and London, 1966) p.20
15) S.Furukawa, H.Matsumura and H.Ishiwara, Jap.J.Appl.Phys, 11 134 (1972)
16) K.Yasuda, H.Masuda, M.Takeda and T.Wada, Jap.J.Appl.Phys. 16

387 (1977)
17) W.Frank, U.Gösele and A.Seeger, Proc. of Int.Conf. Radiation Phys. in Semiconductors and Related Materials, Tbilisi, Sep. (1979)
18) L.Y.Wei, J.Phys.Chem.Solids, 18 162 (1961)
19) T.Wada, K.Yasuda and E.Matsumoto, 6th Int.Cong. of Rad.Res. Tokyo, c-17-6, (1979)
20) T.Wada, E.Matsumoto and T.Endo, Proc of 11th Int.Conf. on Defects and Rad. Effects in Semi. Oiso. Sep. 11. (1980)

APPLICATION OF NTD SILICON FOR RADIATION DETECTOR OF SURFACE BARRIER TYPE

A. Yusa and D. Itoh

Komatsu Electronic Metals Co., Hiratsuka, Japan

C. Kim and H. Kim

Dept. of Electronics Eng., Korea Univ., Kodaira, Tokyo

K. Husimi and S. Ohkawa

Inst. for Nuclear Study, Univ. of Tokyo, Tanashi, Tokyo

ABSTRACT

N-type extrinsic silicons higher than 10 kΩ-cm were produced from ultra-high pure P-type silicons by the neutron transmutation doping technique. Materials produced by the method has properties in dopant homogeneity and controllability better than conventionally doped ones. This has enabled us to fabricate a high performance detector with a large depletion depth. However, some problems were pointed out as to reverse current-voltage characteristics, i.e., a reverse current level was so high to use the detector at room temperature. In order to improve this, a study of starting materials, neutron irradiation and annealing conditions are carried out. Significant improvements in detector performance have been obtained by using a material which is annealed at the temperature of 800°C for 10 minutes in an infrared lamp furnace.

1. INTRODUCTION

The N-type silicon crystal has been used for producing the radiation detector of the surface barrier type. In the case to obtain the detector with a large depletion depth, a good silicon crystal which has a high resistivity and at the same time has an uniform impurity distribution is required. The conventional floating zone (FZ) method is, however, not suitable for producing the silicon crystal which has the resistivity higher than 30 kΩ-cm economically. This is due to the fact that the distribution of impurity phosphorus in the FZ silicon is not uniform caused by a temperature distribution of the crystal and an evaporation of phosphorus from the surface in the FZ process. On the other hand, it has been expected that the neutron transmutation doping (NTD) technique which converts a high purity P-type silicon into N-type is promising for producing the high quality silicon crystal for detector use.[1]

It is widely recognized that mono-silane SiH_4 is suitable as the source gas for producing the raw material of the high purity silicon compared with the other source gases such as $SiHCl_3$ and $SiCl_4$, because the decomposition temperature of SiH_4 is lower than that of the other source gases.[2] In the process of producing SiH_4, one of the by-product boron hydride is completely removed by an associated chemical reaction. As the result, the silicon crystal produced by means of the SiH_4 method contains a very small quantity of boron, and this fact is very favorable for producing the NTD silicon with a controlled resistivity.

However, the characteristics of the detector produced were a little worse than those of the FZ silicon detector. This was supposed to be due to the crystallographic imperfections caused by irradiation of neutrons and an improper thermal annealing. In fact, the minority carrier lifetime of the NTD silicon crystal was 180 to 200 μsec after thermal annealing and was considerably shorter than that of the FZ silicon crystal.

The purpose of this work is to obtain the NTD silicon crystal which has a long lifetime by improving the method of thermal annealing. Several NTD silicon crystals with different lifetime were obtained in the same way but only by changing the thermal annealing condition. By using these NTD silicon crystals, the reverse current and the energy resolution of the surface barrier detector produced are also measured as a function of the lifetime.

2. EXPERIMENTS

1. Preparation of Test Samples

The N-type polycrystalline silicon obtained by the thermal decomposition of SiH_4 was refined by the FZ method once or twice in a high vacuum. The impurity phosphorus in the crystal was vaporized out of the surface almost completely, and the crystal turned to a P-type silicon due to a small residual B and Al. In order to produce the dislocation-free single crystal, floating zone refining was performed once more in the same chamber filled with argon gas. The silicon crystals with dislocations (EPD: 4×10^4 /cm^2) were also obtained in the same way butby a control of necking of the seed crystal. In a good dislocation-free crystal, the swirl pattern could not be observed in the as-grown wafer by means of Sirtl etching method.

Table 1. Starting materials used for neutron transmutation doping

sample No	preparation	crystality	resistivity (kohm-cm)	type	life-time (μ-sec)	dimension (mm)
313	FZ 2 pass	dislocationfree swirlless	≃ 100	P	1000-2000	$23^{\phi} \times 60^{\ell}$
314	FZ 2 pass	dislocationfree swirlless	≃ 100	P	1000-2000	$23^{\phi} \times 60^{\ell}$
308	FZ 2 pass	dislocationfree swirl	≃ 100	P	1000	$19^{\phi} \times 30^{\ell}$
207	FZ 3 pass	dislocation EPD $4 \times 10^{4}/cm^{2}$	≃ 80	P	400-500	$19^{\phi} \times 20^{\ell}$

Characteristics and dimensions of the samples prepared are listed in Table 1.

2. Irradiation of Neutrons

The silicon ingot was cleaned with a mixture of HNO_3 and H_2SO_4 and rinsed with deionized water for removing surface contaminations. The cleaned ingot was wrapped in a precleaned polyethylene bag and was fixed in a plastic capsule. The capsule was inserted in a core of a reactor. The nuclear reactors used were TRIGA-MARK 2 of Rikkyo University and JRR-4 of Japan Atomic Energy Research Institute. Thermal to fast neutron ratios of neutron flux in the irradiation holes of these reactors were estimated to be 3 to 4 and 3 to 7, respectively. The neutron irradiation doses of these samples were 1.2 to 2.5×10^{15} n/cm^2.

3. Thermal Annealing

In order to prevent contamination of lifetime killers such as heavy metals and so, the silicon ingot was etched with a mixture of HF and HNO_3, washed with deionized water and then rinsed with ultra-high purity water after neutron irradiation. Each of the sample ingots was placed in a silicon boat immediately and was inserted into a silica tube of the annealing furnace. In the thermal annealing, two types of furnace were used, one was the infrared lamp furnace and the other was the conventional diffusion furnace. The outside wall of the infrared lamp furnace was cooled by a flow of dry nitrogen, thus only the silicon ingot was heated up. The thermal annealing was performed in a flow of dry oxygen for 10 to 60 minutes. This annealing time was so short that heating up and cooling down of the sample were performed as fast as possible. The annealing temperature was 800°C.

4. Fabrication of Surface Barrier Detector

The NTD silicon rod thus obtained was cutted down to disk-type wafers. The surface barrier electrode was produced conventionally by evaporation of gold onto the mirror-etched surface of the wafer. The area of the electrode was 11 mm in diameter. The ohmic contact was obtained by evaporation of aluminium onto the other side of the wafer. There is no guard ring nor surface coating applied to the detector for reducing the leakage current.

3. RESULTS

1. Elimination of Crystal Defects Caused by Irradiation of Neutrons

Fig.1 shows the photoluminescence spectra of the starting material and the NTD silicon crystal which are annealed at the temperature of 800°C. As seen from the figure, there is no remarkable difference between two spectra except the phosphorus peak. This result shows that the annealing temperature of 800°C is high enough for eliminating the crystal defects caused by irradiation of neutrons and the

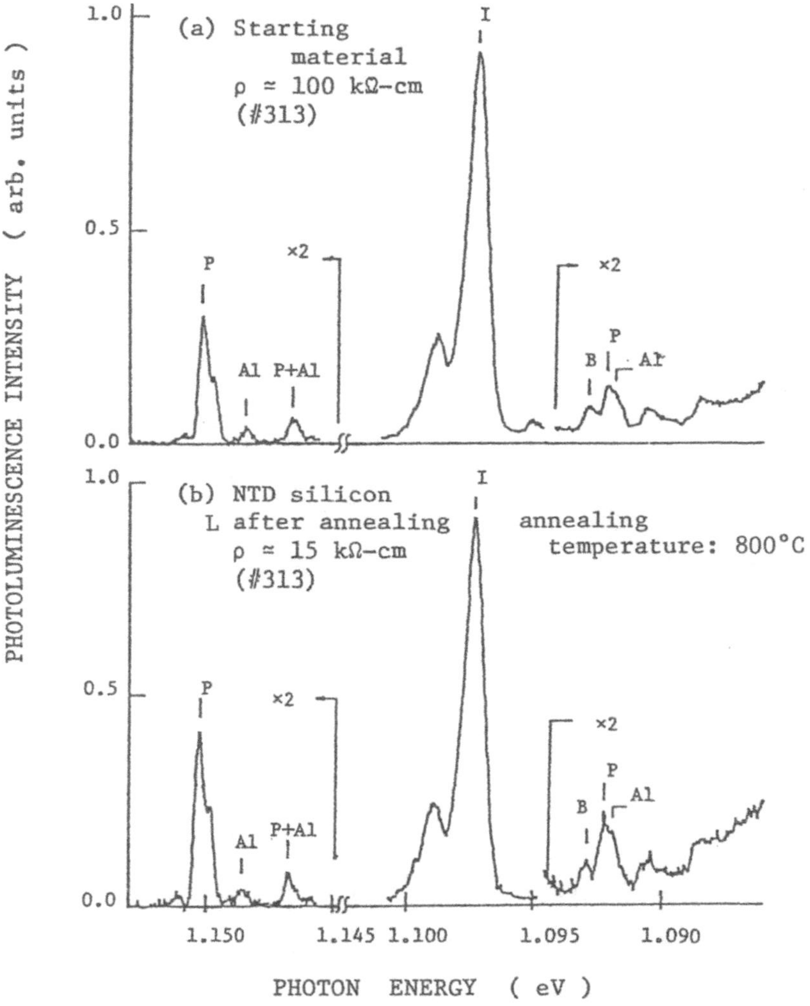

Fig. 1 Photoluminescence spectra of NTD silicons measured at 4.2°K. I is intrinsic luminescence due to the radiative decays of free excitons. The extrinsic luminescences due to the radiative decays of excitons and multiple excitons bounded to impurities are labelled with the chemical symbol of the associated impurity.

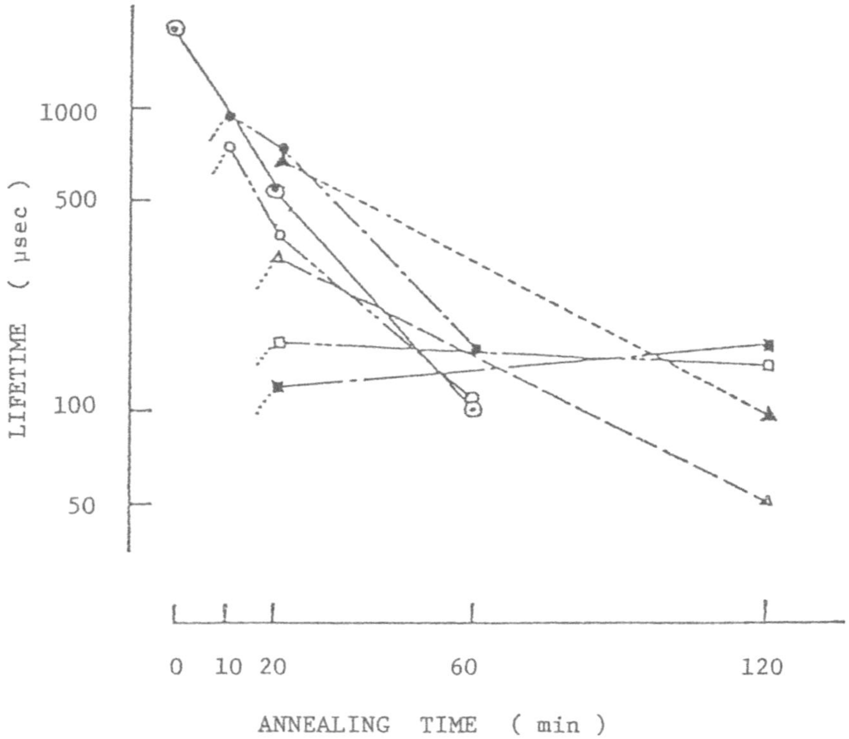

Fig. 2 Minority carrier lifetime of silicon crystal as a function of annealing time

Annealed in infrared lamp furnace

⊙ : dislocationfree as-grown material

• : dislocationfree and swirlless NTD Si (313B)

∘ : dislocationfree and swirlless NTD Si (314B)

▲ : dislocationfree NTD Si with swirl (308)

△ : NTD Si with dislocation (207)

Annealed in diffusion furnace

□ : dislocationfree and swirlless NTD Si (313B)

■ : dislocationfree and swirlless NTD Si (314B)

impurity phosphorus is doped by irradiation of neutrons.

2. Minority Carrier Lifetime and Annealing Time

Fig.2 shows the minority carrier lifetimes of NTD silicon crystals as a function of annealing time. In the case that the infrared lamp furnace is used, the neutron irradiated silicon crystal has recovered the minority carrier lifetime from an unmeasurable small value to nearly a same value of as-grown FZ silicon crystal by the thermal annealing of only 10 minutes. The lifetime becomes short exponentially with an increase of the annealing time and goes down to a value of 100 to 200 μsec after annealing of 60 minutes. In the case of the diffusion furnace, the lifetime is 100 to 200 μsec and is low independent of the annealing time. In this figure, the lifetime decrease of the as-grown FZ silicon crystal caused by the thermal annealing is also shown for comparison. It is found from this experiment that the decrease of lifetime can not be avoidable in the process of thermal annealing. However, with a short annealing time and by using the infrared lamp furnace which can minimize the contamination of the crystal in the furnace, the NTD silicon crystal which has nearly a same lifetime compared with that of the as-grown FZ crystal has been obtained.

3. Characteristics of NTD Silicon Detectors

1). Minority Carrier Lifetime and Detector Reverse Current

The reverse current of the surface barrier detector is composed of a surface current and a bulk current. The bulk current consists of a generation current and a diffusion current. The generation current i_g is given as a function of the lifetime of holes τ by

$$i_g = q \times n_i \times W / (2\tau) \qquad (1)$$

where n_i is the intrinsic carrier density, W is the depletion depth and q is the electronic charge.[3] The diffusion current i_d is also given by

$$i_d = q \times p_n \times (D_p / \tau)^{0.5} \qquad (2)$$

where p_n is the hole density and D_p is the diffusion constant of holes in the N-type silicon layer.[4]

The generation current is proportional to the depletion depth, therefore this becomes dominant compared with the diffusion current when the applied voltage to the detector becomes high. Fig. 3 shows the reverse current of detectors as a function of the minority carrier lifetime for applied voltages of 50 and 450 V. The bulk current of the detector with the resistivity of 10 kΩ-cm calculated by using Eqs. (1) and (2) are also shown in the figure for applied voltages of 50 and 450 V. It is interesting to see from the result that, though the most part of the reverse current is the surface current, this surface current also has a lifetime dependance similar to that of the bulk current.

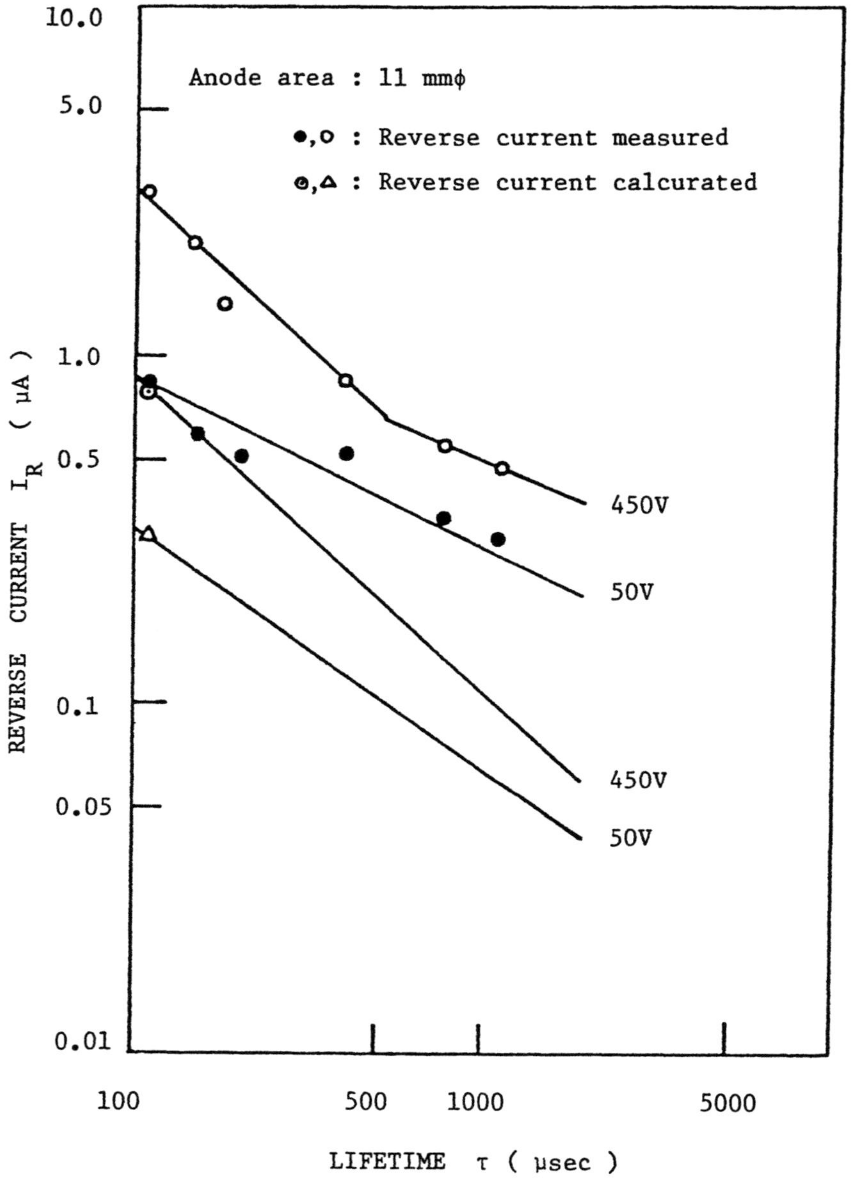

Fig. 3 Reverse current of detectors as a function of lifetime

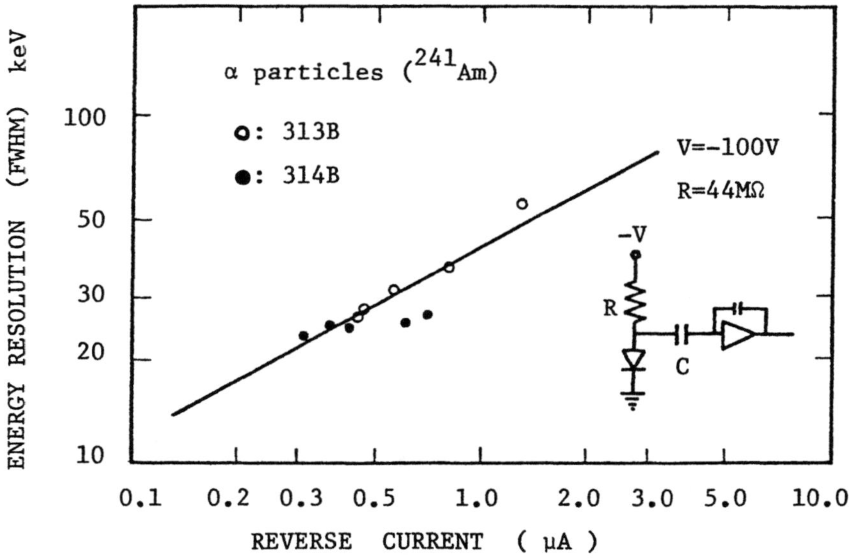

a) α particles from ^{241}Am

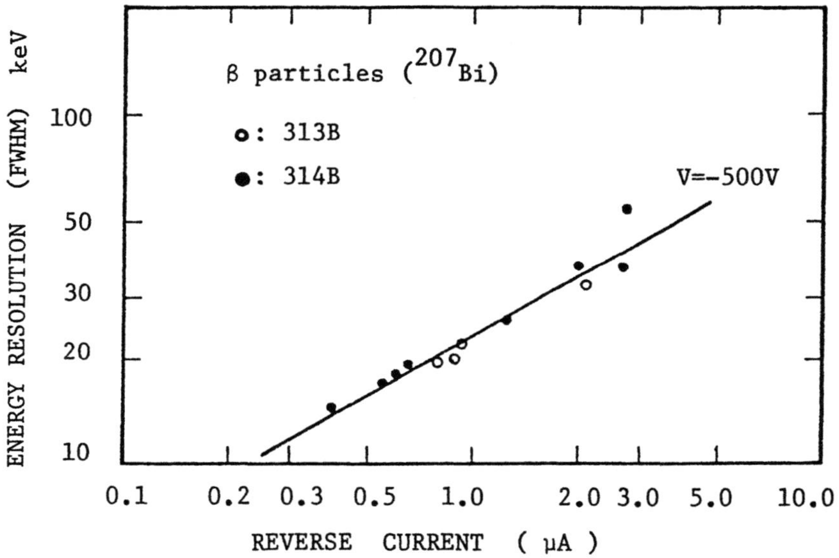

b) β particles from ^{207}Bi

Fig. 4 Energy resolutions (FWHM) as a function of reverse current

Table 2. Sirtl etching patterns and characteristics of NTD silicon detectors

sample No.		314	308	207
Sirtl etch pattern of wafer surface		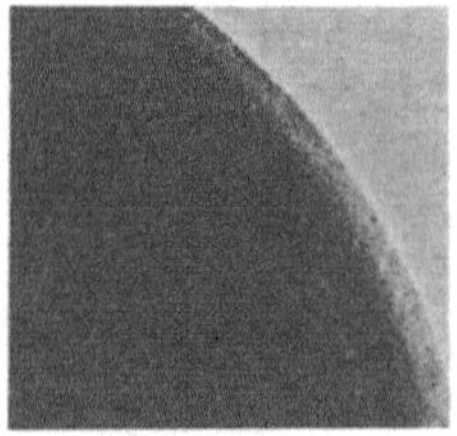		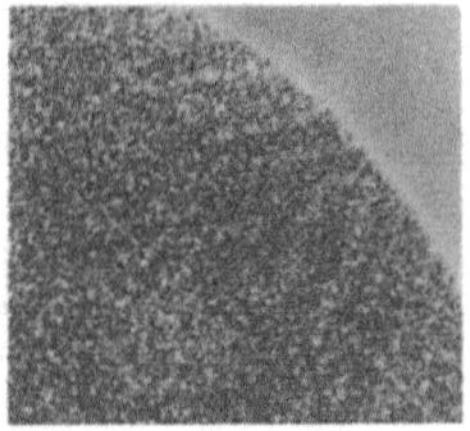
lifetime (μ-sec)		1100	800	400
detector characteristics	bias voltage (V)	50	50	50
	reverse current (μA)	0.4	1.0	2.0
	FWHM for α rays (keV)	15	43	400

2). Energy Resolution and Reverse Current

The energy resolution of detectors expressed by the full width at the half maximum (FWHM) of the spectrum peak are shown in Fig.4 (a) and (b) as a function of the reverse current for applied source voltages of 100 and 500 V. As seen from the result, the energy resolution of the detector becomes worse being inversely proportional to the square root of the reverse current which has been pointed out by Goulding and Hansen.[5]

3). Crystallographic Defects of the Starting Materials and Energy Resolution

The silicon crystals with crystallographic defects are also tested in the same way in order to compare the characteristics of the detectors produced by dislocation-free and swirlless crystals described above. The sample #314 in Table 2 is the dislocation-free and swirlless crystal, on the other hand, samples #308 and #207 are crystals with swirl and dislocation respectively. The minority carrier lifetime of the crystal with crystallographic defects is shorter than that of the defect-free crystal in general. However, it has been found that the reverse current of detectors produced by the NTD silicon crystals with crystallographic defects is considerably larger than that of detectors produced by the defect-free silicon crystals, even if these have nearly a same minority carrier lifetime.

The typical energy spectrum of α particles from ^{241}Am ever obtained by using the surface barrier detector produced by the defect-free NTD silicon crystal is shown in Fig.5. The energy resolution is nearly a same value of that of the FZ silicon detector.

4. CONCLUSION

It has become possible to obtain the NTD silicon crystal which has the minority carrier lifetime as long as that of the FZ silicon crystal. Improvement of the lifetime of the NTD silicon crystal has been made by using the infrared lamp furnace. The annealing time of about 10 minutes at the annealing temperature of 800°C is sufficient enough for removing the damage caused by irradiation of neutrons. With the annealing time longer than this, the lifetime becomes short again.

The reason that the lifetime of the NTD silicon crystal reported in the previous parer was short is found to be due to a long time thermal annealing (16 to 24 hours) in the diffusion furnace.

It has been observed that the reverse current of the surface barrier detector has a distinct correlation with the minority carier lifetime of the crystal used. It has also been found that the good result can not be obtained if a source material which contains crystallographic defects is used.

It is of cause desirable that the fast neutron content in the neutron flux is small. However, no remarkable influence of the type

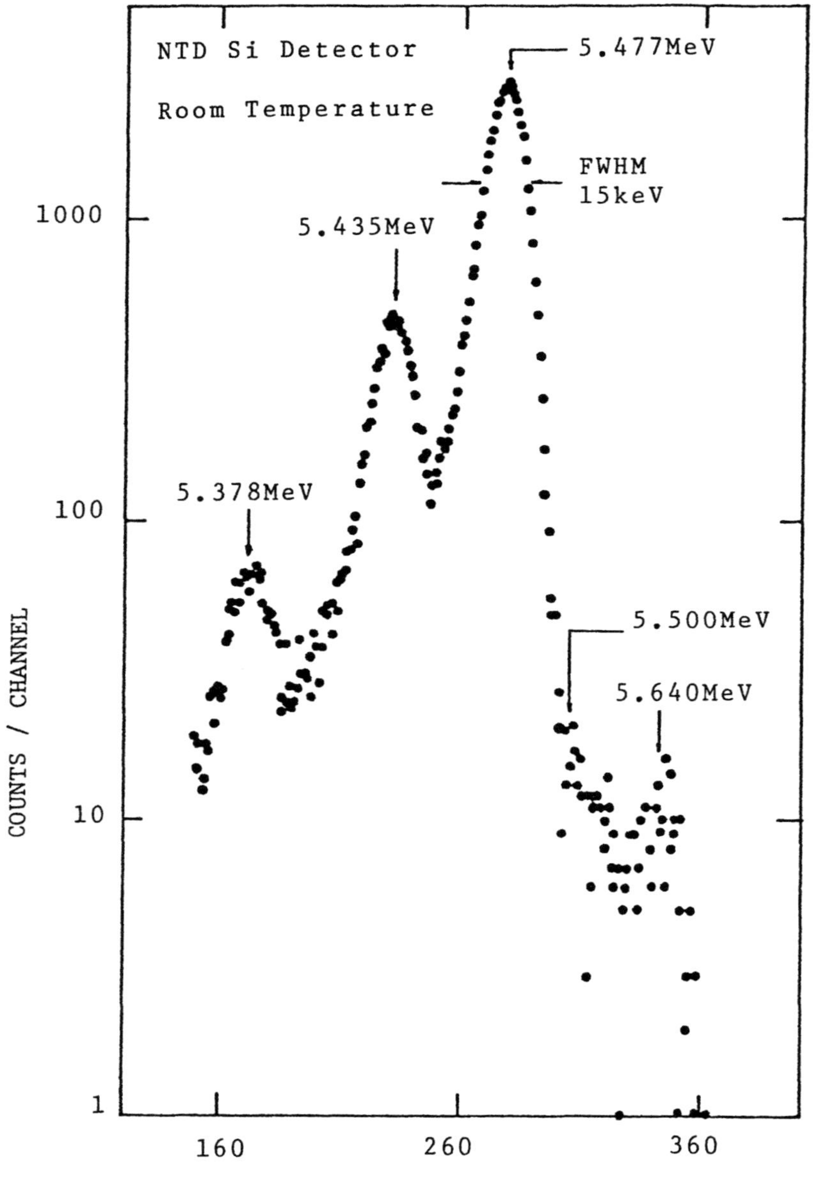

Fig. 5 Spectrum of α particles from ^{241}Am obtained by NTD silicon detector

of nuclear reactor upon the characteristics of the NTD silicon detector has been observed in this experiment.

ACKNOWLEDGEMENTS

The authors would like to express their sincere thanks to Dr. M. Tajima of Electrotechnical Laboratory for his kind supply of the photoluminescence spectra of the NTD silicon and to Mr. F. Shiraishi of Rikkyo University for his helpful collaboration. Thanks are also due to Mr. Y. Yatsurugi of Komatsu Electronic Metals Co. for his encouragement throughout of this work.

REFERENCES

1) C. Kim, H. Kim, A. Yusa, S. Miki, K. Husimi, S. Ohkawa and Y. Fuchi, IEEE Trans. Nuclear Science, NS-26, No.1, p.292 (1979)
2) A. Yusa, Y. Yatsurugi and T. Takaishi, J. Electrochem. Soc., 122, p.1700 (1975)
3) C. T. Sah, R. N. Noyce and W. Shockley, Proc. I.R.E., 45, p.1228 (1957)
4) W. Shockley, Electrons and Holes in Semiconductor D van Nostrand (1950)
5) F. S. Goulding and W. L. Hansen, Nucl. Instr. & Meth., 12, p.249 (1961)

NEUTRON TRANSMUTATION DOPING OF GaAs*

M. A. Vesaghi and H. Fritzsche

James Franck Institute
The University of Chicago
Chicago, IL 60637

INTRODUCTION

Thirty years ago, Cleland, Lark-Horovitz and Pigg[1] carried out the first transmutation doping and showed that it is a convenient and highly reproducible method for introducing a homogneous distribution of dopants into certain semiconductors. This work was continued by Fritzsche et al.,[2] Cuevas,[3] and others[4] who made a detailed investigation of the electronic transport properties of transmutation doped Ge down to He temperatures. The method was applied to Te by Kuehnel et al.[5] and is now widely used in the manufacture of Si devices.

Besides a brief study by Mirianashvili et al.,[6] no detailed investigation of transmutation doping has been carried out on GaAs. This is quite surprising in view of the importance of GaAs as device material and the difficulties encountered with dopants in III-V compounds with regard to impurity and vacancy complexes and the question whether Ge or other impurities are on Ga or As sites.

We report here a study of the resistivity ρ and the Hall coefficient R of GaAs exposed to thermal neutron fluences between 10^{17} and 2×10^{18} n/cm^2. Three kinds of GaAs crystals were used, undoped and semi-insulating (Cr-doped) crystals, and a high purity epitaxial layer.[7] R, ρ, and the Hall mobility R/ρ were measured between 1.4 and 450K.

*Supported by the U.S. Air Force Office for Scientific Research Grant No. AFOSR76-30440.

DOPING EFFICIENCY

In contrast to Si and Ge, all naturally occurring isotopes of Ga and As participate in transmutation doping with the reactions

$$^{69}Ga(n,\gamma)^{70}Ga \longrightarrow {}^{70}Ge + \beta^- \quad (21m) \qquad (1)$$

$$^{71}Ga(n,\gamma)^{72}Ga \longrightarrow {}^{72}Ge + \beta^- \quad (14h) \qquad (2)$$

$$^{75}As(n,\gamma)^{76}As \longrightarrow {}^{76}Se + \beta^- \quad (26h) \qquad (3)$$

The capture cross sections for thermal neutrons are 1.68, 4.86 and 4.3 barn and the natural abundances 60, 40, and 100% for the Ga and As isotopes in the reactions (1), (2), and (3), respectively. If the resultant Ge atoms are located on Ga sites and the Se atoms on As sites, then all end products will act as shallow donors. In this case one expects to find a nuclear transmutation doping efficiency of

$$N_D = 0.16\, \phi t \qquad (4)$$

which is three orders of magnitude higher than that in Si. Here ϕ is the thermal neutron flux ($n/cm^2 sec$) and t the irradiation time (sec).

EXPERIMENTAL DETAILS

The undoped and Cr-doped, semi-insulating crystals were melt-grown and purchased from Laser Diode Lab., Inc. Bridge-shaped samples with three electrode arms on each side and enlarged pads for current contacts were ultrasonically cut from 0.1 cm thick wafers. The undoped crystal had an excess electron concentration of $2\text{-}3 \times 10^{16}$ cm^{-3}. The Cr-doped crystal had $\rho > 10^7$ ohm cm at 300K. The epitaxially grown crystal[7] was a square platelet 0.5 X 0.5 X 0.02 cm. The samples were irradiated at the Research Reactor Facility at the University of Missouri in a thermal neutron flux of $\phi = 5 \times 10^{11}$ $n/cm^2 sec$. Annealing of the recoil and radiation damage was carried out at 800°C for 4.5-10.5 hours in an inert gas atmosphere after pyrolytic Si_3N_4 encapsulation to prevent As evaporation. The epitaxial layer was annealed at 600°C.

RESULTS

Table I lists the values of the Hall coefficient R, resistivity ρ as well as of the electron density $n = 1/Re$ and the Hall mobility $\mu_H = R/\rho$ measured at 300K before and after transmutation doping for the undoped GaAs samples. From measurements at 450K we estimate that errors due to carrier freeze-out are less than 10 percent.

Table I. Characteristics at 300K of Undoped GaAs Samples Before and After Transmutation Doping

	Sample No.:	0	1	3	4	5
Before irradiation	$R(cm^3/C)$	263	216	254	249	245
	ρ(ohm-cm)	0.07	0.06	0.072	0.07	0.068
	$n(10^{16} cm^{-3})$	2.37	2.89	2.46	2.5	2.55
	$\mu_H(10^3 cm^2/Vs)$	3.78	3.60	3.53	3.56	3.60
After irradiation + annealing	$\phi t(10^{17} n/cm^2)$	0	1.0	3.125	9.37	18.75
	anneal time (h)	10.5	5	10.5	4.5	4.5
	$R(cm^3/C)$	193	131	101	37.5	19.2
	ρ(ohm-cm)	0.044	0.035	0.024	0.0094	0.0055
	$n(10^{16} cm^{-3})$	3.23	4.75	6.2	16.7	32.5
	$\mu_H(10^3 cm^2/V_s)$	4.35	3.71	4.19	3.97	3.47
	$\Delta n(10^{16} cm^{-3})$	0.86	1.86	3.8	14.2	30
	$0.16\ \phi t(10^{16} cm^{-3})$	0	1.6	5.0	15	30

In order to test the quality of the Si_3N_4 encapsulation, sample No. 0, which received no irradiation was measured before and after annealing at 800°C for 10.5 hours. Evaporation of As atoms is known to produce deep electron traps and would therefore result in a reduction in n and μ_H. Instead we observe a small increase in carrier concentration and an improvement in mobility. It appears that annealing removes some low lying acceptor states initially present in the material. Unfortunately, this was discovered after the other samples had been irradiated. One therefore may have to correct the final Δn results by subtracting a value which probably is proportional to the anneal time. The last two rows in Table I compare the measured change in carrier concentration Δn with that expected from Eq. (4) and the irradiations ϕt (listed in row 5).

The close agreement between the measured and calculated values of the donor concentration introduced by transmutation doping shows that most Ge atoms are indeed on Ga sites.

A comparison of the last two rows of Table II which lists the data of Cr-doped samples after transmutation doping is less conclusive. However, since the difference 0.16 $\phi t - n$ increases monotonically with ϕt, we believe that the major cause for this difference is that Cr-doping delays complete annealing or creates As vacancies which can capture Ge atoms which have been displaced from their normal lattice sites by the recoil energy.

Table III shows preliminary results on a high purity epitaxial layer. The values for the donor and acceptor concentrations were calculated from the mobility at 77K using the master curve of Wolfe et al.[8] One finds that N_D has increased by 1.1 X $10^{15} cm^{-3}$ which agrees within 10% with the value expected from Eq. (4) even though the sample was annealed for only one hour at 600°C. In impure samples, a much higher anneal temperature of 800°C was needed because the impurities tend to trap the diffusing species. Further anneal studies on this epitaxial sample are needed to see whether the increased N_A value is caused by remaining radiation or by Ge on As sites.

Figs. 1-3 show the temperature dependencies of R, ρ and μ_H, respectively, for the samples listed in Table I and II. The non-irradiated sample is labelled (00) before annealing and (0) after annealing. Attention is drawn to the following features in the figures. The maxima in the Hall curves near 80K mark the transition from dominant band conduction at higher temperatures to impurity conduction in a band of states formed by the shallow donors, at lower temperatures. The critical donor concentration N_c at which the

Table II. Characteristics at 300K of Cr-Doped GaAs ($\sim$9 X $10^{16} Cr/cm^3$) Samples After Transmutation Doping and Annealing at 800°C for 4.5 hours.

Sample No.:	1C	2C	3C	4C
$R(cm^3/C)$	560	323	111	40.2
ρ(ohm-cm)	0.235	0.143	0.043	0.014
$\mu_H(10^3 cm^2/V_s)$	2.38	2.25	2.56	2.97
$n(10^{16} cm^{-3})$	1.10	1.72	5.63	15.5
$0.16\ \phi t(10^{16} cm^{-3})$	7.5	9.0	15.0	30.0

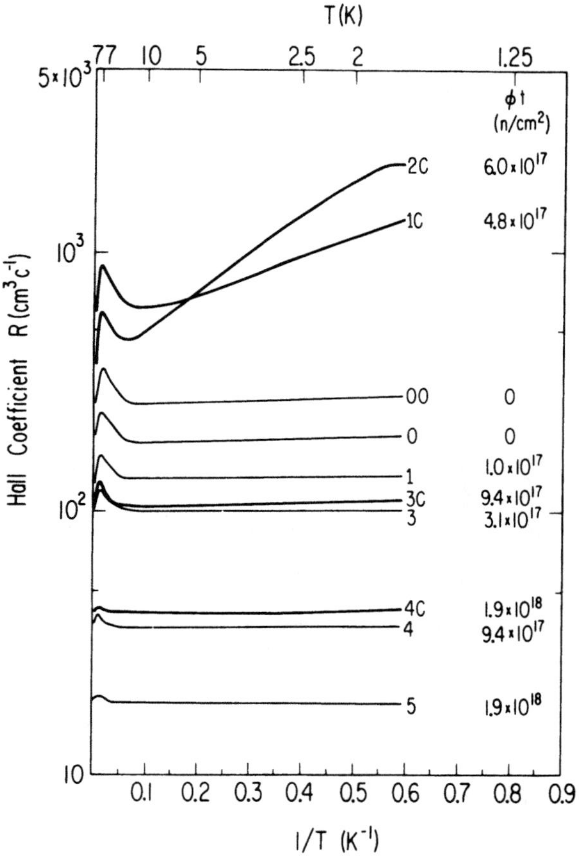

Fig. 1. Temperature dependence of the Hall coefficient of GaAs samples after various amounts of neutron transmutation doping. The characteristics of these samples are listed in Tables I and II.

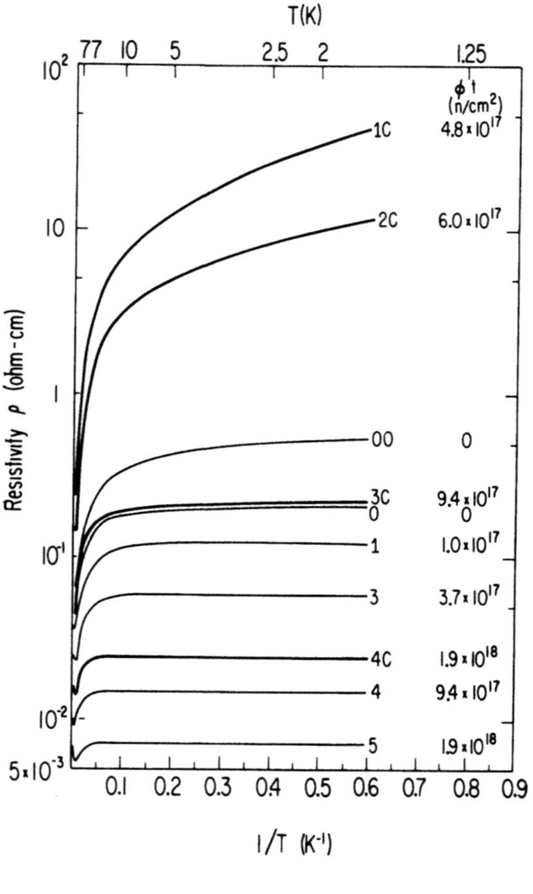

Fig. 2. Temperature dependence of the resistivity of the samples of Figure 1.

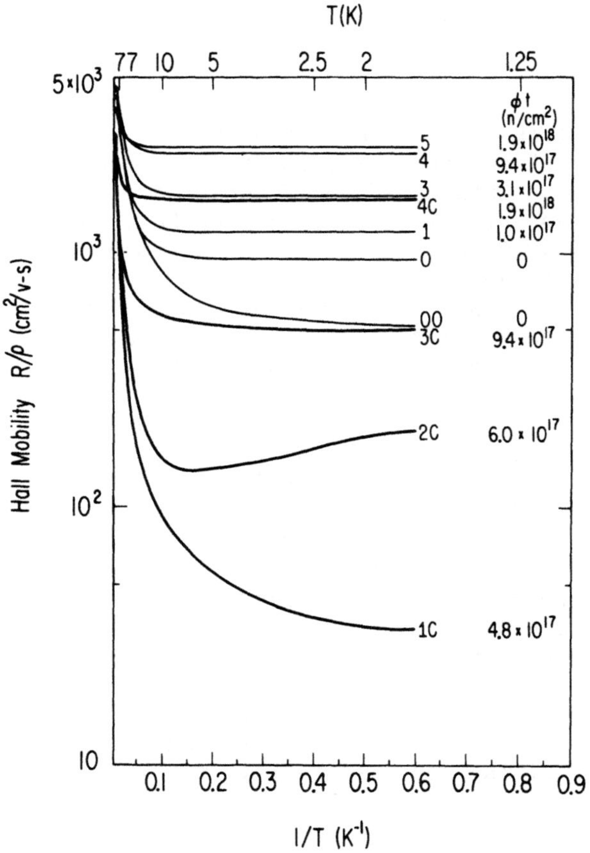

Fig. 3. Temperature dependence of the Hall mobility of the samples shown in the previous two figures.

Table III. Characteristics at 300K of Epitaxial GaAs Before and After Transmutation Doping. 0.16 $\phi t = 1.2 \times 10^{15} cm^{-3}$. Anneal: 1 hour at 600°C.

	R (cm^3/C)	ρ (ohmcm)	μ_H (cm^2/V_s)	n (cm^{-3})	N_D (cm^{-3})	N_A (cm^{-3})
Before	39,800	5.18	7,690	$1.57\ 10^{14}$	$2.1\ 10^{14}$	$5.7\ 10^{13}$
After	7,820	1.01	7,770	$7.98\ 10^{14}$	$1.3\ 10^{15}$	$4.7\ 10^{14}$

nonmetal-metal transition occurs is about $3 \times 10^{16} cm^{-3}$ for the undoped samples.[9] R of these samples is therefore temperature independent at low T. Samples 1C and 2C which are partially compensated by Cr-doping have a temperature activated Hall coefficient at low T. The nonmetal-metal transition occurs at a somewhat higher concentration than that of sample 3C. We estimate $n_c = 6 \times 10^{16} cm^{-3}$ and $N_c = 1.5 \times 10^{17} cm^{-3}$ for this sample series. Such increase of the critical concentration with compensation has also been observed in Ge.[10] It is at variance with Mott's criterion for N_c at the metal-nonmetal transition.[11]

The resistivity curves of Fig. 2 show two activation regimes. The larger activation ε_1 at higher T is due to excitation of carriers into the conduction band. The smaller low temperature activation energy ε_2 decreases to zero as the nonmetal-metal transition is reached. For the metallic samples ε_1 is still finite which means electronic transport takes place in a band of donor states below the conduction band. This is supported by the fact that the Hall mobility increases with donor concentration even for the metal-like samples (Fig. 3). Similar results have been observed in Ge above the nonmetal-metal transition.[12] We believe the increase of mobility with donor concentration is caused by the increased overlap of the donor state wave functions. The low temperature Hall mobility of the partially compensated samples is lower than that of the uncompensated ones because of the random disorder potentials of the compensating acceptor ions.

REFERENCES

1. J. W. Cleland, K. Lark-Horovitz, and J. C. Pigg, Phys. Rev. 78: 814 (1950).
2. H. Fritzsche and M. Cuevas, Phys. Rev. 119: 1238 (1960).
3. M. Cuevas, Phys. Rev. 164: 1021 (1967).
4. S. Golin, Phys. Rev. 132: 178 (1963).
5. A. Kuehnel, H. Siethoff, and G. Landwehr, Phys. Stat. Solidi A29: 387 (1975).

6. Sh. M. Mirianashvili and D. I. Nanobashvili, Soviet Physics, Semiconductors 4: 1612 (1971).
7. We are very grateful to Prof. C. M. Wolfe of Washington University, St. Louis, for making this sample available to us.
8. C. M. Wolfe, G. E. Stillman, and J. P. Dimmock, J. Appl. Phys. 41: 504 (1970).
9. H. Fritzsche in:"Metal Non-Metal Transition in Disordered Systems," edited by L. R. Friedman and D. P. Tunstall (Scottish Universities Summer School in Physics, 1978) p. 193.
10. H. Fritzsche, Phil. Mag. in press.
11. N. F. Mott, "Metal-Insulator Transitions," Taylor and Francis, London (1974).
12. M. Cuevas and H. Fritzsche, Phys. Rev. 137: A1847 (1965); ibid 139: A1628 (1965).

PARTICIPANTS

Alm, A.	ASEA AB, Fack, S-721 83 Västeraes, Sweden
Andresen, K.	Isotope Division, Risoe National Laboratory 4000 Roskilde, Denmark
Badham, K.A.	Atomic Energy of Canada Ltd., P.O. Box 6300, Ottawa, Ontario K2A 3W3, Canada
Barraclough, K.G.	RSRE Malvern, St. Andrews Rd., Gt. Malvern Works, Malvern, England
Blanc, F.	CEA-CENG-LETI/LIR, 85x 38041 Grenoble Cedex, Grenoble, France
Braggins, T.	Westinghouse R&D Center, 1310 Beulah Road, Pittsburgh, PA 15235
Cao, Y.-Z.	Institute of Atomic Energy, P.O. Box 275, China
Cherruau, F.	Services de Piles de Saclay, B.P. 2, 91190 Gif sur Yvette, France
Chu, C.	Westinghouse Electric Corp., Semiconductor Div., Youngwood, PA
Cleland, J.W.	Oak Ridge National Lab., P.O. Box X, Oak Ridge, Tennessee 37830
Corelli, J.	Rensselaer Polytechnic Institute, Dept. of Nuclear Engineering JEC 5050, Troy, NY 12181
Crees, D.	AEI Semiconductors Ltd., Carholme Road Lincoln, England
Crick, N.	AERE Harwell, Oxfordshire OX11 ORA, England

Domenici, M.	Smiel, P.O. Box 255, Merano, Italy
Ewaldt, H.	VALVO GmbH, Stresemannallee 101, 2000 Hamburg 54, Federal Republic of Germany
Genthon, J.P.	Services des Piles de Saclay, B.P. 2, 91190 Gif sur Yvette, France
Guldberg, J.	Topsil A/S, P.O. Box 93 3600 Frederikssund, Denmark
Gunn, S.	University of Missouri, Research Reactor Columbia, Missouri 65211
Hansen, K.	Risoe National Laboratory, 4000 Roskilde, Denmark
Hee, T.	Litex A/S, Risingevej 1, 2660 Broendby Strand, Denmark
Henry, K.	AERE Harwell, Oxfordshire OX11 0RA, England
Herzer, H.	Wacker Chemitronic GmbH, P.O. Box 1140 D-8263 Burghausen, Federal Republic of Germany
Heydorn, K.	Isotope Division, Risoe National Laboratory, 4000 Roskilde, Denmark
Hohmuth, K.	Central Inst. of Nuclear Research, Rossendorf, 8051 Dresden, Postfach 19, German Democratic Republic
Husimi, K.	Institute for Nuclear Study, University of Tokyo, Midoricho, Tanashi, Tokyo 188, Japan
Hyder, A.	EG&G ORTEC, 100 Midland Road, Oak Ridge, Tennessee 37830
Kaltenborn, N.	Institute for Energy Technology, P.O. Box 40, N-2007, Kjeller Norway
Kenworthy, I.	Mullard Ltd., 6200 Dept., Millbrook Industrial Est., Southampton, England
Kirk, M.A.	Argonne National Laboratory, University of Oxford, Oxford OX1 3PH, England
Kramer, H.	Monsanto Company, P.O. Box 8, St. Peters, Mo. 63376

Larrabee, R.	National Bureau of Standards, Building 225, Room A-327, Washington, D.C. 20234
Larsen, F.	Topsil A/S, P.O. Box 93, 3600 Frederikssund, Denmark
Leon, N. de	Topsil A/S, P.O. Box 93, 3600 Frederikssund, Denmark
Li, Y.-X.	Institute of Atomic Energy, P.O. Box 275 Beijing, China
Lin, C.-K.	Institute of Atomic Energy, P.O. Box 275 Beijing, China
Lindstrom, L.	Forsvarets Forskningsanstalt Dep. 3, P.O. Box 1165, Linköping, Sweden
Lu, C.G.	Institute of Atomic Energy, P.O. Box 275 Beijing, China
Lund, K.	Topsil A/S, P.O. Box 93, 3600 Frederikssund Denmark
Malmros, O.	Topsil A/S, P.O. Box 93, 3600 Frederikssund, Denmark
Marsh, O.	Hughes Research Laboratories, 3011 Malibu Canyon Rd., Malibu, Ca. 90265
Martin, G.	Monsanto Ltd., 10-18 Victoria Street, London SW1, England
Meese, J.	University of Missouri, Research Reactor, Columbia, Missouri 65211
Molinari, M.A.	Recherches Nuclaires, Rue de Loess, Strasbourg, France
Newman, R.C.	University of Reading, J.J. Thomson Physical Laboratory, Whiteknights, Reading RG6 2AF England
Nielsen, V.	Topsil A/S, P.O. Box 93, 3600 Frederikssund, Denmark
Nielsen, K. Falcon	Risoe National Laboratory, 4000 Roskilde Denmark

Okamura, M.	Hitachi Research Lab., Hitachi Ltd., 4026 Kuji, Hitachi, Japan
Pajot, B.	Université Paris 7, 75221 Paris Cedex 05, France
Paxman, D.H.	Philips Research Laboratory, Cross Oak Lane, Salfords, Redhill, Surrey RH1 5HA, England
Popp, G.	BBC, D-6840 Lampertheim, Postfach 200, Federal Republic of Germany
Raabe, G.	VALVO GmbH, Stresemannallee 101, 2000 Hamburg 54, Federal Republic of Germany
Roche, D.	CEA-CENG-LETI-NCE, 85X - 38041 Grenoble Cedex, Grenoble, France
Ross, R.	Central Inst. of Nuclear Research, Rossendorf, 8051 Dresden, Postfach 19, German Democratic Republic
Rowlands, V.	Topsil A/S, 61 Quarry Street, Guilford, Surrey GU1 3UA, England
Rundquist, H.	Studsvik Energiteknik AB, S-611 82 Nyköping, Sweden
Salling, J.	Topsil A/S, P.O. Box 93, 3600 Frederikssund, Denmark
Schöller, M.	Siemens AB, Frankfurter Ring 152, 8000 Munich, Federal Republic of Germany
Schulz, E.	AEG-Telefunken, Theresienstrasse 2, 71 Heilbronn, Federal Republic of Germany
Senes, A.	Thomson-CSF, SSC. Rue P.M. Curie 37100, Tours, France
Sigfridsson, B.	Försvarets Forskningsanstalt Dep. 3 P.O. Box 1165, Linkoping, Sweden
Skarstam, B.	Dept. of Physics & Measurement Tech., FOA-3, Box 1165, S-581 11 Linkoping, Sweden
Smith, T.	AERE Harwell, Oxfordshire OX11 0RA, England
Sommer, K.H.	AEG-Telefunken, Postfach 2160, 4788 Warstein 2, Federal Republic of Germany

Sonntag, A.	AEG-Belecke, 4788 AEG Belecki-Warstein 2, Postfach 2160, Warstein, Federal Republic of Germany
Souza, A. de	International Rectifiers, Hurst Green Oxted, Surrey, England
Stendal, K.	Risoe National Laboratory, 4000 Roskilde, Denmark
Stone, B.D.	Monsanto Company, P.O. Box 8, St. Peters, Mo. 63376
Tajima, M.	Electrotechnical Laboratory, 1-1-4 Umezono Sakura-Mura, Ibaraki 305, Japan
Teszner, J.-L.	D.R.E.T. 26 Boulevard Victor, 75996 Paris Armees, Paris, France
Tole, J.	AEI Semiconductors Ltd. Carholme Road, Lincoln, England
Vanthienen, R.	Monsanto NV Europe, Ottersemse Stw 707, B-9000 Gent, Belgium
Veyrat, J.-F.	CEA-CENG, 85X - 38041 Grenoble Cedex, Grenoble, France
Vieweg-Gutberlet, F.G.	Wacker Chemitronic GmbH, P.O. Box 1140, D-8263 Burghausen, Federal Republic of Germany
Vlasak, T.	Brown, Boveri & Cie, 5242 Birr, Switzerland
Wada, T.	Mie University, Faculty of Engineering, Kamihama, Tsu, Japan
Washburn, J.	Motorola Inc., 5005 East McDowell, Phoenix, Arizona 85008
Wilkes, J.G.	Mullard Ltd., 6200 Dept., Millbrook Industrial Est., Southampton, England

INDEX

Batch number: 10838062

Printed by Printforce, the Netherlands